Smart Cities in Asia

CITIES SERIES

Series Editor: John Rennie Short, *Department of Public Policy, University of Maryland, Baltimore County, USA*

As we move into a more urban future, cities are the main setting for social change, economic transformations, political challenges and ecological concerns.

This series aims to capture some of the excitement and challenges of understanding cities. It provides a forum for interdisciplinary and transdisciplinary scholarship. International in scope, it will embrace empirical and theoretical studies, comparative and case study approaches. The series will provide a discussion site and theoretical platform for cutting edge research by publishing innovative and high quality authored, co-authored and edited works at the frontier of contemporary urban scholarship.

Titles in the series include:

Smart Cities in Asia

Governing Development in the Era of
Hyper-Connectivity

Edited by

Yu-Min Joo

KDI School of Public Policy and Management, South Korea

Teck-Boon Tan

*S. Rajaratnam School of International Studies,
Nanyang Technological University, Singapore*

CITIES SERIES

 Edward Elgar
PUBLISHING

Cheltenham, UK • Northampton, MA, USA

Published by
Edward Elgar Publishing Limited
The Lypiatts
15 Lansdown Road
Cheltenham
Glos GL50 2JA
UK

Edward Elgar Publishing, Inc.
William Pratt House
9 Dewey Court
Northampton
Massachusetts 01060
USA

A catalogue record for this book
is available from the British Library

Library of Congress Control Number: 2019954435

This book is available electronically in the **Elgar**online
Social and Political Science subject collection
DOI 10.4337/9781788972888

ISBN 978 1 78897 287 1 (cased)
ISBN 978 1 78897 288 8 (eBook)

Printed and bound by CPI Group (UK) Ltd, Croydon, CR0 4YY

Contents

v

Tables

Contributors

Tathagata Chatterji is Professor of Urban Management and Governance at Xavier University, Bhubaneswar. His research interests are urban economic development, global cities and political-economy of urbanisation. He has published two books: *Local Mediation of Global Forces in Transformation of the Urban Fringe* (Lambert Academic Publishing 2014) and *Citadels of Glass – India's New Suburban Landscape* (Tata-Westland Press 2015). He received the Gerd Albers Award in 2016 from the ISOCARP for his research on comparative governance. He holds a PhD in Urban Governance and Planning from the University of Queensland.

Ming-Yee Foo holds a MPP (Public Policy) from the Lee Kuan Yew School of Public Policy (LKYSPP), National University of Singapore. During her graduate studies (2017–2019), she focused on urban studies and international affairs as part of her curriculum. Her research areas are in the realms of smart cities, technology adoption and sustainable development. Prior to joining LKYSPP, she had interned and worked in various departments of the Singapore Public Service.

Jong-Sung Hwang is a lead researcher at the National Information Society Agency in Korea. He leads research projects on smart city and future government and is a master planner of the Busan EDC National Pilot Smart City. From 2011 to 2013, he was a chief information officer of the Seoul Metropolitan Government. He was awarded the Order of Civil Merit, Camellia Medal by the Korean Government in 2016. He holds a PhD in Political Science from Yonsei University in Korea.

Satria Aji Imawan is a member of the research staff at the Centre for Population and Policy Studies (PSKK) Universitas Gadjah Mada, Indonesia. His research interests are behavioural public administration, public service innovation and regulatory impact assessment.

Yu-Min Joo is an associate professor at the KDI School of Public Policy and Management in Korea. She holds a PhD in Urban and Regional Planning from the Massachusetts Institute of Technology. From 2012 to 2019, she was an assistant professor at the Lee Kuan Yew Policy, National University of Singapore. She teaches and publishes on urban development and policy issues in Asia, particularly on the topics of urban governance, smart cities, city branding, mega-projects and mega-events.

Chee Keong Khoo is a doctoral candidate at the City University of Hong Kong under the Hong Kong PhD Fellowship Scheme. He earned his bachelor's degree in Building Surveying with Distinction from the University of Malaya, Malaysia. His dissertation deals with cost–benefit analysis of green building certification and occupant behaviour of green public housing in Hong Kong.

Kian Cheng Lee is an assistant professor at the Faculty of Political Science and Public Administration and School of Public Policy, Chiang Mai University, Thailand. He holds a PhD (Social Sciences), MA (Southeast Asian Studies), MTh (Asian Christianity), MDiv, MA (Biblical Studies) and BSc (Physics). He is also a board director of Asian Pastoral Institute Ltd. Kian Cheng's research interests include Chinese transnationalism (tourism, entrepreneurship and social embeddedness), cultural diplomacy and religious networks.

Xin Li is an assistant professor in the Department of Architecture and Civil Engineering at the City University of Hong Kong. She has a PhD in Urban and Regional Studies from the Massachusetts Institute of Technology and a Master of Urban Planning from the University of Michigan at Ann Arbor. Her recent research focuses on cultural and creative industry and urban land policy.

Michael Manning is a doctoral candidate in Public Administration and lecturer for the Institute of General Education at Incheon National University. He has published in the *Asian Journal of Cultural Policy*, *Songdo Journal of Social Sciences*, and *Korea TESOL Journal*. His research interests include city planning, environmental policy and comparative cultural policy.

Kyung-Min Nam is an assistant professor at the University of Hong Kong and holds a PhD in International Development and Regional Planning from the Massachusetts Institute of Technology. His research interests focus on the institutional and policy dimensions of economic and environmental sustainability, and his scholarly work has been widely published in the major peer-reviewed journals in this field. He has also served on the editorial boards of several academic journals, including the *Annals of Regional Science* (Springer).

Jae-in Noh is a doctoral candidate in Public Administration at Incheon National University. She has worked for the Korea Local Information and Research Development Institute as a researcher and has published articles in Korean, on the topics of Jeongbohwa Jeongchaek (Informatisation Policy), and in English (forthcoming). Her research interests lie in e-government, local politics and administration.

Ora-orn Poocharoen is the Director of Chiang Mai University School of Public Policy. She is also a nominated member of the Committee of Experts on Public Administration (CEPA) of the United Nations from 2018 to 2021. She obtained her PhD in Public Administration from Syracuse University, New York. Her research focuses on comparative public policy and administration in Asia, smart city, deliberative processes and innovation in the public sector. She has provided consultation to many government agencies in Thailand and multilateral organisations.

Arif Budy Pratama is a lecturer at the Public Administration Programme, Universitas Tidar, Indonesia. He is currently undertaking his PhD at the Bonn International Graduate School for Development Research (ZEF), University of Bonn, Germany. He holds an MPA (with Distinction) from University of Exeter, UK. His research interests include bureaucracy, local governance and development. Prior to joining academia, he worked as the head of Program Formulation in the Ministry for Political, Legal, and Security Affairs, Republic of Indonesia.

Souvanic Roy is a professor and head of School of Ecology, Infrastructure and Human Settlement Management at the Indian Institute of Engineering Science and Technology (IIEST), Shibpur. His areas of interest include urban planning, urban policy for inclusive cities and housing strategies for the urban poor. He is a member of the Working Group for National Civil Society Consultation and has contributed to the Delhi Declaration on 'India's Urban Future: Choice Not Chance – Civil Society Contribution to Habitat III'.

Teck-Boon Tan is a research fellow and coordinator of the Science and Technology Studies Programme in the Office of the Executive Deputy Chairman, S. Rajaratnam School of International Studies (RSIS), Nanyang Technological University, Singapore. His research covers science and technology policy, smart urbanism and disruptive technology. He received his PhD from the Lee Kuan Yew School of Public Policy, National University of Singapore.

Jill L. Tao is the first international professor and chair of the Department of Public Administration at Incheon National University, Korea. She has published widely in public administration, local government and public policy journals. She currently works with a research team at the United Nations University, Operating Unit on Policy-Driven Electronic Governance (EGOV) in Guimaraes, Portugal, examining the effects of culture on how local officials utilize social media platforms, and whether such differences have demonstrable effects on transparency, trust and participation.

Poon Thiengburanathum is an assistant professor and Director of the Excellence Center for Urban Study and Public Policy (ECUP), Chiang Mai University. His expertise lies in the field of complex system analysis for planning and policy recommendations. The complex system analysis has been adopted for a variety of research in regional development, urban planning and environmental studies, such as urban mobility, haze, supply chain and logistic analysis for food products.

Shang-su Wu is a research fellow at S. Rajaratnam School of International Studies (RSIS), Nanyang Technological University, Singapore. He has a PhD from the University of New South Wales, Australia. Before joining the RSIS, he taught at the National Defense University in Taiwan. He is the author of *The Defence Capabilities of Small States: Singapore and Taiwan's Responses to Strategic Desperation* (Palgrave 2016). His research interests are military modernisation, Taiwan issues, railway and international relations.

Xinhui Yang is a postgraduate at the School of Government, Sun Yat-sen University (SYSU) in Guangzhou, China. His primary research interests are in the field of smart city, urban governance and rural education. His expertise is in building econometric models and in data analysis including Excel, Stata, Python, R and CiteSpace.

Masaru Yarime is an associate professor in the Division of Public Policy at the Hong Kong University of Science and Technology. He has appointments as an honorary reader at University College London and as a visiting associate professor at the University of Tokyo. He has received a BEng and MS in Chemical Engineering from the University of Tokyo and the California Institute of Technology respectively and a PhD in Economics and Policy Studies of Innovation and Technological Change from Maastricht University in the Netherlands.

Lin Ye is a professor at the Center for Chinese Public Administration Research and School of Government, Sun Yat-sen University (SYSU) in Guangzhou, China. His primary research interests are in the field of urban policy and regional governance.

Acknowledgements

We would like to thank Ms Eileen Tong for her editorial assistance. This research was financially supported by the Lee Kuan Yew School of Public Policy, National University of Singapore, and partially funded by the KDI School of Public Policy and Management, South Korea. We would also like to thank the S. Rajaratnam School of International Studies, NanyangTechnological University, for its support.

1. Smart cities in Asia: an introduction

Yu-Min Joo and Teck-Boon Tan

INTRODUCTION

The smart city has become a trendy policy concept in recent years within the context of two notable trends: (1) the rapid development of novel Fourth Industrial Revolution technologies, such as the Internet of Things (IoT), big data and artificial intelligence (AI); and (2) a rapidly urbanising planet, with more than half the population now living in cities. By providing off-the-shelf technological solutions to urban challenges and proposing to change the way in which we manage and live in cities, smart cities have captured the imagination of policymakers and politicians worldwide. Against this backdrop, we are witnessing an upsurge in the number of smart city developments being launched in Asia. Take these examples: India announced its plan to build 100 smart cities in 2015; Singapore launched its Smart Nation initiative in 2014; South Korea and Japan have been developing a series of ubiquitous- and eco-cities since the early 2000s; and China has recently been receiving attention for eagerly adopting the latest smart technology for urban management. As these Asian economic powerhouses deeply engage in their smart city projects, other Asian countries are also quick to follow suit by putting into motion their own smart city plans. In 2018, the Association of Southeast Asian Nations (ASEAN) launched the ASEAN Smart Cities Network (ASCN), with 26 pilot cities across all ten countries in the regional intergovernmental organisation.

This edited volume brings together a collection of studies on the latest smart city developments in Asia. The focus on Asian smart cities was prompted by our observation that a rapidly growing number of smart city projects is developing in the region, but, nevertheless, a relative dearth of proper attention has been given to them in smart city literature. Having undergone (or currently undergoing) a rapid catch-up in economic development and equally fast-paced urbanisation, Asian countries in general appear to exhibit a much more favourable attitude towards the development of smart cities in their quest for the prestige – symbolic or otherwise – of being able to adopt the latest information and communication technology (ICT) for their urban management systems.

This stands in stark contrast to the more sceptical perceptions and privacy concerns found in the West.

In fact, we may look upon the phenomenon of smart city developments in Asia as a unique trend emerging in a region that is quickly urbanising and growing in the global economy, rather than interpreting them as merely 'catching up' with the smart city models and experiences of the West. For sure, many Asian smart cities embody globally shared key concepts, such as living labs, sustainability and, increasingly, the city-as-a-platform. Yet, the kind of challenges that Asian governments seek to address via their smart city plans and, in particular, the conditions under which these projects are being rolled out are quite different from those faced by American or European cities. After all, many Asian countries and cities have different governance structures, national institutions, and development histories and challenges. Hence, it is not surprising that the motivations behind their smart city projects would differ as well. The majority of smart city research, however, has thus far been carried out based on cases in the West, with but a few exceptions. Even when smart cities in Asia are examined, only a handful – such as Singapore and Songdo – are found in the literature. Given the relentless rise of smart cities in the region and their growing significance in the global economy, it is necessary to examine the phenomenon through a uniquely Asian perspective to achieve a richer and more accurate interpretation. Indeed, while politicians and policymakers in Asia launch their smart city initiatives with much fanfare and enthusiasm, exactly what these projects are and their policy relevance to development remain rather ambiguous in the concurrent discussion on smart cities in the literature.

The collection of chapters herein attempts to offer a more comprehensive and closer look at how and why various governments in Asia pursue smart cities in their specific locales. The goal is to examine the real-world applications of contemporary smart cities, including what motivates their developments, processes and concerns situated within their specific local contexts. As it seeks to draw attention to contextually grounded, specific case studies depicting actual planning experiences of smart cities in various parts of Asia, this edited volume also addresses ongoing calls in smart city literature for more empirically based, international, comparative research to understand the complex reality of smart city development (see, e.g., Carvalho 2015; Glasmeier & Christopherson 2015; Kitchin 2015; Luque-Ayala & Marvin 2015). The idea is that cities are not merely passive backdrops for globally trending policy; instead, local actors actively shape and reshape the smart city idea around their own priorities and objectives, situating it within their local-specific political economy and institutions. Hence, the 'lack of comparative analysis and a dearth of knowledge about the range of urban contexts within which [smart urbanism] is emerging' is an issue (Luque-Ayala & Marvin 2015, p. 2108).

Karvonen et al. (2018) recently published an edited volume, stating that it is the 'first attempt to reflect on how smart city initiatives are being realised in different locales' (p. 295) by studying 23 cities in both the Global North and the Global South. More of such international, comparative research to 'reveal the discursive and material realities of actually existing smart city developments' remains warranted (Glasmeier & Christopherson 2015, p. 9), which this book responds to by focusing exclusively on Asia.

How is the globally promoted smart city idea situated, and how does it become reshaped within the Asian context? What are the substantive differences between smart city initiatives in Asia and in the West? What motivates their developments, which processes are involved and what kind of concerns might there be? One of our key objectives is to understand the true nature of Asian smart cities as they are, rather than viewing them through the critical lens of theories and frameworks mostly based on cases from the West. For this reason, we have made it a point to bring together local scholars and experts on Asian smart cities, most of whom came from or are based in the country or city of their expertise. Being local experts not only on smart cities but also on other related local policies, governance and urban issues, their candid examination of contemporary smart city projects is expected to result in a meaningful and timely collection that will help unravel the globally shared idea of smart cities being reproduced in Asia through local knowledge.

Consequently, our book also addresses Robinson's critique (2002, 2006) of urban scholarship as being focused on only a small handful of cities, particularly in the West, for the purpose of theory building and analyses. Following her call to 'de-colonize' urban scholarship and expand its geographic focus, this edited volume brings hitherto understudied cases of smart city development in Asia into the limelight. In addition to expanding the knowledge and analysis of smart cities as contextualised within diverse urban and national contexts, we hope it can be a starting point for triggering a discussion on how the field can develop appropriate narratives and ground-up theories for smart cities in Asia.

CONTEXTUALISING SMART CITIES IN ASIA

An Overview of Existing Smart City Narratives

Before we begin our discussions on situating the concept of the smart city in Asia, it is worthwhile to first unpack the existing understanding and narratives of smart cities. Indeed, the globally promulgated concept of the smart city has had quite a few definitions and conceptualisations (e.g., Caragliu et al. 2009; Nam & Pardo 2011; Deakin & Al Waer 2012; Angelidou 2015; Yigitcanlar et al. 2018) and has also been criticised for being somewhat ambiguous and elusive at times (Vanolo 2014; Carvalho 2015). The lack of one clear and concrete definition, however, does not hinder our study on the actual practices of smart cities in Asia. Moreover, we note that there are emerging features associated with smart cities that allow us to start with some understanding of what they really are. For instance, smart cities involve actively using advanced, off-the-shelf technologies to *solve* urban problems (Glasmeier & Christopherson 2015). They also entail developing the 'city-as-a-platform', where silo-based services and systems are interconnected to produce a collaborative and integrated model (Anttiroiko 2016; Hwang 2016). A smart city can be summed up as a 'shift from innovation to application, from the back-office to front-line services' (Allwinkle & Cruickshank 2011, p. 9), based on a technical platform supported by cutting-edge technologies, including IoT, big data and AI. Its emphasis on the application of existing and available technologies seems to be opening doors for a diverse assortment of cities to not only aspire towards but also realise smart urbanism. Now, becoming smart is no longer the prerogative of a small handful of knowledge-intensive cities. Additionally, with the end goal emphasis on *cities*, rather than technology itself, multi-dimensional aspects are highlighted. The famous European Union (EU) smart city framework comprises multiple dimensions, including smart economy, people, governance, mobility, environment and living (EU 2014). Likewise, Yigitcanlar et al. (2018) highlight the desired economic, societal, environmental and governance outcomes in smart cities, which are then further developed into a multi-dimensional framework that includes the smart city's assets, drivers and outcomes.

Given the multi-dimensional aspects and goals inherent in smart cities, who the key actors are and their motivations behind the push for smart cities become important in relation to the kind of directions these urban projects might take. The most apparent ones identified in the literature are high-tech companies, which began to see 'cities' as huge untapped sources of revenue amid steep competition in their traditional markets (Townsend 2013). The popularisation of the term 'smart city' itself took off after IBM Chairman Sam Palmisano's speech 'A Smarter Planet: The Next Leadership Agenda' in November 2008, following the financial crisis of 2007–2008 (Söderström et al. 2014). Framing the trend as a 'corporate smart city', the literature began to explain the smart city as an outcome of corporate vision and the market creation strategy of profit-driven multinational corporations (e.g., IBM, Cisco, Intel, Accenture and Siemens) (Hollands 2015; Kitchin 2015) backed by international consulting firms (Glasmeier & Nebiolo 2016).

The emphasis placed on the role of global companies does not marginalise local actors in smart city pursuits, not least because their goals appear to be aligned. In fact, global companies have deliberately designed smart city concepts to court local governments. IBM (2012), for example, has marketed its smarter city solutions by claiming that 'cities compete globally to attract both citizens and businesses [and] a city's attractiveness is directly related to its ability to offer the basic services that support growth opportunities, build economic value and create competitive differentiation'. Such local economic development motivation present in the smart city narrative has attracted quite a few entrepreneurial governments in post-industrial cities to come on board, urging them to keep up with the global circulation of smart city discourse.

This pursuit of a growth agenda under smart city development is analogous to Harvey's (1989) neoliberal urban entrepreneurialism. Indeed, becoming a smart city is often considered another form of entrepreneurial project, serving as an important global signal for competitiveness in today's tech-driven society and a means of attracting footloose capital and talent. It is not a surprise then that smart cities have often been criticised as neoliberal artefacts, improving little – if not aggravating – the digital divide and social inequality issues afflicting post-industrial cities (Hollands 2008, 2015; Greenfield 2013; Vanolo 2014).

In reaction to these concerns, the concept of the smart city has since evolved to emphasise participatory orientation and citizen empowerment, using IT-enhanced platforms to tackle urban problems with civic innovation (e.g., see Allwinkle & Cruickshank 2011; Komninos et al. 2013; Townsend 2013; Deakin 2014; Kitchin 2015) and even to promote a 'new citizenship regime' (Joss et al. 2017). In this view, people and communities are encouraged to take ownership and initiative, as well as to use technology to create social and public value (Hollands 2015; Foth 2016). Indeed, with smart cities' open data and innovation platforms, an entirely new technologically enhanced opportunity now exists to build not just participatory platforms but also citizen-driven, open and inclusive models to shape cities and innovations (Komninos et al. 2013; Deakin 2014; Anttiroiko 2016). The idea is to create an ecosystem or 'living lab' in which citizens can take part in co-designing and co-creating solutions with the government and other stakeholders. Building smart cities as participatory innovation platforms thus requires 'sufficient political will and a culture of citizen empowerment and co-creation' (Anttiroiko 2016, p. 7). Apparently, most cases discussed in this light are about Western cities with deeply rooted local democracy.

In fact, all smart city narratives introduced thus far arise from the Western context, with minimum contributions from the Asian archetype. At most, the latter has been viewed (somewhat critically) as being government-centred, hardware-driven projects that use cutting-edge technologies principally for economic competitiveness (Anttiroiko 2013; Yigitcanlar & Lee 2014). Needless to say, they serve as less than ideal examples of smart cities, especially when juxtaposed against the Western conception of the smart city as a new urban governance future based on digital democracy and a modern ecosystem where local citizens actively develop urban solutions and economic opportunities with user-focused technologies. Given the substantively distinct economic and political contexts and development experiences in Asia, however, we argue that the smart city narrative situated in this region should be given a more nuanced treatment.

Setting the Scene: Government-led Smart Cities in Asia

Our edited volume focuses on both national and local government-led smart city initiatives in Asia, following the noticeably observable trend in the region where governments are not only announcing ambitious smart city initiatives but are spearheading them as well. The following chapters in the book will illustrate how the smart city concept is being used, developed and implemented to meet the policy objectives of these governments in Asia within their local contexts. To be sure, quite a few issues raised in the existing smart city narratives are also pertinent to these Asian cases, such as the nebulous role of multinational companies, widening inequality and the need to underline citizen-centricity in smart city projects. However, for government-led smart city initiatives in Asia, we must also draw attention to the following major differences.

Most notably, while national governments have seldom been discussed in smart city literature, they are often the main actor, actively promoting and planning smart cities in many cases in Asia. In the case of East Asian countries, national governments had a strong hand in their countries' industrialisation and development in the latter half of the twentieth century, and state-led smart city initiatives could indicate a lingering legacy of this state-led developmentalism. This trend may also be interpreted as national governments – whether in the East or in other parts of Asia – seeing the smart city as an opportunity to pursue their national development goals amid greater accessibility to modern technology in today's fast-paced society. What is evident is that economic development remains a high priority for many Asian countries, after having achieved rapid industrialisation relatively recently or currently undergoing one.

The role of local governments should also not be overlooked, as the concept is a smart *city*, not a smart *nation* – except in the case of the city-state of Singapore. Rather than local governments' entrepreneurial pursuits to become globally competitive as identified in the literature, in Asia, the motivations and capacity of local governments in developing smart cities are expected to vary widely. Some local governments of Asian global cities are vying for a leading position in the global smart city ranking, but these examples would be comparatively few amid the widespread boom of smart city projects in the region. More participating local governments are likely to be challenged by either underdevelopment or being unable to catch up with their population growth. While 'development' is the key driver behind smart city endeavours, the level of development goals and the kinds of communication technology become subject to local interpretation and capacity.

For sure, this would not be the very first book to examine Asian smart cities. There are, for example, *Smart Eco-City Development in Europe and China* (Caprotti et al. 2017); *Singapore: Smart City, Smart State* (Calder 2016); and *Smart Cities and Urban Development in India* (Mani 2016). However, our edited volume stands out in the literature, as it surveys the latest smart city projects initiated by governments across varying geographical, economic, political, social and institutional settings in Asia. As implied earlier, although we attempt to highlight notable characteristics and motivations of smart city developments in Asia in this book, there are considerable variations among the countries involved. Each chapter herein studies contextually situated smart cities as they travel within unique local institutions, political dynamics, socio-economic conditions and urban development challenges.

SURVEY OF SMART CITIES IN ASIA: BOOK OVERVIEW

A major contribution of this book is that it surveys countries in Asia, examining the similarities as well as the differences in their development and implementation of the globally adopted smart city concept. To highlight Asia's diverging local contexts in which smart cities are situated and to provide a clearer understanding of them, this book is divided into three parts. In an attempt to group those countries or cases with key shared commonalities, the first part brings together smart cities of the developed-economies of Japan and the Four Asian Tigers (i.e., Singapore, Hong Kong, Taiwan and South Korea); the second part examines the smart cities of the two giants in Asia – China and India – and the third and final part specifically focuses on the smart city development and plans of second-tier cities in Indonesia, Thailand and South Korea. While we do arrive at some conclusions in the subsequent section of this chapter, we also invite the readers to draw their own conclusions and meaningful lessons from this survey.

Smart Cities of the Four Asian Tigers and Japan

Japan and the Four Asian Tigers are leading economies in Asia, with a keen interest in technological developments and a strong state that guided their rapid economic growth throughout the latter half of the twentieth century. The developmental state theory has explained the industrialisation and economic successes of Japan (Johnson 1982) and the Four Asian Tigers (see, e.g., Amsden 1989; Evans 1989), highlighting the states' strong intervention in the economy and society with their forward-looking plans, albeit to varying degrees (Castells 1992). What these successful economies share is an interventionist state, followed by rapid industrialisation based primarily on the strategy of catching up. Given their initial condition of rather limited natural resources for economic development, these countries had to rely almost exclusively on human capital while placing heavy emphasis on technological developments. Hence, it is not a surprise that a number of these countries have been early adopters of technologies in their urban landscapes, in addition to promoting high-tech industries as part of their economic development strategy. Regardless of whether the 'smart city' title is attached to them, the major cities of these five countries are known for their extensively wired urban environments. In fact, TechRepublic ranked Seoul, Hong Kong, Tokyo and Singapore as the top four 'most connected, innovative cities' in the world (Forrest 2015). These cities are also well-acknowledged 'global cities', with the following Global Cities Index ranking by AT Kearney (2019): Tokyo fourth, Hong Kong fifth, Singapore sixth and Seoul thirteenth.

For these Asian economies – already boasting their high-tech global cities – what are the main motivations behind their latest smart city pushes? What policy ends are they trying to achieve? What are their key strategies and challenges? The first part of this book begins with Singapore, which has received much attention from policymakers and scholars across both the East and West for its latest flagship Smart Nation initiative. Yu-Min Joo, Teck-Boon Tan and Ming-Yee Foo (Chapter 2) unpack this mega digitalisation push, launched in 2014, including its development motivations, governance structure and notable characteristics of related projects that span across diverse aspects of the Singaporean economy and society. The authors argue that the Smart Nation is the Singapore Government's next big idea – after the city-state's successful economic transformation under the global-city development goals – to be pursued in the coming decades amid an impending wave of technological change and disruption. The chapter underscores the 'going global' thinking associated with the Smart Nation as a continuation of Singapore's development strategy, and it highlights the international networks and urban solutions (to be exported) sought after under the flagship project. It also introduces the case of an autonomous vehicle living lab as an example of the developmental state's targeting of sunrise industries, *inter alia*. From this calculated state-led push, the authors then draw attention to how difficult it has been to get citizens on board.

Hong Kong and Singapore have often been placed side-by-side in comparative studies. When it comes to smart city development, Hong Kong can be considered a latecomer, having announced its Smart City Blueprint only in late 2017. It also lacks the well-acclaimed position and branding of Singapore's Smart Nation. However, this does not automatically imply that the city lags behind smart city solutions and their implementations in the urban landscape. As Xin Li, Kyung-Min Nam and Chee Keong Khoo (Chapter 3) explain, Hong Kong had long embraced ICT in various domains of urban development and management. What differentiates its latest smart city vision is how it attempts to bring disparate technology applications together in a comprehensive, system-level and integrated manner, as well as its ostensibly people-centric (as opposed to its previously techno-centric) approach to smart city development. The plan promises to deliver smart mobility, environment, people, economy, government and living to cater to the city's contemporary challenges and needs. As a notable case, the chapter introduces the Kowloon East smart city pilot project, which has been spurred by urban renewal goals of building Hong Kong's second central business district.

Next, Shang-su Wu (Chapter 4) takes us to Taipei, Taiwan. As the author points out, few studies exist on Taipei as a smart city in the literature, even though Taiwan has long been known for its ICT industry. The chapter explains the government's ambition to turn Taiwan into a digital smart island and how decision makers in both the public and private sectors are eager to promote smart Taipei as the way for its next development path by leveraging Taiwan's strength in ICT. Under the policy to develop innovative industries, the chapter discusses how Taipei develops and implements comprehensive smart city plans that meet local needs and challenges. A particularly notable point to highlight in this chapter is how Taipei seizes the smart city movement as an opportunity for city diplomacy to overcome the constraints and pressure it faces. In addition to forming a platform for international cooperation under the smart city banner, the goal is for local companies to showcase and export their smart city technologies.

The case of smart cities in South Korea, especially exemplified by the well-known case of Songdo, has often been portrayed as top-down, infrastructure-driven developments (see, e.g., Carvalho 2015; Anttiroiko 2013). This, however, is an old version of the country's smart city plan, according to Jong-Sung Hwang (Chapter 5). South Korea has been one of the early pioneers of smart cities, beginning in the early 2000s, even before the smart city concept became popular. They were then locally referred to as ubiquitous-cities, and due to lacklustre outcomes in some of these smart city projects, South Korea underwent what Hwang termed the 'smart city winter'. With this experience in mind, the national government recently launched its latest smart city plan as a new development model based on a holistic city-as-a-platform strategy. The chapter provides a comprehensive overview of the evolution of smart city developments in South Korea as well as detailed explanations behind each phase.

Part I ends with the case of Japan – the frontrunner of East Asian economic development. Japan's smart city aspirations place heavy emphasis on *inter alia* energy issues, rather than the all-encompassing aspects of urban life – which, in that regard, sets it apart from the other four smart city cases discussed earlier. Using the innovation systems framework, Masaru Yarime (Chapter 6) introduces the key actors and policies behind Japan's smart city development to tackle the country's persistent concerns over efficient and sustainable energy supply and uses. While such concerns can be traced back to the oil crisis of the 1970s, the agenda received a shot in the arm when the Fukushima accident occurred in 2011. In the case of Japan, smart cities are demonstrations or showcase projects constructed by the Japanese Government and the private sector (i.e., electric power companies and electronic companies). As a form of 'societal experimentation', their purpose is to localise and test out new energy-related technologies and to build business models with multiple stake-holders, while at the same time raising energy and climate change awareness among the public. The chapter also provides some details of policies designed to facilitate innovation in the energy industries and new smart technology adoption by end-users, which could offer useful insights to smart city policy-makers elsewhere.

Smart City Initiatives of Two Asian Giants

China and India are by far the most populous countries in the world. Both have a population of more than 1.3 billion, while the world's third most populous country – the United States – only has a population of roughly 330 million. In terms of size, China and India rank third and seventh in the world, respectively (World Population Review 2019). With their sheer population and land size, China and India no doubt stand out from other countries in Asia. They both share great economic potentials – with China already wielding its economic power in the global economy. At the same time, they are often compared with one another for their different political systems and institutions. This begs the question: how are these two giants going to develop and govern their latest smart city initiatives, and equally important, what are their key motivations?

Souvanic Roy and Tathagata Chatterji (Chapter 7) explore the Smart City Mission launched by India in 2015, also widely known as India's ambitious plan to build 100 smart cities. The authors aptly capture the challenges of India's smart city initiative, describing it as 'implementing big data and a knowledge-intensive urban agenda in a developing country with a large, mostly poor, digitally divided population across diverse geographies and varied institutional landscapes'. The chapter explains how the smart city initiative signals the Indian government's focus on cities as the development engine in the twenty-first century, departing from its decades of rural bias. Seeking to embark on leapfrog development, smart city projects are expected to deliver 'world-class' urban services and trigger further developments in other parts of the city. In addition to introducing India's smart city initiative – including its key agencies, selection processes, implementation and financing – the chapter analyses the policy through the 'good governance framework' and makes a compelling argument that the initiative is resulting in corporatised urban governance in India. In other words, the authors raise the concern over a centralised technocratic governance model that is being reinforced by top-down and outcome-oriented smart city development plans, implemented in localities by newly institutionalised corporation-like agencies called special purpose vehicles.

Equally ambitious, China has also launched hundreds of smart city pilot projects – a scale that is unimaginable for most countries. While China's smart city development aspirations might not have received as much attention and exposure as India's Smart City Mission, its speedy integration of advanced technologies in managing cities and urban life (e.g., the social credit system, cashless payments, AI-enabled transportation network and face-recognition system) is beginning to attract international scrutiny (Ng & Tanu 2019). Xinhui Yang and Lin Ye (Chapter 8) introduce China's smart city endeavours, including a series of supporting policies and pilot projects launched by various ministries, as well as China's latest push towards developing an evaluation index system for its many smart cities. While the chapter points out that a number of current evaluation index systems stress 'economic development and infrastructure construction', it also emphasises Chinese smart cities as a key remedy for solving urban woes and management challenges, reflecting the country's rapid and massive urbanisation. In particular, it explores the traffic congestion problem, introducing the development of the smart transportation system in the Sino–Singapore Suzhou Industrial Park. The chapter thus showcases one of the examples of Singapore 'going global' with its smart city expertise and branding.

Second-tier Cities and Smart City Development

In the last part of the book, we turn our attention to the smart city aspirations of second-tier cities in Asia. Glasmeier and Nebiolo (2016) note that the top smart cities are often national capitals or major metropolitan areas. In Asia – often manifesting in the concentrated development of the national capital or a few key cities – smart cities indeed tend to be national capitals or representative metropolitan cities. However, second-tier cities are not left out of the picture. For example, there are cases in which the smart city concept is applied to a new city (such as Songdo in South Korea). There are also cases in which existing second-tier cities actively seek to get on board the national smart city agenda to boost local development or become selected as pilot cases. Part III highlights some of the issues and local perspectives that arise from these cases.

Two of the three chapters in Part III are on ASEAN countries. There are a total of ten countries in ASEAN, and Indonesia and Thailand (examined herein) are its two largest economies. Besides Singapore, ASEAN comprises developing countries currently undergoing rapid urbanisation with relatively high economic growth rates. At various stages of development, how are smart city policies interpreted, developed and implemented? In addition, these countries differ from East Asian countries in that they comparatively lack the experience of strong national governments guiding and leading their economic and technological developments. In the case of Indonesia, Arif Budy Pratama and Satria Aji Imawan (Chapter 9) explain that it is the local governments that initiate their own smart city developments, amid the absence of a clear conceptualisation of the smart city at the national government level. Accordingly, 'every city has its own understanding of the "smart city" concept, which affects their agenda setting and smart city policymaking', and the authors argue that 'bureaucracy readiness' at the local level is thus a critical factor for smart city projects in Indonesia. Based on interviews with Yogyakarta's key civil servants, they elaborate on the four dimensions of readiness in bureaucracy that have allowed the small secondary city of education and tourism to successfully pursue its smart city agenda, centred on smart culture.

The chapter on Thailand (Chapter 10) by Ora-orn Poocharoen, Poon Thiengburanathum and Kian Cheng Lee highlights the case of another second-tier city, Chiang Mai, which happens to be one of the first three smart city projects in Thailand. (The other two are Phuket and Khon Kaen.) According to the authors, 'the smart city outlook … is viewed as critical for secondary cities with a burgeoning urban population' and to 'help redistribute the unbalanced wealth generation in Thailand, which has been concentrated in Bangkok'. Based on the field experiences of the authors – who have been involved in developing the smart city plan of Chiang Mai – the chapter mainly discusses the city as a complex adaptive system and introduces the multiplicity of urban systems as a framework for smart city projects. While this framework is far from being specific to Thailand, it nevertheless offers a rare glimpse into the developing country's local visions and viewpoints vis-à-vis its smart city development.

The book ends with the case of Songdo – a smart city built from scratch. In addition to criticisms of being a top-down smart city development (Carvalho 2015; Anttiroiko 2013), Songdo has been referred to as a by-product of the South Korean government and companies' attempt to develop a new growth engine and exportable development model (Shwayri 2013; Mullins & Shwayri 2016), combined with the development aspiration of the local government of Incheon (Shin 2016; Joo 2019). Michael Manning, Jill L. Tao and Jae-in Noh (Chapter 11) shed light on yet another angle of analysis. Focusing on the overlap between 'smart' and 'green' cities in the project, the authors carried out a survey with the city's residents and visitors in order to explore their perceptions of the concept of 'green'. The authors find that ordinary citizens tend to perceive 'green spaces' and 'cleanness' as the key attributes of a green city. This differs from the concept envisioned by the national government and how it markets the Songdo project abroad. In fact, the Korean Government's own perception of a 'green city' – which mixes 'sustainability elements and economic development' as a growth model – is in itself catering to 'Korean sensibilities', rather than identifying with general international viewpoints. The authors thus argue that Songdo represents uniquely local renditions of the green and smart city, especially those of the residents, by being 'modern and convenient'. In that regard, it may then warrant a more nuanced evaluation than concurrent critiques of Songdo.

CONCLUSION: GOVERNING DEVELOPMENT IN AN ERA OF HYPER-CONNECTIVITY

The collection of chapters in this edited volume are empirical case studies, with an eye to drawing out relevant narratives and findings of smart cities in the Asian context and policymaking space. In particular, the chapters showcase local academics' views and interpretations of their smart cities and what they consider to be the key issues, which may or may not always complement views from the West. What stands out from our collection of studies is that many smart cities in Asia are part of the governments' new development strategy. Whether governments are searching for the next big development push or just trying not to fall behind in this hyper-connected era, they are embarking on the smart city journey as a key development strategy. At the same time, many national and local governments also appear to be taking it as an opportunity to test out new governance modes or ways in which development is pursued in their locales, leveraging on the ideas being promoted by the smart city concept. Examples such as an emphasis on citizens as co-creators, living labs and system-level integration that have come up in a number of chapters illustrate such efforts. We also note that while there are certainly globally shared aspects of the smart city, what is particularly noteworthy from this collection is local specificity. Just to highlight a few examples: Taipei's attempt to forge international connections via smart city networking; Japan's focus on energy issues; India's use of smart cities for leapfrog development; and Singapore's emphasis on building international cooperation networks and promoting urban solution exports, all reflect goals and approaches that are closely associated with specific local challenges and contexts. According to our cases, the smart city is more of a development strategy in an era of hyper-connectivity than the pursuit of a tech-enabled utopia.

Admittedly, this book is by no means comprehensive, but we do hope that it serves as an important starting point to explore a contextually grounded examination of smart cities, inviting diverse viewpoints and analyses. Although this book touches on a range of smart city development projects in Asia, it could have benefited from more extensive coverage of ASEAN in particular. The contextually grounded study of smart cities in ASEAN, which is rapidly growing in terms of economy and urbanisation, would be meaningful, as their conceptualisation and development of smart cities can differ significantly from those not only in the West but also the tech-savvy frontrunner economies of East Asia. What kind of smart cities these developing countries envision and how they approach their developments definitely merit more detailed examination. Especially with all ten countries of ASEAN having at least one (if not a few) pilot smart cities as part of their ASCN, the dearth of knowledge on these rising smart cities suggests much room for further investigation down the road.

Our book also does not address the outcomes and achievements of smart city projects. How do smart city policies fare as a development strategy? What positive and negative impacts do they have on our societies? Many of these policies have been introduced only recently, and it is yet too early for an accurate evaluation. As they engender more concrete outcomes in the future, it would be wise to revisit and analyse whether they have achieved their development goals and how they have changed our cities. Nevertheless, there is still value in currently studying the local perceptions and approaches of smart cities in order for us to understand the diverse range of conceptualisations and evolutions of the globally popular trend taking place on the ground. We hope that the featured cases, as well as some of the commonly shared themes and insights distilled from the chapters, will benefit not only those readers interested in smart cities and contemporary Asian urbanisation but also practitioners and policymakers looking to gain insights into the specific and actual practices of our much talked about smart cities.

REFERENCES

Allwinkle, S & Cruickshank, P 2011, 'Creating smart-er cities: An overview', *Journal of Urban Technology*, vol. 18, no. 2, pp. 1–16.

Amsden, AH 1989, *Asia's next giant: South Korea and late industrialization*. New York: Oxford University Press.

Angelidou, M 2015, 'Smart cities: A conjuncture of four forces', *Cities*, vol. 47, pp. 95–106.

Anttiroiko, AV 2013, 'U-cities reshaping our future: Reflections on ubiquitous infrastructure as an enabler of smart urban development', *AI & Society*, vol. 28, no. 4, pp. 491–507.

Anttiroiko, AV 2016, 'City-as-a-platform: The rise of participatory innovation platforms in Finnish cities', *Sustainability*, vol. 8, no. 9, article 922.

AT Kearney 2019, 'A question of talent: How human capital will determine the next global leaders', *2019 Global Cities Report*, viewed 28 June 2019, https://www.atkearney.com/global-cities/2019.

Calder, KE 2016, *Singapore: Smart city, smart state*. Washington, DC: Brookings Institution Press.

Caprotti, F, Cowley, R, Bailey, I, Joss, S, Sengers, F, Raven, R, Spaeth, P, Jolivet, E, Tan-Mullins, M, Cheshmehzangi, A & Xie, L 2017, *Smart Eco-city development in Europe and China: Opportunities, drivers and challenges*. Exeter: University of Exeter.

Caragliu, A, Del Bo, C & Nijkamp, P 2009, 'Smart cities in Europe', *Series Research Memoranda 0048*. Amsterdam: Free University Amsterdam, Faculty of Economics, Business Administration and Econometrics.

Carvalho, L 2015, 'Smart cities from scratch? A socio-technical perspective', *Cambridge Journal of Regions, Economy and Society*, vol. 8, no. 1, pp. 43–60.

Castells, M 1992, 'Four Asian tigers with a dragon head: A comparative analysis of the state, economy, and society in the Asian Pacific Rim', in Appelbaum, R & Henderson, J (eds.), *States and development in the Asia Pacific Rim*, pp. 33–70. London: Sage.

Deakin, M 2014, 'Smart cities: The state-of-the-art and governance challenge', *Triple Helix*, vol. 1, no. 7, pp. 1–16.

Deakin, M & Al Waer, H 2012, *From intelligent to smart cities*. London: Routledge.

EU 2014, *Mapping smart cities in the EU*. Brussels: European Union Directorate General for Internal Policies.

Evans, P 1989, 'Predatory, developmental and other apparatuses: A comparative political economy perspective on the third world stat', *Sociological Forum*, vol. 4, no. 4, pp. 561–587.

Forrest, C 2015, 'Photos: 20 of the world's most connected, innovative cities', TechRepublic, 9 April, viewed 28 June 2019, https://www.techrepublic.com/pictures/photos-20-of-the-worlds-most-connected-innovative-cities/.

Foth, M 2016, 'Early experiments show a smart city plan should start with people first', *The Conversation*, 1 June, viewed 11 August 2017, http://theconversation.com/early-experiments-show-a-smart-city-plan-should-start-with-people-first-60174.

Glasmeier, A & Christopherson, S 2015, 'Thinking about smart cities', *Cambridge Journal of Regions, Economy and Society*, vol. 8, no. 1, pp. 3–12.

Glasmeier, A & Nebiolo, M 2016, 'Thinking about smart cities: The travels of a policy idea that promises a great deal, but so far has delivered modest results', *Sustainability*, vol. 8, no. 11, article 1122.

Greenfield, A 2013, *Against the smart city*. New York: DO projects.

Harvey, D 1989, 'From managerialism to entrepreneurialism: The transformation in urban governance in late capitalism', *Geografiska Annaler, Series B: Human Geography*, vol. 71, no. 1, pp. 3–17.

Hollands, RG 2008, 'Will the real smart city please stand up? Intelligent, progressive or entrepreneurial?', *City*, vol. 12, no. 3, pp. 303–320.

Hollands, RG 2015, 'Critical interventions into the corporate smart city', *Cambridge Journal of Regions, Economy and Society*, vol. 8, no. 1, pp. 61–77.

Hwang, JS 2016, 'Seumateusiti baljeonjeonmang-gwa hangug-ui gyeongjaenglyeog', *IT & Future Strategy*, NIA, vol. 6, pp. 1–42.

IBM 2012, 'Smarter more competitive cities', *IBM Smarter Cities: Point of view*, viewed 2 June 2017, http://smartcitiescouncil.com/system/tdf/public_resources/ Smarter,%20more%20competitive%20cities.pdf?file=1&type=node&id=156.

Johnson, CA 1982, *MITI and the Japanese miracle: The growth of industrial policy, 1925–1975*. Stanford, CA: Stanford University Press.

Joo, YM 2019, *Megacity Seoul: Urbanization and the development of modern South Korea*. Oxford and New York: Routledge.

Joss, S, Cook, M & Dayot, Y 2017, 'Smart cities: Towards a new citizenship regime? A discourse analysis of the British smart city standard', *Journal of Urban Technology*, vol. 24, no. 4, pp. 29–49.

Karvonen, A, Cugurullo F & Caprotti, F 2018, *Inside smart cities: Place, politics and urban innovation*. London: Routledge.

Kitchin, R 2015, 'Making sense of smart cities: Addressing present shortcomings', *Cambridge Journal of Regions, Economy and Society*, vol. 8, no. 1, pp. 131–136.

Komninos, N, Pallot, M & Schaffers, H 2013, 'Special issue on smart cities and the future Internet in Europe', *Journal of the Knowledge Economy*, vol. 4, pp. 119–134.

Luque-Ayala, A & Marvin, S 2015, 'Developing a critical understanding of smart urbanism?', *Urban Studies*, vol. 52, no. 12, pp. 2105–2116.

Mani, N 2016, *Smart cities and urban development in India*. New Delhi: New Century Publications.

Mullins, PD & Shwayri, ST 2016, 'Green cities and "IT839": A new paradigm for economic growth in South Korea', *Journal of Urban Technology*, vol. 23, no. 2, pp. 47–64.

Nam, T & Pardo, T 2011, 'Conceptualizing smart city with dimensions of technology, people, and institutions', in *The Proceedings of the 12th Annual International Conference on Digital Government Research*, pp. 282–291. New York: ACM.

Ng, D & Tanu, E 2019, 'From dispensing toilet paper to shaming jaywalkers, China powers up on facial recognition', *Channel NewsAsia*, 2 February, viewed 28 June 2019, https://www.channelnewsasia.com/news/cnainsider/shaming-jaywalkers -china-facial-recognition-technology-privacy-11196684.

Robinson, J 2002, 'Global and world cities: a view from off the map', *International Journal of Urban and Regional Research*, vol. 26, no. 3, pp. 531–554.

Robinson, J 2006, *Ordinary cities: Between modernity and development*. London: Routledge.

Shin, HB 2016, 'Envisioned by the state: Entrepreneurial urbanism and the making of Songdo City, South Korea', in Datta, A & Shaban, A (eds.), *Mega-urbanization in the Global South: Fast cities and new urban utopias of the postcolonial state*, pp. 83–100. Oxon: Routledge.

Shwayri, ST 2013, 'A model Korean ubiquitous eco-city? The politics of making Songdo', *Journal of Urban Technology*, vol. 20, no. 1, pp. 39–55.

Söderström, O, Paasche, T & Klauser, F 2014, 'Smart cities as corporate storytelling', *City*, vol. 18, no. 3, pp. 307–320.

Townsend, A 2013, *Smart cities: Big data, civic hackers, and the quest for a new utopia*. New York: WW Norton & Co.

Vanolo, A 2014, 'Smartmentality: The smart city as disciplinary strategy', *Urban Studies*, vol. 51, no. 5, pp. 883–898.

World Population Review 2019, 'Largest countries in the world by area 2019', viewed 28 June 2019, http://worldpopulationreview.com/countries/countries-in-world-by -area/.

Yigitcanlar, T & Lee, SH 2014, 'Korean ubiquitous-eco-city: A smart-sustainable urban form or a branding hoax?', *Technological Forecasting and Social Change*, vol. 89, pp. 100–114.

Yigitcanlar, T, Kamruzzaman, LB, Ioppolo, G, Sabatini-Marques, J, da Costa, EM & Yun, JH 2018, 'Understanding "smart cities": Intertwining development drivers with desired outcomes in a multidimensional framework', *Cities*, vol. 81, pp. 145–160.

PART I

Smart cities of the four Asian Tigers and Japan

2. The Smart Nation: unpacking Singapore's latest mega-digitalisation push

Yu-Min Joo, Teck-Boon Tan and Ming-Yee Foo

INTRODUCTION

In November 2014, Singapore unveiled the multibillion-dollar Smart Nation initiative – a mega-digitalisation project that aimed to infuse cutting-edge information and computing technology into almost every aspect of citizens' lives. Ever since, the city-state has come to be known internationally as a leading smart city. The reality is that Singapore has always been an early and avid adopter of information and communication or infocomm technologies to transform its economy, society and government. Since the early 1990s, it has launched plans (e.g., Intelligent Island and Intelligent Nation) that have laid a strong foundation for the country's development into a smart nation today. In that regard, the Smart Nation initiative is more than just a trendy policy dovetailing on the global smart city boom. Launched by Prime Minister Lee Hsien Loong, it is an ambitious national development strategy of the Singapore Government – one that aims to apply a new set of technology-oriented development strategies after Singapore successfully achieved its development status as a global city.

In this chapter, we underline why the Smart Nation initiative should be understood from the perspective of a developmental state, of which Singapore has always been one. This contextually grounded perspective allows for a more accurate understanding of the country's latest mega-digitalisation push – a critically important lens, not least because the smart city concept has often been criticised as a form of neoliberalism.[1] As a case in point, the literature has repeatedly referred to smart city projects as a neoliberal strategy – an urban stratagem that is 'sold, resold, modified or augmented to make money' (Glasmeier & Christopherson 2015, p. 6), profiting global corporations and capitalists rather than citizens (Greenfield 2013; Hollands 2008; Vanolo

2014). Such criticisms fail to note the fact that, although global neoliberalism has spread to the newly industrialised economies of East Asia, it has assumed a variety of forms, combining with state developmentalism – an ideology associated with, among other things, active state intervention, a determined governing elite and a singular focus on economic development (Huff 1999; Leftwich 1994; Park et al. 2012).

In Singapore, state intervention has played a leading role in economic and social development since its independence in 1965. To this day, the government continues to intervene extensively in the country's development, albeit in the hybridised form of a 'neoliberal-developmental state' (Liow 2012). With that in mind, we argue that it is this developmental state mentality that has led the Singapore Government to again intervene in the nation's development through the Smart Nation initiative. This view fits in squarely within the East Asian tradition of 'big pushes' – investing heavily in overhead capital to accelerate national development (Shin 2005).

For a more nuanced understanding of the Smart Nation as a state-driven development strategy, we first provide a brief overview of Singapore's development and its past digitalisation policies for context. We then unpack the Smart Nation initiative, including the government's key motivations, its governing structure and the notable characteristics of related projects. In the subsequent section, we look into the autonomous vehicle (AV) living lab to illustrate what the government's visions for Smart Nation are and, more importantly, how it plans to achieve them. We also briefly explore the major hurdle for the Smart Nation initiative before concluding this chapter with a discussion of some policy implications for Singapore and challenges for other governments looking to replicate the model.

SINGAPORE'S DEVELOPMENT AND TECHNOLOGICAL PURSUITS: A BRIEF OVERVIEW

It can be argued that Singapore's economic development in the latter half of the twentieth century was closely tied to an ideology of national survival (Loh 1998). Since Singapore earned self-rule from the British in 1959 and its merger attempts with Malaysia failed in 1965, the government, led by the People's Action Party (PAP) – which continues to dominate Singaporean politics today – has been preoccupied with ensuring the economic survival of the island-state. With an economy revolving around entrepôt trade and few industries to speak of (Lim 2002), Singapore recognised very early on the need to be open to foreign direct investments and adopted a non-discriminatory policy towards multinational corporations (MNCs; see Goh 2016). Its leaders also recognised that technology is a key enabler in industrialisation and modernisation (Lim 2001) and that it could bring managerial and technical 'know-how' into the country by attracting MNCs. With the goal of achieving long-term economic progress, the government assumed an active role in this strategy, whether it was courting MNCs with tax and investment incentives, revamping the education system to produce better-trained workers, providing a generous infrastructure to support industrial activities or taming testy relations between labour unions and employers (Goh 2016; Huff 1999; Sebastian 1997). When other countries started replicating this low-cost strategy, the Singapore Government switched gears and began promoting high-tech industries to create more value-added jobs and services (Pereira 2008). It also pushed for a regionalisation pro-gramme to capitalise on the rapid economic growth in other Southeast Asian countries (Yeung 2000). This developmental state model was instrumental to Singapore's economic and national development, and much of it can still be seen in post-industrialised Singapore today (Castells 1992; Yeung 2000).

In a nutshell, Singapore's survival as a nation-state hinged on its ability to adopt technology and 'go global'. The idea was promoted in a 1972 speech by then Minister for Culture, Sinnathamby Rajaratnam, in which he highlighted Singapore as a 'global city' and described it as a 'child of modern technology' (Rajaratnam 1972, p. 230). 'Going global' for Singapore not only entailed opening doors to attract MNCs. It implied more proactive international rela-tions and solidifying Singapore's position in the global arena. According to Rajaratnam (1972), a future could be envisioned in which cities become the nexus of interaction between nation-states and they 'reach out to one another through the tentacles of technology' (p. 231).

During the 1980s, the government realised that Singapore could no longer rely on low-cost manufacturing industries to sustain the country's economic competitiveness and so technology began to play an even more important role in Singapore's development. It was then that the Singapore Government sought to establish the city-state as a 'regional centre for computer software and services' (Tan 1983, p. 1) by launching the National Computerisation Plan in 1981 and then the National IT Plan in 1986 (see Table 2.1 for a brief summary of major technology initiatives launched in Singapore).

A more ambitious project – the Intelligent Island – was launched in 1992, following the release of the IT2000 report. The Intelligent Island aimed to expand the use of information technology (IT) across Singapore with a goal to ultimately improve the quality of life for Singaporeans. It focused on four key areas: transportation, government and business services, cash transactions, and culture and leisure (Teo & Lim 1999). Based on this initiative, in 1999 – long before the term 'smart city' was used – Mahizhnan published a journal article titled 'Smart cities: The Singapore case' (Mahizhnan 1999), showcasing the advanced IT infrastructure covering the entire city-state. The image of a technology-enabled utopian future for Singapore in the new millennium was

Table 2.1   Summary of Singapore's major technological initiatives prior to Smart Nation

Year	Major initiative	Key directions and policies
1981	National Computerisation Plan	To improve public administration and public service through computerisation.
1986	National IT Plan	To extend computerisation to the private sector to ramp up efficiency and competitiveness. Developed pipeline for the national information and broadband infrastructure. Identified IT as a new growth industry.
1992	A Vision of an Intelligent Island Report: IT2000	A more comprehensive plan to expand the use of IT in transportation, government and business services, cash transactions, and culture and leisure. Launched Singapore ONE (national multimedia broadband network). Other key projects included Electronic Road Pricing, Singapore Infomap and TradeNet.
2006	Intelligent Nation 2015	To develop Singapore as a global leader in infocomm and e-governance, with an emphasis on innovation, integration and internationalisation. Pushed for the island-wide installation of broadband networks.

Source: Mahizhnan & Yap 2000; Hornidge 2010; Teo & Lim 1999.

captured in a narrative where Singapore was framed as the 'home of the future' (Lim 2001, p. 187).

Extending the concept of Intelligent Island, Intelligent Nation 2015 was launched in 2006 as a ten-year strategy to help Singapore fully transform into a 'global city' (Hornidge 2010), in tandem with the rise of the ubiquitous Internet during the global digital revolution. The plan's roadmap was guided by the three goals of 'innovation, integration and internationalisation', which included plans to increase infocomm exports, create jobs, increase the value of the infocomm industry and further expand its broadband infrastructure coverage (IMDA 2006).

This brief examination of Singapore's past digitalisation and IT-related plans suggests that Singapore was already a smart city, even before the Smart Nation initiative was launched in 2014. In fact, academic literature, even recent publications, on the topic of smart cities has mentioned or discussed the Intelligent Nation 2015 (e.g., Angelidou 2017; Glasmeier & Christopherson 2015) and the Intelligent Island (or IT2000 plan) (e.g., Hollands 2008; Mahizhnan 1999; Vanolo 2014) as Singapore's case. Singapore, based on its previous masterplans, has already been framed by the scholars as a smart city. The Smart Nation initiative can thus be considered an extended version of the Singapore Government's ongoing plans to stay on top of advances in digital technology and achieve a techno-utopian vision for the city-state. In other words, the Singapore Government – in addition to its goals of attracting foreign capital and building an efficient and politically stable society conducive to global businesses – has long been planning to leverage on technology to propel the country to the next stage of its development. The Smart Nation initiative is a cumulation of such efforts.

At the same time, the Smart Nation stands out from the previous digitalisation plans, as it is a reflection of the developmental state's attempt to ride the waves of the Fourth Industrial Revolution – a major technology revolution that will see the convergence of cutting-edge technology like artificial intelligence, biotechnology and robotics. The Singapore Government has revised the country's economic development strategies several times in keeping with international trends and domestic needs, and the Smart Nation initiative is the country's latest agenda. Rather than simply being the next digitalisation plan for Singapore, the Smart Nation is the next big push for economic and national development in today's fast-changing, technology-driven world. The following section explains in more detail the Smart Nation initiative as Singapore's latest and the most important development agenda.

UNPACKING THE SMART NATION INITIATIVE

Key Motivations

Singapore's multibillion-dollar Smart Nation initiative has two primary goals. One is to provide a transformational solution to a complex set of policy challenges confronting the country, including a rapidly aging population,[2] land scarcity, climate change and international terrorism. The idea is to leverage recent advances in technology, such as the Internet-of-Things, artificial intelligence, virtual reality, robotics and big data, to tackle this broad spectrum of issues. As such, the Smart Nation goes beyond improving 'connectivity' and 'efficiency' – the buzzwords of Singapore's past digitalisation policies. It feeds off the wave known as the Fourth Industrial Revolution, in which hyper-connectivity and technological advancements make it possible for data to be mined much faster and deeper than previously imagined, thus allowing for predictive analytics to further customise public services at an individualised level (Centre for Public Impact 2017; Chang & Kannan 2008).

The second goal of the Smart Nation is to develop the national economy. As a government press release makes clear, in addition to improving the quality of life, the Smart Nation initiative is 'most importantly ... about transforming Singapore's economy, and ensuring that there will be jobs and opportunities' (Smart Nation Digital Government Office (SNDGO) 2018a). An excerpt from

Prime Minister Lee Hsien Loong's 2014 Smart Nation speech clearly show-cases the Singapore Government's motivation:

> We are a leading city today but other leading cities like San Francisco, New York, London, Sydney [and] Shanghai ... are attracting capital, talent [and] ideas. They are building outstanding urban environments. They are pulling ahead of the rest of the pack ... We have to move ahead with them and stay up there amongst the leading cities of the world. (Lee 2014)

In 2017, the prime minister further commented that the Smart Nation is about Singapore taking full advantage of IT 'to create new jobs [and] new business opportunities, [and] to make our economy more productive' (Lee 2017). The Smart Nation thus reflects the Singapore Government's goal for the country to stay relevant and competitive in the global economy and generate economic growth as it keeps pace with the latest technological advancements.

Related to this economic development goal, the Smart Nation also aims to increase Singapore's visibility on the global stage, particularly amid the worldwide boom in smart city development. As a small nation with limited resources, Singapore is well aware of the importance of being a useful ally of international and regional powers, a central tenet of the country's foreign policy (Balakrishnan 2016). Reinforcing its 'going global' mind-set, the Singapore Government expects to expand the mindshare of its policies, promote goodwill and foster more economic partnerships with relevant coun-tries and cities through the Smart Nation initiative. In addition to a number of memoranda of understanding signed to help build smart city projects in China and India, Singapore led the formation of the ASEAN Smart Cities Network (ASCN) in 2018. All these strategic moves suggest that Singapore is seeking to cultivate its image and reputation as a regional (if not global) leader in smart city development and to actively play a role in the global society, while at the same time seizing economic opportunities in rapidly urbanising countries in the region. Reflecting the importance of 'going global' with the Smart Nation initiative, the Minister for Foreign Affairs, Dr Vivian Balakrishnan, leads the mega-digitalisation project.

Centralised Government Structure

With key developmental goals on the line, the Smart Nation is directed by a special body – a governing elite, so to speak – from within the government. When the initiative was first unveiled in 2014, the office created to oversee it – the Smart Nation Programme Office – was part of the Prime Minister's Office (PMO). In 2017, it was decided that the initiative was not progressing quickly enough (Tham 2017), and to speed things up, a new office – the Smart

Nation Digital Government Group (SNDGG) – was established to consolidate the planning and implementation of the Smart Nation initiative (see Figure 2.1). The SNDGG is comprised of two organisations – the SNDGO and the Government Technology Agency of Singapore (GovTech). Comprised of staff from various government ministries and offices, the SNDGO is responsible for planning and prioritising key Smart Nation projects, whereas GovTech is tasked with implementing them (SNDGO 2019).

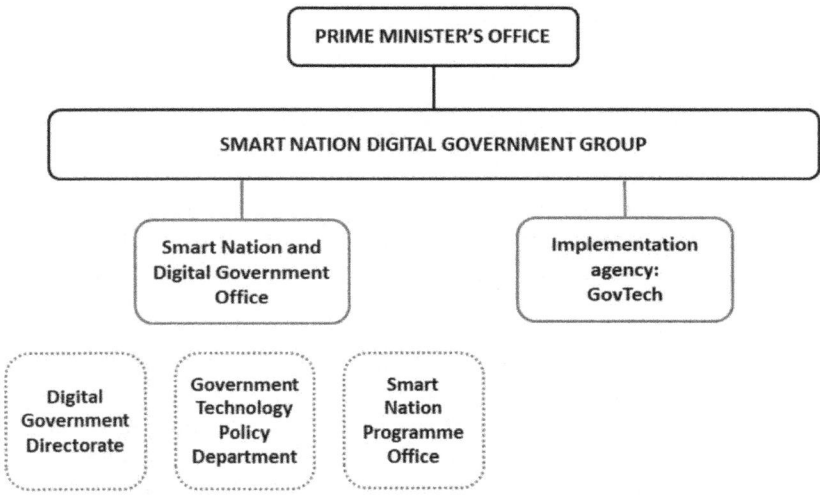

Source: Prime Minister's Office 2017.

Figure 2.1   Governance structure of the SNDGG

Placing the SNDGG under the PMO means that the Smart Nation initiative remains under the direct aegis of the prime minister. More significantly, the move is also strikingly different from the approach taken in previous plans (i.e., Intelligent Island and Intelligent Nation). Further strengthening the government's oversight of the Smart Nation, the SNDGG is overseen by a powerful ministerial committee consisting of various senior ministers (Nasir 2017). By having top ministers with key appointments in charge, the hope is that Smart Nation projects will be accelerated within their respective ministries. In many respects, this new powerful body overseeing the Smart Nation initiative is reminiscent of the governing elite in the developmental state during the country's earlier push for industrialisation and subsequent economic restructuring. In today's context, however, the move solidifies the Smart Nation's position as the government's latest policy endeavour and, equally important, is illustrative of the government's commitment to economic growth through long-term planning.

Key Characteristics of Smart Nation Projects

In addition to its highly centralised governing structure, the comprehensive list of Smart Nation projects further demonstrates the Smart Nation's position as a major government-driven development agenda. What particularly stands out in the Smart Nation initiative is the scale and depth of its technological development and adoption. Unlike previous digitalisation masterplans, which mostly focused on improving efficiency in the public service and economy, the Smart Nation covers a broad spectrum of urban life, going so far as to cater to an individual's life-cycle and needs.[3] Its key initiatives include strategic national projects that build nationwide platforms for delivering improved and enhanced services to citizens and businesses as well as for urban planning processes. There are also more specific initiatives to support urban living, transportation, health, digital government services, and startups (Table 2.2).

Despite the centralised governing structure of the Smart Nation, there is an accompanying desire on the part of the Singapore Government to co-design and co-create solutions with citizens and other stakeholders. This is evident in the government's provision of open datasets and apps – a practice that, one can argue, is novel for Singapore. Hoe (2016) thus defined the Smart Nation as 'an effort by the government to co-create innovative people-centric solutions with industr[ies] and citizens' (p. 327). In short, there is a conscious effort to invite citizens and businesses to actively engage and participate in Smart Nation projects, not only in terms of finding urban planning and management solutions together but also in growing high-tech industries targeted by the government.

Indeed, quite a number of Smart Nation projects are geared towards preparing Singapore for a future of improved urban living assisted by technology. The Singapore Government has not only rolled out nationwide infrastructure-related platforms such as Virtual Singapore (a 3D-modelling platform) and the Smart Nation Sensor Platform (an integrated platform to improve urban services and security), but also launched projects such as TeleHealth (a project to deliver healthcare remotely) and AVs that could transform the way Singaporeans live. The futuristic visions encompassed in the Smart Nation are not a surprise considering the Singapore Government's forward-looking mentality. In the words of Prime Minister Lee Hsien Loong, 'the ethos of our society is rational, technological and forward-looking. We have always looked ahead' (Lee 2015). Both this mentality of looking ahead and the targeting of sunrise industries under the Smart Nation are part of the

Table 2.2　　　*List of Smart Nation initiatives and key projects*

Initiatives	Projects
Strategic national projects	Core Operations Development Environment and eXchange (CODEX)
	E-Payments
	Moments of Life
	National Digital Identity (NDI)
	Smart Nation Sensor Platform
	Smart Urban Mobility
Urban living	Automated meter reading
	Drones to survey dengue hotspots
	MyEnv app
	OneService app
	Planning for our people and businesses
	Smart Elderly Alert Systems
	Smart Towns
	Virtual Singapore
Transport	AVs
	Contactless fare payment for public transport
	On demand autonomous shuttles
	Open data for analytics and urban transportation
	Spearheading research in standards for self-driving vehicles (SDVs)
Health	Assistive technology and robotics in healthcare
	Health hub
	National Steps Challenge and the Healthy 365 app
	TeleHealth
Digital Government Services	Business Grants portal and LicenseOne portal
	CentEx
	HDB Resale portal
	Multilingual Digital Services
	Parents Gateway app
Start-ups and Businesses	CorpPass
	Data Innovation Programme Office (DIPO)
	FinTec Sandbox
	Networked Trade Platform
	Punggol Digital District (PDD)

Note: Please see www.smartnation.sg for the details of each project.
Source: Data from the Smart Nation website (Smart Nation Singapore 2019).

Singapore Government's national development plans. The AV living lab is a case in point and is discussed in more detail next.

AUTONOMOUS VEHICLE LIVING LAB CASE

The AV living lab is a prime example of what the Singapore Government is aiming to achieve through the Smart Nation initiative. Exemplifying how the Smart Nation is ultimately a concerted attempt to recalibrate and upgrade the economy by targeting sunrise industries, the Singapore Government has transformed the entire country into a living lab where novel smart city technologies can be tested, evaluated and fine-tuned in a realistic setting. The world is a complex and dynamic place, and for that reason novel technologies tested in simulated environments often do not meet the rigors and standards of real-world scenarios. Using the living lab concept, the Singapore Government seeks to attract promising companies to establish operation in the city-state. The selling point of Singapore is that in this real-life laboratory, technologies under development can be put through a series of rigorous tests to prepare them for wider implementation down the road. Under this living lab concept, Singapore is, in a way, a petri dish for incubating innovative technologies – a test setting in which emerging systems can be tested and refined under complex and realistic conditions (Lee 2013).

One of the most striking examples of this living lab in Singapore is nuTonomy. A US-based software startup founded in 2013, nuTonomy is the first private company to secure the Singapore Government's approval to conduct road tests for AVs in the country (SNDGO 2018b). Although much smaller than well-known AV giants such as Uber and Waymo, nuTonomy nevertheless has big plans. According to its website, nuTonomy is aiming to ease traffic conditions not just in the US and Singapore but across the world by developing a globally applicable AV (nuTonomy 2018). With the AV industry projected to be worth more than US$550 billion by 2026, nuTonomy stands to gain considerably if its vision of an AV that can be used worldwide is fulfilled (Garsten 2018).

With many of its engineers, software developers and research scientists located in Singapore, nuTonomy's value to Singapore goes beyond that of a commercial self-driving taxi service in the city-state – a plan that nuTonomy has already set in motion. Similar to how foreign MNCs had supported Singapore's industrialisation, nuTonomy's strong presence in Singapore today serves to upgrade the economy, just as leading-edge technologies such as artificial intelligence, big data and advanced robotics are defining a new global economic order. By hosting high-tech companies such as nuTonomy, Singapore stands to gain the kind of technical know-how it needs to compete in an increasingly tech-driven world. At the same time the Singapore

Government can also cultivate an image of Singapore as a global leader in smart city solutions.

Although this calculated strategy of positioning Singapore as a living lab is somewhat of a departure from earlier periods of industrialisation led by the developmental state, the idea is fundamentally similar: the assumption is that the government must take on an active role to support the economy by working closely with industries and companies deemed strategically beneficial to the country. In that regard, the government has offered strong political support and the right regulatory environment for nuTonomy to thrive. Meanwhile, at a more granular level, the government has provided generous infrastructure support so nuTonomy could test its fleet of AVs. Aside from designating the sizable One-North Research and Business Park as a test site, the government has also installed special closed-circuit televisions to help monitor the AVs while they are on the road, and high-tech beacons to direct them (EDB 2017). It is worth noting that these incentives came on top of funding support from EDBI, which is a dedicated corporate investment arm of the Singapore Economic Development Board (nuTonomy undated). Whereas private companies drive similar AV projects in the West, what is fundamentally different in the case of Singapore is that the government is firmly in the driver's seat.

In addition to the goal of incubating a sunrise industry, the promotion of AVs helps build a next-generation transportation system for Singapore. As revealed in the Smart Nation initiative, the idea is that, in the next decade or so, one will be able to hail an AV with a smartphone, use it to get to the nearest train station and then take another AV to get to the final destination after getting off the train (Sim 2018). In addition to solving this 'first and last mile' problem, a fleet of shared AVs available on demand can, in theory, reduce road congestion, optimise public transit use and cut down on pollution. Indeed, a fleet of shared AVs is expected to reduce the number of vehicles in Singapore by as much as two-thirds (Matheson 2016), turning it into a car-light society. In land-scarce Singapore, that will also reduce pressure on the government to build more roads and parking lots. For these reasons, even when doubts over the technology emerged after a nuTonomy AV hit a truck during a routine road test in 2016, the government's plan to incorporate AVs into the country's transportation ecosystem was not derailed (Lin 2016).

Another important reason for incorporating AVs into the transportation ecosystem of the Smart Nation is that the technology can be adapted to meet a range of urban needs. For example, if a fleet of autonomous cargo trucks operates only at night, the streets will become less congested during the day for commuters and drivers. Beyond cargo trucks, Singapore is also aiming to apply AV technology to buses. Without having to stop for the driver to take breaks, autonomous buses can be optimised to reduce waiting times and overcrowding. Likewise, autonomous taxis, which nuTonomy is currently testing on the road, will become an integral part of this next-generation transportation ecosystem. Aside from being safer, autonomous taxis are also likely to be cheaper than human-driven ones, as driver compensation typically accounts for a sizable portion of the fare. Going further, there are plans to tap AV technology for public sanitation works such as waste collection and road sweeping in the Smart Nation (SNDGO 2018b).

However, this high-tech urban vision of the Smart Nation also implies a certain degree of social disruption: specifically, many taxi, truck and bus drivers will no longer be needed. An argument could be made that new jobs will be created for displaced workers due to improvement in productivity associated with AV technology. However, the fact is that many drivers face losing their livelihoods with uncertain prospects for transitioning into new high-quality jobs. Indeed, a survey by Randstad – an MNC specialising in human resources services – found that about one in five Singaporeans were fearful of losing their jobs due to technological displacement (Williams 2017). For the Smart Nation to be embraced by the country, it needs to be inclusive of all citizens regardless of age, income and education. The Smart Nation cannot be inclusive, if certain segments of the population are severely disadvantaged by it. Indeed, it would defeat the whole idea of the Smart Nation as an initiative to improve the overall quality of life in the city-state.

To be sure, many legal and technical hurdles remain before AVs can be deployed in any considerable number in the Smart Nation. Yet, because the Singapore Government is in a position to plan long-term and has been actively intervening in a broad spectrum of policy issues in the country, it is well-placed to introduce a coherent set of policies to tackle difficult problems and fallouts. The ability to engage in long-term planning is especially crucial in the case of AVs because the gestation period before these systems become widely applicable can be lengthy and costly. Singapore, as a developmental state, is in a relatively good position to realise this vision of a techno-utopian society, because of the government's relative freedom to plan ahead and, when necessary, provide strong policy backing.

A ROADBLOCK FOR THE SMART NATION: BRINGING CITIZENS ON BOARD

Despite the various advantages the Singapore Government has when it comes to forging ahead with the Smart Nation as its next big development agenda, it faces one major hurdle: bringing citizens on board and deepening their participation. As mentioned earlier, the Singapore Government now considers citizens as co-creators of the Smart Nation (Balakrishnan 2016). In that regard, it has been releasing data to the public in the hope that Singaporeans will contribute meaningful ideas to Smart Nation development. The most striking example is the publication of transportation data to help promote the development of third-party transportation apps for reducing traffic congestion (SNDGO 2018c). This is quite a departure from the past, when the population was idealised as a source of labour to feed the country's state-planned industrialisation (Pereira 2008). The dynamic between the government and its people has been turned on its head, with citizens now considered as joint creators in the Smart Nation initiative. Despite the government's attempt at citizen engagement, however, ordinary Singaporeans appear to be somewhat unenthusiastic. Deepening citizen participation as Singapore embarks on the journey to become a Smart Nation remains a challenge at this point.

The bottom line is that for a nationwide project like the Smart Nation to become a reality, full support from the population is important, maybe even critical. Although the Smart Nation has solid backing from the government, there is anecdotal evidence indicating that Singaporeans are not very excited about it. As a case in point, a 2016 pilot project to introduce smart homes to about 3,000 households in a mature public housing estate saw only about 50 sign-ups (Tham 2017).[4] One reason for this low level of citizen participation may be that the Smart Nation – as a vision – is too abstract and, as a result, has failed to resonate with Singaporeans. Such apathy on the part of citizens only holds back government plans for wider implementation of Smart Nation solutions since quite a few related projects are meant to improve the lives of citizens (Kong & Woods 2018). If citizens do not come on board, then these Smart Nation projects will have little meaning.

Insofar as the Smart Nation initiative is a worthwhile pursuit for the country, its practical relevance to ordinary citizens has neither been properly communicated nor explained – at least not to a degree that resonates with them. This point is especially salient today, when technology is increasingly being perceived as disruptive. As automation threatens jobs and fake online content erodes trust in governments, active citizen engagement is even more important to underline the benefits of modern technology (see, e.g., Hio 2018; Williams 2017). Through the survival ideology, Singapore's industrialisation was concretised in a way that citizens could readily comprehend the dire consequences if the Singapore economy were to falter. With the population behind the plan to industrialise the economy, the government was able to focus on delivering economic growth. The Smart Nation similarly captures the state's inclination to plan ahead and keep Singapore relevant in the global economy amid rapid technological advancements. Yet, boasting one of the highest living standards in the world today, Singapore is already a developed global city-state with a gross domestic product (GDP) per capita close to US$64,000 in 2018 (Singapore Department of Statistics 2019). In that context, the survival ideology no longer seems to resonate strongly with the citizens of this highly developed economy and, despite the emphasis on job creation made by the Singapore Government, some citizens remain apprehensive of ongoing changes (Cheng 2019).

Local enthusiasm for Smart Nation is critical, not only to facilitate policy implementation but also because active citizen engagement leads to technologies that meet the needs of users more adequately. It is useful to bear in mind that the main beneficiaries of the Smart Nation are the citizens, so it is vitally important that the Smart Nation solutions implemented meet their needs. This is why tech manufacturers conduct extensive market studies at the design phase to ascertain the specific requirements of potential users for their products. In many respects, technology implementation in the absence of active user engagement during the design phase is tantamount to taking a shot in the dark – an attempt to push for technology adoption under great uncertainty and without adequate knowledge. For that reason, user interaction at the design phase reduces the risk of offering something to those who have no need for it (Carvalho 2015). Without knowing what citizens want, building a Smart Nation that is inclusive and welcomed by its intended beneficiaries will not be easy.

Under these circumstances, the Singapore Government can give more attention to rebranding the Smart Nation development agenda for citizens, now that the government's rebranding efforts have been largely successful internationally. This can involve more frequent policy communications on the ways in which the Smart Nation can benefit the daily lives of Singaporeans from different walks and stages of life, including specific stories and evidence of citizens becoming more empowered through the use of technology. When backed by carefully designed policies that showcase the Smart Nation as, ultimately, a public good, such efforts will increase local enthusiasm for the mega-digitalisation project.

CONCLUSION

This chapter examined Singapore's latest mega-digitalisation push – the Smart Nation initiative. Amid the global smart city boom, the Smart Nation is undoubtedly a major national development strategy of the Singapore Government, as Singapore looks for a new development agenda after achieving global city status. In many ways, the Smart Nation is the next 'big idea' that Singapore can work towards in the coming decade. By discussing the AV living lab concept, this chapter also illustrated how the PAP government sought to meet development goals through active state intervention and long-term planning – all hallmarks of a developmental state.

In that regard, the Smart Nation should be understood not as one of many trendy policies to build a smart city, but as a dedicated nationwide policy goal of a thriving developmental state that is endowed with decades of accumulated policy know-how and expertise. Furthermore, as explained in this chapter, the Smart Nation is built on the foundation of years of technological innovations and digitalisation masterplans set in motion long before the smart city concept became popular. For that reason alone, the Smart Nation initiative may not be easily replicated by other governments looking to jump-start their economic development with smart city projects. There are certainly lessons to be learned from Singapore's experience, but a more accurate understanding of the institutional setting behind the Smart Nation initiative and the history on which it is built should come first before embarking on any mega-smart city projects.

It is also useful to note that the Smart Nation has its own set of challenges that may not be easily resolved. Distinguishing itself from the country's past economic development policies, the Smart Nation now underscores citizens as co-creators. Ordinary Singaporeans, however, appear to be disconnected from the Smart Nation initiative. If the goal is to build a nation that is not only 'smart' but at the same time inclusive and sustainable, deepening citizens' engagement is imperative. Indeed, a lesson that we can draw from Singapore's industrialisation is that it was implemented in a way that struck a chord with the population at that time. Singaporeans rallied behind the country's industrialisation, as the narrative resonated with them. Regrettably, it appears that what makes the Smart Nation appealing to Singaporeans today remains somewhat nebulous at this point.

NOTES

1. At the risk of over-simplifying, neoliberal governments are generally aligned with the logic of a free market and competition to facilitate the interests of private capital – oftentimes at the expense of the public (Ostry et al. 2016).
2. Singapore is rapidly aging and will see a near-doubling of the number of seniors – aged 65 and above – from about 516,692 (13 per cent of population) to an estimated 900,000 (25 per cent of population) between 2017 and 2030 (Singapore Department of Statistics 2018; Population.sg 2016).
3. For example, Moments of Life is a platform that seeks to provide integrated and continuous services to citizens at different stages of their life.
4. Regarding the monitoring system for the elderly as part of smart home technology, it was reported that many seniors rejected the system because they 'did not want to be seen as vulnerable and useless' and feared 'losing their privacy' (Tham 2017).

REFERENCES

Angelidou, M 2017, 'The role of smart city characteristics in the plans of fifteen cities', *Journal of Urban Technology*, vol. 24, no. 4, pp. 1–28.
Balakrishnan, V 2016, 'Innovative solutions for a smart city', *Smart Nation Singapore*, 13 July, viewed 28 May 2019, https://www.smartnation.sg/whats-new/speeches/innovative-solutions-for-a-smart-city-conference-at-world-cities-summit-2016.

Carvalho, L 2015, 'Smart cities from scratch? A socio-technical perspective', *Cambridge Journal of Regions, Economy and Society*, vol. 8, pp. 43–60.

Castells, M 1992, 'Four Asian tigers with a dragon head: A comparative analysis of the state, economy, and society in the Asian Pacific Rim', in Appelbaum, R & Henderson, J (eds.), *States and development in the Asia Pacific Rim*, pp. 33–70. London: Sage.

Centre for Public Impact 2017, 'Destination unknown: Exploring the impact of Artificial Intelligence on government', *Working Paper: Artificial Intelligence and the Future of Governance*, viewed 28 May 2019, https://publicimpact.blob.core.windows.net/production/2017/09/Destination-Unknown-AI-and-government.pdf.

Chang, AM & Kannan, PK 2008, 'Leveraging Web 2.0 in government', *The IBM Center for the Business of Government*, viewed 28 May 2019, http://www.businessofgovernment.org/report/leveraging-web-20-government.

Cheng, K 2019, 'In the pursuit of technology, what happens to workers left behind?', *Channel News Asia*, 5 March, viewed 18 June 2019, https://www.channelnewsasia.com/news/singapore/pursuit-of-technology-what-happens-to-workers-skills-11309094.

Department of Statistics, Singapore 2018, *Population, Annual*, viewed 28 May 2019, from: http://www.tablebuilder.singstat.gov.sg/publicfacing/createDataTable.action?refId=14911.

EDB 2017, 'Asia's testbed for smart cities', *Economic Development Board*, viewed 28 May 2019, https://www.edb.gov.sg/en/news-and-resources/insights/innovation/asia-s-testbed-for-smart-cities.html.

Garsten, E 2018, 'Sharp growth in autonomous car market value predicted but may be stalled by rise in consumer fear', *Forbes*, 13 August, viewed 28 May 2019, https://www.forbes.com/sites/edgarsten/2018/08/13/sharp-growth-in-autonomous-car-market-value-predicted-but-may-be-stalled-by-rise-in-consumer-fear/#4b7a4f2b617c.

Glasmeier, A & Christopherson, S 2015, 'Thinking about smart cities', *Cambridge Journal of Regions, Economy and Society*, vol. 8, pp. 3–12.

Goh, CB 2016, *From traders to innovators: Science and technology in Singapore since 1965*. Singapore: ISEAS Yusof Ishak Institute.

Greenfield, A 2013, *Against the smart city*. New York: DO projects.

Hio, L 2018, 'Information from 65k Singapore users in hands of Cambridge Analytica', *The Straits Times*, 6 April, viewed 28 May 2019, https://www.straitstimes.com/singapore/information-from-65k-spore-users-in-hands-of-cambridge-analytica.

Hoe, SL 2016, 'Defining a smart nation: the case of Singapore', *Journal of Information, Communication and Ethics in Society*, vol. 14, no. 4, pp. 323–333.

Hollands, RG 2008, 'Will the real smart city please stand up? Intelligent, progressive or entrepreneurial?', *City*, vol. 12, no. 3, pp. 303–320.

Hornidge, A 2010, 'An uncertain future: Singapore's search for a new focal point of collective identity and its drive towards "knowledge society"', *Asian Journal of Social Science*, vol. 38, pp. 785–818.

Huff, WG 1999, 'Turning the corner in Singapore's developmental state', *Asian Survey*, vol. 29, no. 2, pp. 214–242.

IMDA 2006, 'iN2015 Masterplan offers a digital future for everyone', *Infocomm Media Development Authority*, last updated 3 November 2017, viewed 28 May 2019, https://www.imda.gov.sg/infocomm-and-media-news/buzz-central/2006/9/in2015-masterplan-offers-a-digital-future-for-everyone.

Kong, L & Woods, O 2018, 'The ideological alignment of smart urbanism in Singapore: Critical reflections on a political paradox', *Urban Studies*, vol. 55, no. 4, pp. 679–701.

Lee, HS 2014, 'Prime Minister Lee Hsien Loong's National Day rally 2014 speech (English)', Prime Minister's Office, 17 August, viewed 28 May 2019, http://www.pmo.gov.sg/newsroom/prime-minister-lee-hsien-loongs-national-day-rally-2014-speech-english.

Lee, HS 2015, 'Speech by Prime Minister Lee Hsien Loong at Founders Forum Smart Nation Singapore Reception on 20 April 2015', *Smart Nation Singapore*, 20 April, viewed 28 May 2019, https://www.smartnation.sg/whats-new/speeches/founders-forum-smart-nation-singapore-reception-2015.

Lee, HS 2017, 'Prime Minister Lee's National Day rally speech', *Prime Minister's Office*, 20 August, viewed 30 May 2019, https://www.pmo.gov.sg/Newsroom/national-day-rally-2017.

Lee, T 2013, 'NUS's Living Lab makes university a petri dish for tech startups and researchers', *Tech in Asia*, 15 May, viewed 30 May 2019, https://www.techinasia.com/nuss-living-lab-makes-university-a-petri-dish-for-tech-startups-and-researchers.

Leftwich, A 1994, 'Governance, the state and the politics of development', *Development and Change*, vol. 25, no. 2, pp. 378–380.

Lim, A 2001, 'Intelligent Island discourse: Singapore's discursive negotiation with technology', *Bulletin of Science, Technology & Society*, vol. 21, no. 3, pp. 175–192.

Lim, A 2002, 'The culture of technology in Singapore', *Asian Journal of Social Science*, vol. 30, no. 2, pp. 271–286.

Lin, M 2016, 'Driverless car hits lorry during test drive', *The Straits Times*, 19 October, viewed 28 May 2019, https://www.straitstimes.com/singapore/driverless-car-hits-lorry-during-test-drive.

Liow, ED 2012, 'The neoliberal-developmental state: Singapore as case study', *Critical Sociology*, vol. 38, pp. 241–264.

Loh, KS 1998, 'Within the Singapore story: The use and narrative of history in Singapore', *Crossroads: An Interdisciplinary Journal of Southeast Asia Studies*, vol. 12, no. 2, pp. 1–21.

Mahizhnan, A 1999, 'Smart cities: The Singapore case', *Cities*, vol. 16, pp. 13–18.

Mahizhnan, A & Yap, MT 2000, 'Singapore: The development of an Intelligent Island and social dividends of information technology', *Urban Studies*, vol. 37, no. 10, pp. 1749–1756.

Matheson, R 2016, 'Startup bringing driverless taxi service to Singapore', *MIT News*, 24 March, viewed 28 May 2019, http://news.mit.edu/2016/startup-nutonomy-driverless-taxi-service-singapore-0324.

Nasir, N 2017, 'Smart Nation and Digital Government Group Office to be formed under PMO', *PMO*, 20 March, viewed 14 June 2019, https://www.gov.sg/news/content/smart-nation-and-digital-government-group-office-to-be-formed-under-pmo.

nuTonomy 2018, 'Tomorrow's car, today. Software for driverless fleets', *nuTonomy*, viewed 28 May 2019, https://www.nutonomy.com/.

nuTonomy (undated), 'nuTonomy completes $16M Series A funding to accelerate delivery of self-driving car software', *nuTonomy*, viewed 28 May 2019, https://www.nutonomy.com/press-release/nutonomy-series-a/.

Ostry, JD Loungani, P & Furceri, D 2016, 'Neoliberalism: Oversold?', *Finance and Development*, vol. 53. no. 2, pp. 38–41.

Park, BG, Hill, RC & Saito, A 2012, *Locating neoliberalism in East Asia: Neoliberalizing spaces in developmental states*. Chichester: Wiley-Blackwell.

Pereira, AA 2008, 'Whither the developmental state? Explaining Singapore's continued developmentalism', *Third World Quarterly*, vol. 29, no. 6, pp. 1189–1203.

Population.sg 2016, 'Older Singaporeans to double by 2030', *Population.sg*, 22 August, viewed 28 May 2019, https://www.population.sg/articles/older-singaporeans-to -double-by-2030.

Prime Minister's Office 2017, 'Formation of the Smart Nation and Digital Government Group in the Prime Minister's Office', *Prime Minister's Office*, 20 March, viewed 28 May 2019, https://www.pmo.gov.sg/newsroom/formation-smart-nation-and -digital-government-group-prime-minister%E2%80%99s-office.

Rajaratnam S 1972, 'Singapore: Global City, text of Address by Mr S. Rajaratnam, Minister for Foreign Affairs to the Singapore Press Club on February 6, 1972', *National Archives of Singapore*, viewed 28 May 2019, http://www.nas.gov.sg/ archivesonline/data/pdfdoc/PressR19720206a.pdf.

Sebastian, LC 1997, 'The logic of the guardian state: governance in Singapore's development experience', *Southeast Asian Affairs*, pp. 278–298.

Shin, JS 2005, 'The future of development economics: A methodological agenda', *Cambridge Journal of Economics*, vol. 29, pp. 1111–1128.

Sim, Y 2018, 'How driverless car startup nuTonomy is building the future of transport', *TECHINASIA*, 2 February, viewed 28 May 2019, https://www.techinasia.com/ driverless-car-startup-nutonomy-future-transport.

Singapore Department of Statistics 2018, 'Population trends 2018', viewed 13 November 2019, https://www.singstat.gov.sg/-/media/files/publications/population/ population2018.pdf.

Singapore Department of Statistics 2019, 'Singapore economy', viewed 24 July 2019, https://www.singstat.gov.sg/modules/infographics/economy.

Smart Nation Singapore 2019, 'Initiatives', viewed 13 November 2019, https://www .smartnation.sg/what-is-smart-nation/initiatives.

SNDGO 2018a, 'Strategic national projects to build a Smart Nation', *Smart Nation Singapore*, 5 December, viewed 28 May 2019, https://www.smartnation.sg/whats -new/press-releases/strategic-national-projects-to-build-a-smart-nation.

SNDGO 2018b, 'Autonomous vehicles', *Smart Nation Singapore*, 12 December, viewed 28 May 2019, https://www.smartnation.sg/what-is-smart-nation/initiatives/ Transport/autonomous-vehicles.

SNDGO 2018c, 'Open data & analytics for urban transportation', *Smart Nation Singapore*, 12 December, viewed 28 May 2019, https://www.smartnation.sg/what-is -smart-nation/initiatives/Transport/open-data-and-analytics-for-urban-transportation-1.

SNDGO 2019, 'Smart Nation progress', *Smart Nation Singapore*, viewed 14 June 2019, https://www.smartnation.sg/why-Smart-Nation/smart-nation-progress.

Tan, T 1983, 'Speech by Dr Tony Tan Keng Yam, Minister for Trade and Industry at the opening ceremony of Applications 83/Computa 83 at the World Trade Centre Exhibition Hall 3 on Wednesday 11 May 1983 at 9.15am', National Archives of Singapore, Singapore.

Teo, TSH & Lim, VKG 1999, 'Singapore – an 'intelligent island': Moving from vision to reality with information technology', *Science and Public Policy*, vol. 26, no. 1, pp. 27–36.

Tham, I 2017, 'Untangling the way to a Smart Nation', *The Straits Times*, 26 March, viewed 28 May 2019, https://www.straitstimes.com/singapore/untangling-the-way -to-a-smart-nation.

Vanolo, A 2014, 'Smartmentality: The smart city as disciplinary strategy', *Urban Studies*, vol. 51, no. 5, pp. 881–896.

Williams, A 2017, 'Nearly 1 in 5 Singapore employees fears losing jobs to automation', *The Straits Times*, 5 June, viewed 28 May 2019, https://www.straitstimes.com/

business/economy/nearly-1-in-5-singapore-employees-fear-losing-their-jobs-to
-automation.

Yeung, H 2000, 'State intervention and neoliberalism in the globalizing world
economy: Lessons from Singapore's regionalization programme', *The Pacific
Review*, vol. 13, no. 1, pp. 133–162.

3. Smart-city vision and strategy in Hong Kong

Xin Li, Kyung-Min Nam[1] and Chee Keong Khoo

INTRODUCTION

The smart city is a concept built on the application of information and communication technology (ICT) to urban development and management (Townsend 2013). Its ultimate goal is to enhance urban life quality and sustainability by adding an ICT layer to the city, which can be considered a platform on which various human activities are based (Komninos 2015). In contemporary urban development, the initial concept of the smart city as a completed urban product – where economic agents are offered a choice only from a ready-made, fixed set of functions – has increasingly given a way to the smart city as a flexible urban platform – where it is open to change, and its offerings evolve with public needs (Hwang 2020). Given the goal and the flexibility embedded in the urban platform, a smart-city model can embrace other alternative city-development concepts, such as the eco-city, low-carbon city and resilient city, or be jointly applicable to a single city.

The "smart city" concept has gained popularity since IBM first used the term in 2009, but its precursors date back to the mid-1990s, when the "digital city" idea was first introduced (Dirks & Keeling 2009; Söderström et al. 2014) (Figure 3.1). Many digital-city projects initiated in an earlier period (1996–2002) were led by the private sector, with ICT application to a limited scope of smart-city domains. In the following growth stage (2003–2011), alternative technology-driven smart-city models, such as the ubiquitous city, were proposed. During this period, the government teamed up with private firms for ICT industry and infrastructure development, and focused on city-wide installation of hardware such as sensors, Internet, servers and controllers. Such hardware upgrade projects, however, often lacked a thorough understanding of how cities work and operate, and the scope of potential ICT application was narrowly defined under technological determinism. Since 2012, smart-city development has entered a new phase. Smart-city initiatives have spread on

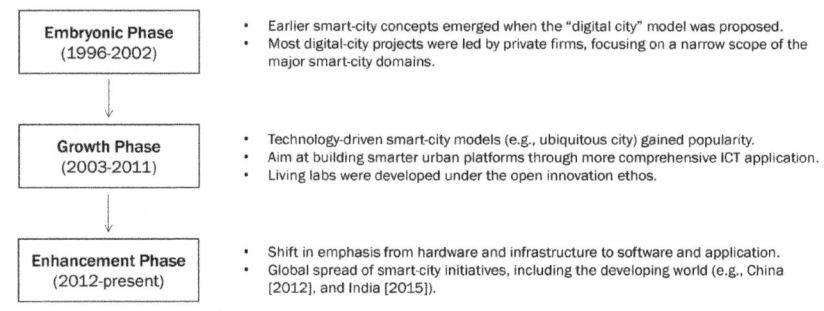

Source: Created by the authors.

Figure 3.1   Three stages of smart-city development

a global scale, and software aspects have received greater attention, with comprehensive application domains identified, such as smart mobility, government, economy and environment.

Hong Kong is a latecomer in joining the burgeoning smart-city initiatives (Figure 3.2). In Hong Kong, the *Digital 21 Strategy*, which was first announced in 1998 and updated in 2001, 2004 and 2008, may represent the city's earlier vision for ICT-based urban development (Tang 2017). In some aspects of the smart-city domains, such as e-commerce, Hong Kong was among the first movers, as exemplified by the Octopus electronic transaction system introduced in 1997. The strategy, however, leaned toward ICT sector promotion and infrastructure development, and was implemented without a comprehensive smart-city vision and guidelines. In the absence of strong public leadership, the private sector tended to take the initiative for narrow-scoped ICT projects in a piecemeal manner. This partly explains why Hong Kong is overall ranked lower in smart-city indices than other rival cities (e.g., Tokyo, Seoul and Singapore), failing to take full advantage of its world-class business environments and infrastructure (EasyPark 2017; IESE Business School 2018; Schwab 2016; Yeung & Singh 2017). The Hong Kong Government announced its first comprehensive smart-city vision in late 2017, three to eleven years behind other Asian competitors.

Despite its late start, Hong Kong currently has an ambitious plan to employ smart technologies to improve its long-lasting and emerging urban challenges. The *Hong Kong Smart City Blueprint*, the city's smart-city vision statement announced in late 2017, declares a people-centric approach to ICT application, where a smart city is viewed as a key means to enhance urban sustainability and liveability. The blueprint covers comprehensive smart-city application

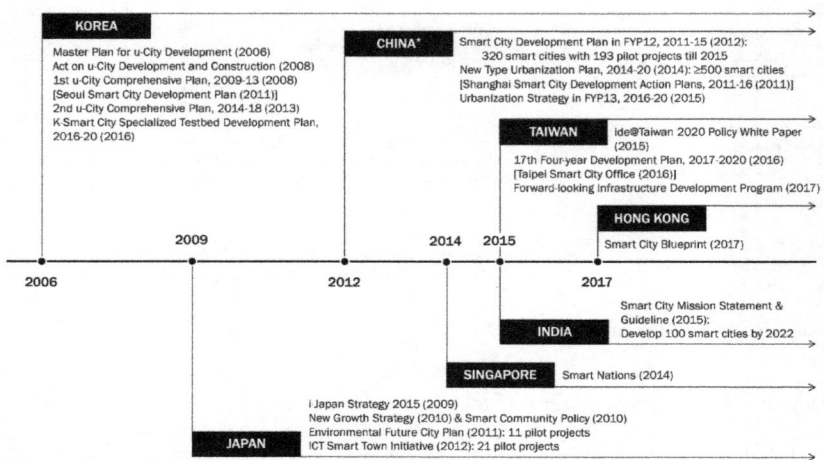

Notes: * Mainland only; FYP12: 12th Five Year Plan; FYP13: 13th Five Year Plan.

Source: Created by the authors.

Figure 3.2   Smart-city development landmarks in Asia

areas with both short-run and longer-term strategies. The six main smart-city domains identified in the blueprint are Smart Mobility, Smart Living, Smart Environment, Smart People, Smart Government and Smart Economy. In this chapter, we offer a review of Hong Kong's vision and strategy in each domain and introduce several ongoing pilot and showcase projects.

HONG KONG'S SMART-CITY INITIATIVES

Although some smart-city application domains have long received substantial attention within Hong Kong's policy circle, early smart-technology adoption projects tended to be patchy and target only narrow aspects involving public facility and infrastructure operations. Most notably, the *Digital 21 Strategy*, a precursor of the *Hong Kong Smart City Blueprint*, presents a strong bias toward an economic development domain (Legislative Council 2013). With the goal of smooth and timely transition toward a digital economy, the *Digital 21 Strategy* gives high policy priority to ICT industry promotion and startup incubation, as well as construction of backbone Internet infrastructure. Cyberport and the Science Park – two of the city's largest business park projects for high-technology firms and startups – were developed during the strategy's implementation period.

Despite lacking a comprehensive vision, the scope of the ICT application areas was gradually extended to cover other smart-city domains, such as mobility, government and environment. The Octopus transaction system introduced in 1997 and the new resident-identity card introduced in 2003 represent Hong Kong's existing smart-card systems adopted for efficient e-commerce and public administration. Real-time transit and flight information tracking systems, coupled with increased public wi-fi hotspots, were another early smart-technology application field in the city, aimed at improving the quality of local and international transport services. T-Park in Tuen Mun, a waste-to-energy plant in operation since 2015, is a showcase project in the areas of smart environment and energy. Such earlier projects, however, were rather independent, and the lack of cross-sectoral coordination called for a more comprehensive and integrated vision to create system-level efficiency and synergy.

It is only very recently that Hong Kong added a full-scale smart-city development initiative to its priority policy agenda. In late 2015, the Hong Kong Government announced that it would use Kowloon East as a testbed for upcoming smart-city projects (Office of the Chief Executive 2015), with an internal research report on the smart city (Central Policy Unit 2015). This was a critical turning point which signalled an upcoming change in the city's then piecemeal approach to smart-city development. As exemplified in *Hong Kong 2030+*, the smart-city concept has been incorporated into the city's strategic territorial plan as one of its tri-pillar (Smart, Green, Resilient) urban development framework, with particular emphasis on smart mobility and environment areas (Development Bureau and Planning Department 2016). A follow-up public consultation report (PricewaterhouseCoopers (PwC) Advisory Services 2017) synthesises what emerging challenges Hong Kong is facing and how smart-city initiatives can offer potential solutions, demanding stronger government leadership in the six major application domains.

In late 2017, Hong Kong's smart-city vision was further concretised with a shortlist of proposed projects and dedicated budgets. In October 2017, Hong Kong's Chief Executive devoted one section of her Policy Address (items 80–83) to "smart city", committing public investment of HK$700 million to key infrastructure projects, including electronic identification (eID) and e-government systems, and a pilot multi-functional smart lampposts scheme (Office of the Chief Executive 2017). In two months, the launch of Hong Kong's full-scale smart-city initiatives was formalised with the public release of the *Hong Kong Smart City Blueprint*, the city's first strategic plan dedicated to smart-city development (Innovation and Technology Bureau (ITB) 2017). The blueprint proposes a people-centric smart city, where an extensive ICT-based platform upgrading ultimately aims to enhance public welfare and liveability. Various city-wide smart-city initiatives are to be kicked off in six main areas – smart mobility, living, environment, people, government and economy – which are well aligned with the six-dimensional wheel concept introduced by Cohen (2012).

The strategic importance of the blueprint lies in the fact that it integrates a number of quasi-independent but related projects implemented under various public initiatives into a single framework. In Hong Kong, earlier efforts at smart-technology adoption tended to be dispersed throughout various specialised plans and policies, which different government bureaus and departments were in charge of. For example, multiple smart-environment projects were implemented under the leadership of the Environment Bureau as part of its environmental protection and resources management plans (Environment Bureau 2013, 2017), and many smart-economy and mobility projects were orchestrated by the Development and the Transport and Housing Bureaus in line with the city's economic and spatial development visions (Planning Department 2018). The main weakness of this practice was that the absence of a clear, comprehensive vision coupled with the project-optimised implementation structure tended to limit the scope of potential cross-project synergy toward broader platform upgrading. In this context, the blueprint has significance as a first step toward addressing the cross-sectoral coordination problem.

The ITB, created under the Financial Secretary in November 2015, is set to take primary responsibility for overall coordination and monitoring of the proposed smart-city projects. A three-layered governance structure has been proposed as an implementation model (ITB 2017; PwC 2017). At the top of the structure is the Steering Committee on Innovation and Technology, a ministerial committee under the Chief Executive. This committee will bear the main responsibility for setting a clear development scope, steering strategic policy and directions, prioritising key projects, and monitoring overall progress with key performance indices. At the next level is the Smart City Office, in charge of programme management under the ITB. The Smart City Office functions as a secretariat to the steering committee, coordinates the directions, manages the programmes approved by the committee, coordinates cross-bureau/department collaborations within the government, and monitors each programme's implementation status. The Smart City Office includes the Energizing Kowloon East Office (EKEO) responsible for Kowloon East smart-city pilot projects, and the Office of the Government Chief Information Officer, which will offer technical support to various bureaus and departments involved in smart-city projects. At the third layer is an execution agency, which consists of project teams and is under the direction of the EKEO. The project teams are primarily responsible for resource management and project planning/execution at the programme level, as well as staff training and regular implementation outcome reporting.

THE *HONG KONG SMART CITY BLUEPRINT*

Hong Kong faces various urban challenges, and the Hong Kong Government expects that its smart-city initiatives will help deal with them and create new opportunities for citizens and businesses. In the *Hong Kong Smart City Blueprint*, the main challenges Hong Kong faces are identified under each of the six smart-city domains and are matched with various ongoing or proposed smart-city projects and initiatives (Table 3.1).

Weakening environmental sustainability, compromised to support intense human activities on limited built-up land, is one of the most serious urban challenges in Hong Kong. The city's soaring ecological footprints are primarily driven by the carbon footprint component, associated with energy consumption (World Wide Fund for Nature (WWF) & Global Footprint Network (GFN) 2013). In the local context, the electricity and building sectors are particularly important in curbing greenhouse gas (GHG) emissions; the local power sector alone accounts for over two-thirds of Hong Kong's GHG emissions, and >90 per cent of the electricity generated is consumed by buildings (Environment Bureau 2017). That is, a supply-side mitigation strategy needs to pay primary attention to regulating the local power sector, and at the same time it should go hand in hand with demand-side public actions promoting energy savings in the building sector.

In this context, meeting green building standards has been encouraged as a demand-side response, placing high priority on public office buildings and residential apartments. Since 1996, Hong Kong has operated an independent green building certification system – the Building Environmental Assessment Method (BEAM) Plus or its predecessor HK-BEAM. As of December 2018, the number of certified green buildings reached 2,182 units or 5.2 per cent of Hong Kong's entire building stock (BEAM Society 2019, Hong Kong Green Building Council 2018). Supply-side strategies include stringent regulations of the local power sector, enforcing a phase-out of the existing coal-fired thermal plants and strengthened market incentives in favour of renewables. In compliance with China's commitment to the Paris Agreement, the Hong Kong Government has announced binding targets for 2030 to reduce carbon intensity (or absolute emissions) by ≥ 65 per cent (or ≥ 26 per cent), compared with the 2005 levels, and to mitigate per capita carbon emissions from 6.2 tonnes in 2014 to ≤ 3.8 tonnes in 2030. Besides such GHG mitigation efforts, Hong Kong has also placed at the top of its smart-environment agenda improving the current solid-waste management and pollution monitoring systems through an extensive application of enhanced sensor, drone and remote-sensing technologies.

Table 3.1   *Proposed smart-city initiatives for key issues and challenges in Hong Kong*

	Current status	Proposed strategy and initiatives
Smart Environment	Buildings account for 90% of local electricity consumption.	Green and intelligent buildings and energy efficiency: - Building retro-commission with smart/IT tech. - LED lighting in both new and retrofitted public residential projects. - Stricter green building requirements in Kowloon East Smart City pilot site.
	Electricity generation accounts for two-thirds of local carbon emissions.	Climate Action Plan 2030+: - Reduce carbon intensity by ≥ 65 per cent by 2030 (vs. 2005). - Increase proportion of natural gas and renewable energy on electricity generation.
	A third of municipal solid waste (MSW) is recovered for recycling.	Waste Management & Pollution Monitoring: - Charging scheme for MSW. - Remote sensing for monitoring.
Smart Mobility	Extremely high passenger traffic volumes in transit modes. Limited options for digital payment.	All-in-one transport mobile application. New payment system. Smart public transport interchange. Use of smart buses.
	High volumes of air passengers.	Smart Airport: - Facial biometrics tech. - Mobile check-in in off-airport locations. - Driverless vehicles in restricted area.
	Road traffic congestion with >350 vehicles per km of road.	Smart Parking. Installation of in-vehicle units (IVUs). Intelligent signalised junctions and pedestrian lights. Traffic detectors for monitoring real-time traffic conditions. Electronic road pricing system. Automatic tolling system.

	Current status	**Proposed strategy and initiatives**
Smart Government	HK$9 billion allocated to public ICT infrastructure in fiscal year 2017–18.	Smart-city infrastructure: - 5G mobile networks. - Adoption of eID for online government services. - Smart lamppost. - Enhance cybersecurity.
	>220 e-government services via GovHK.	Real-time big-data analytic. Public cloud services. Common Spatial Data Infrastructure (CSDI) via adoption of Building Information Modelling (BIM): - Building plan submission hub. - Asset and facilities management.
	Over 3,100 datasets and 1,000 application programming interfaces (APIs) provided by public sector information portal.	Open data access via one-stop portal. Promote open data for smart-city solutions.
Smart People	Free education up to secondary education. Nearly 100,000 students enrolled in government-funded programmes in academic year 2016–17.	Research and development (R&D) through postdoctoral and internship programmes. Facilitate R&D partnerships between institutions.
	Low entrepreneurial spirit of young generation.	Incubation programme. Venture Capital Fund for entrepreneurship.
	Only 60% of senior secondary students studied in science, technology, engineering and mathematics (STEM)-related elective subjects. Gaps in supply and demand for data science.	Intensive programme for STEM. Enhanced Information Technology (IT) curriculum for secondary schools. Fast-track IT talent scheme for professionals.

	Current status	Proposed strategy and initiatives
Smart Economy	Top three industries in gross domestic product (GDP): Trading and Logistics, Financial Services, Professional Services.	Tax deduction for R&D. Build the Data Technology Hub and the Advanced Manufacturing Centre.
	High volumes of transactions via credit cards and Internet banking services.	Promote sharing economy.
	Licences for trying out Fintech innovations.	Fintech: - Promote Fintech for cross-boundary trade. - API framework for banking. - Virtual banking sector. IT Re-industrialisation: - Government-to-government in science park development. - Government IT-based procurement for technological innovations.
Smart Living	High mobile and household broadband penetration.	Wi-Fi connected city. Faster Payment System (FPS) using mobile number and email address.
	Projected 31% ageing population in 2036.	Adoption of technology for elderly and rehabilitation services.
	Public health services and online health records.	Smart Hospital Services. Expand the electronic health record system for information sharing.

Source: Created by the authors from ITB 2017.

The transport and mobility sector is one of the most active smart-technology application areas in Hong Kong, but it still presents room for further improvement. One aspect is to develop and operate an *integrated* real-time traffic information provision platform. Despite Hong Kong's quality public transport system and high transit ridership levels, a loose network among independent transit operators sets barriers to cross-operator and trans-mode data sharing essential for an integrated system-level service, such as real-time traffic control or all-in-one travel information services. Also, demand for diversified payment options for transit users tends to increase. At present, Hong Kong's transit systems only accept Octopus to charge fares, while various payment options based on mobile applications, as well as smart cards, are widely adopted in many other cities. The parking system can also benefit from smart-technology applications, but data sharing among independent car park operators is a prerequisite. At present, the city-wide consolidated parking information service in Hong Kong is very weak, as the majority of the available parking spaces are under the control of private, estate-level parking service providers. The application area can be extended to the road traffic management. In addition to the automatic toll collection and real-time traffic management systems introduced in 1993 and 2004, respectively, an electronic road pricing system targeting congestion in city centres will soon be implemented for peak-hour traffic control (Transport Department 2019). The local air transport sector also strengthens smart-technology adoption in response to increased congestion. The Airport Authority is currently testing mobile check-in systems and driverless vehicles for onsite baggage transport, and plans to introduce a facial biometrics system for immigration/customs clearance.

The smart-government domain aims to offer upgraded public services through transparent, responsive and user-friendly service platforms, coupled with 5G-class ICT networks. Upgrading of the government itself is a prerequisite for this application domain, and will be spurred by a new big-data analytics platform for real-time data transmission and sharing within the government. The scope and speed of e-government service offerings will substantially increase with the adoption of the eID system and public cloud services as well as the operation of the smart lampposts supporting the city-wide Internet of Things (IoT) networks. Access to a broad range of public data, from published to real-time data covering district to building-level details, will be authorised to the public in a geo-spatial data format through a one-stop open database portal called the CSDI. In contrast to its liberal and open economic environment, Hong Kong has been lagging behind other cities in public data sharing. In 2016, for example, only 20 per cent of government-produced datasets were given open access (24th out of the 94 global economies), while public access to the remaining datasets was highly restricted with case-by-case authorisation granted upon special request (Open Knowledge International 2017). The launch of the CSDI, planned for 2023, is expected to bring significant change to Hong Kong's weak environment for open data sharing and exchange.

A main goal of the smart people domain is to strengthen the stock of human capital and cultivate an innovative and entrepreneurial local milieu along the path to Hong Kong's smart-city development. A weak pool of local talent, particularly in the science and engineering fields, and a feeble local entrepreneurship culture are often seen as a key challenge Hong Kong is facing. Insufficient emphasis on STEM in public education and a lack of a holistic approach to nurturing a skilled labour pool partly underlie the problem (Lam 2018). Evolution toward a knowledge-based and technology-adaptive society increases demand for high-quality human resources in emerging areas like machine learning, big-data analytics, artificial intelligence and robotics, and nurturing young talent in these areas through strengthened STEM curricula is essential. The local climate for innovation and R&D also needs to be improved through extended university–industry collaboration and public–private partnerships as well as strengthened institutional and financial support for business incubation and overseas talent attraction.

The smart economy pillar is intended to strengthen the link between the city as a platform and various economic activities based on it through smart technologies. One direction of the effort is offering new e-commerce platforms to further strengthen the existing business models. In 2018, for example, the Hong Kong Monetary Authority (HKMA) announced open API guidelines and virtual bank authorisation, setting new standards for online banking and trading. The former aims to promote the synergy between the banking sector and third-party ICT firms, and the latter allows firms to offer retail banking services through online or mobile applications without offline service centres. The other direction is to facilitate an active response to new economic opportunities. For example, Hong Kong, which currently operates two business parks for high-technology ventures and startups, has recently extended its regional cooperation with Shenzhen to develop a new world-class science and technology cluster, called the Lok Ma Chau Loop Innovation and Technology Park. The park targets the world's leading ICT enterprises and research/educational institutions as potential tenants through its appealing locational advantage and attractive financial and tax incentives. Together with increased policy support for ICT startups and businesses, the Hong Kong Government also plans an extensive review of the existing legislations to reduce institutional lags and catch up with rapid technological change.

Finally, the smart living domain touches upon the ultimate goal of smart-city development – public welfare gains through enhanced and extended urban services. The Hong Kong Government has committed to creating a secured wi-fi connected city, in support of the FPS just launched in October 2018. The FPS offers a real-time monetary transaction platform, using a mobile number or an email address without additional personal identity requirement. The *Hong Kong Smart City Blueprint* seeks to enhance the living experience and well-being across various citizen groups and to improve public healthcare services for the elderly population. Initiatives proposed include supporting the elderly's use of gerontechnology, teleconsultation and remote health monitoring. The electronic health record sharing system will be expanded for information sharing among public–private health institutions.

Hong Kong's smart-city development in these six domains takes a living lab approach, based on extensive public–private partnerships and citizen participation. At present, smart-technology application is already underway throughout the city, taking various sites as testbeds for a particular segment. A new technology solution, tested in selected sites and improved through trial-and-error and learning-by-doing practices, will eventually be implemented throughout the city. Kowloon East currently functions as Hong Kong's largest smart-city testbed, and is home to various pilot projects, which are simultaneously carried out in close connection with the area's urban redevelopment master plan. In the following section, we briefly introduce the smart-city pilot projects implemented in Kowloon East.

KOWLOON EAST SMART CITY PILOT PROJECTS

In the 2015 Policy Address, Kowloon East was proposed as a pilot project site dedicated to testing the feasibility of Hong Kong's smart-city programmes. The project site includes part of Kwun Tong and Kowloon Bay Business districts, which once functioned as the city's leading industrial district. Kowloon East, however, has lost its economic vibrancy since the city's manufacturing sector declined in the 1980s. Aged buildings/infrastructure and underutilised industrial estates have long symbolised the community. In this context, the Hong Kong Government established the EKEO in 2012, and has spurred urban renewal with an aim of transforming Kowloon East into the city's second central business district (CBD). The role of Kowloon East as a testbed for the city-wide application of smart technologies is seen as essential in creating a new, attractive urban centre, which stands out in four aspects – connectivity, branding, design and diversity (CBD2).[2]

A Brief History of the Kowloon East Area

Kowloon East is located at the heart of the city's former industrial base (Figure 3.3). Kowloon Bay, facing Victoria Harbour, was home to the Kai Tak International Airport, in operation between 1925 and 1998. For a long time, the area functioned as Hong Kong's international transport/logistics hub, granting a substantial locational advantage for Kwun Tong-based exporters. Kwun Tong's positioning as Hong Kong's industrial hub dated back to the early 1960s, when the city experienced rapid population growth. In 1961, when the first post-war census was conducted, Hong Kong's total population reached 3.1 million, an almost four-fold increase from 0.8 million in 1931, when the last pre-war census was conducted (Fan 1974). In response, industrial sector development was initiated to offer citizens jobs on a massive scale. Kwun Tong, located on the outskirts of the old Kowloon, was chosen as home to a cluster of light-manufacturing factories and warehouses, and formed the city's post-war production centre (Chan 1973). Such industrial development was accompanied by large-scale public housing projects. Wo Lok Estate with 1,941 flats, the district's oldest public housing estate, was built during this period, and, as of December 2018, Kwun Tong's public rental housing stock housed 143,200 households or 379,300 residents in total (Housing Department 2019).

China's industrialisation drive after Deng's open-door policy, however, brought a serious external shock to Hong Kong's manufacturing sector. Hong Kong's economy faced an enormous challenge to maintain its labour-intensive industrial bases when exposed to increased competition with mainland cities, and deindustrialisation was accelerated from the mid-1980s. Kwun Tong's economic fate was in the same boat as the city's shrinking manufacturing sector. An increasing number of industrial buildings were abandoned or underutilised and their physical condition was left to deteriorate. Even in 2017, when urban redevelopment projects for the district were under way, vacancy rates for Kwun Tong-based private office/commercial properties and industrial estates were as high as 15.9 per cent and 6.9 per cent, respectively (Rating and Valuation Department 2018).

Despite its long-standing economic decline, Kwun Tong is still one of Hong Kong's most populous districts – 677,300 residents in 2018 with a city population share of 9.2 per cent – and its residents desperately demand improved physical working and living environments. It is in this context that the Hong Kong Government designated Kowloon East as a consolidated redevelopment project site in 2011 with a dedicated government body (EKEO) launched the following year. Kowloon East consists of four main sub-project sites covering 514 hectares of land: Kai Tak, Kwun Tong, Kowloon Bay and San Po Kong (Legislative Council 2017). The Kai Tak Development Area has been largely vacant since the former Kai Tak International Airport ceased operation in 1998 and was replaced by the Chek Lap Kok International Airport located on Lantau Island. The entire 320 hectares within this area is currently designated as a Comprehensive Development Area, a land-use zone set to facilitate large-scale urban development projects. The remaining three sub-project sites – Kwun Tong, Kowloon Bay and San Po Kong – are run-down manufacturing centres housing a number of dilapidated industrial buildings and residential properties.

After releasing a series of revised master plans, the government has confirmed its vision to transform Kowloon East into Hong Kong's CBD2. Given the scale of the site and its proximity to existing urban centres, Kowloon East possesses great potential to become a new economic centre, sharing part of the role as Hong Kong's traditional CBD (CBD1 referring to Central/Wan Chai and Yau Tsim Mong Districts). CBD2 will be located across three districts: Kwun Tong (Kwun Tong and Kowloon Bay), Kowloon City (Kat Tak) and Wong Tai Sin (San Po Kong). These three districts, accounting for one-fifth of Hong Kong's population, form a large local market, and offer a strategic location with a massive amount of urban land available for development. Kowloon East can be particularly appealing to startup companies and finan-

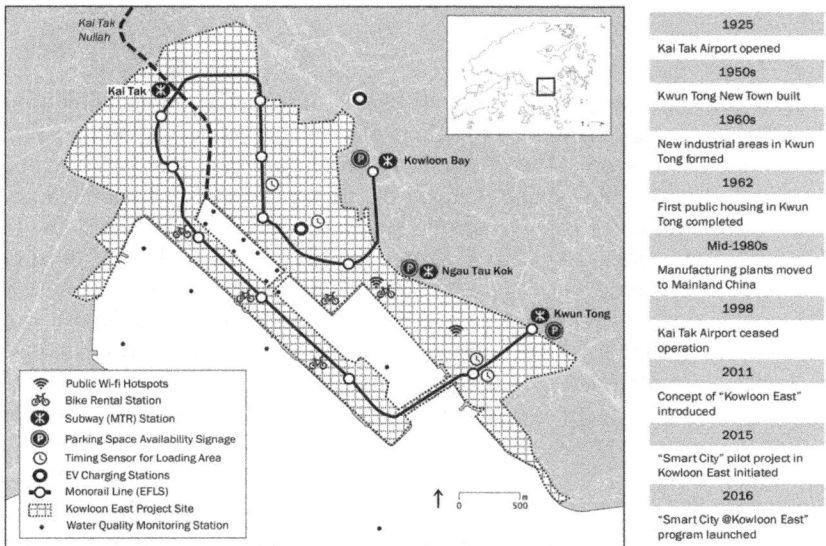

1925	
Kai Tak Airport opened	
1950s	
Kwun Tong New Town built	
1960s	
New industrial areas in Kwun Tong formed	
1962	
First public housing in Kwun Tong completed	
Mid-1980s	
Manufacturing plants moved to Mainland China	
1998	
Kai Tak Airport ceased operation	
2011	
Concept of "Kowloon East" introduced	
2015	
"Smart City" pilot project in Kowloon East initiated	
2016	
"Smart City @Kowloon East" program launched	

Legend:
- Public Wi-fi Hotspots
- Bike Rental Station
- Subway (MTR) Station
- Parking Space Availability Signage
- Timing Sensor for Loading Area
- EV Charging Stations
- Monorail Line (EFLS)
- Kowloon East Project Site
- Water Quality Monitoring Station

Source: Created by the authors.

Figure 3.3   *Kowloon East smart city pilot area*

cial enterprises looking for quality office spaces at competitive market rents (Mingtiandi 2018).

A series of the smart-city pilot projects implemented in Kowloon East since 2015 are a significant addition to a master plan to transform the area into CBD2. In addition to their primary responsibility as a living lab for the city-wide technology adoption, they will also benefit EKEO-led urban redevelopment initiatives by strengthening advanced smart-city features. As shown in Figure 3.3, Kowloon East will be equipped with elements of a green infrastructure, such as bike tracks and rental stations, electric vehicle (EV) charging stations and a monorail system called the Environmentally Friendly Linkage System (EFLS). This intra-district green transport system will complement the inter-district public bus and metro rail systems. Smart technologies based on sensors and real-time data processing will assist the district-wide efficient use of available public resources, such as parking and loading spaces. These smart-city pilot components embrace the basic principles of new urbanism in favour of walkability, connectivity, diversity/mixed land use, blue-green infrastructure and quality public spaces/facilities, and are intended to serve as a significant promoter of urban liveability and environmental sustainability.

Current Progress

Since 2017, the feasibility of various city-wide smart-technology adoption programmes has been tested in Kowloon East under eight Proof of Concept (PoC) trial categories (Figure 3.4). Many of the eight PoC trials fall under the smart-mobility domain. The "persona & preference-based way-finding for pedestrians" sector (PoC Trial 1) intends to promote walkability and mobility, and has experimented with a smart-track network with a synchronised mobile application support in the former airport runway site. A mobile application named MyKE offers pedestrians and drivers key navigation guides, such as optimal pedestrian/driving routes, real-time, route-specific traffic information and parking availability. MyKE also functions as a one-stop information platform for geo-spatial data, including main events held within the region, recommended points of interest and personalised walking/trail routes, integrating virtual or augmented reality technologies.

Four other PoC trial sectors based on surveillance cameras, sensors, and video analytics also focus on enhancing mobility. The smart crowd management system (PoC Trial 2) detects anomalies in pedestrian and vehicular flows, and any abnormal situation, if detected, will automatically call for an immediate response, including the dispatch of an emergency or crisis-management team. The kerbside loading/unloading bay monitoring system (PoC Trial 4) aims to reduce traffic congestion through effective regulations over roadside cargo handling, which often causes serious bottlenecks on Hong Kong's narrow streets and roads. The "real-time road works information" (PoC Trial 7) and "illegal parking monitoring system" (PoC Trial 8) sectors propose to mitigate congestion and improve safety through ICT-enabled effective monitoring, communication and enforcement channels.

Smart environment is another domain that has received much attention in PoC trials. The energy efficiency data system (PoC Trial 3) monitors real-time energy demand with smart electricity sensors and shares the information with each household to redistribute aggregate daily energy consumption away from the peak hours. This system is currently being tested with a group of households selected from Kai Ching and Tak Long Estates, and will be implemented with various incentive/reward schemes for energy savings and real-time energy pricing. The smart waste bin system (PoC Trial 5) operated with fill-level sensors and artificial intelligence monitors the real-time status and forecasts a reasonable fill-up time. A city-wide implementation of the system can help optimise waste collection schedules and workforce distribution, contributing to reducing pressure on the city's solid-waste management capacity.

Regarding its potential contributions, the multi-purpose lamppost project (PoC Trial 6) spans multiple smart-city domains, such as smart environment, living and mobility. The lamppost, which is equipped with security cameras,

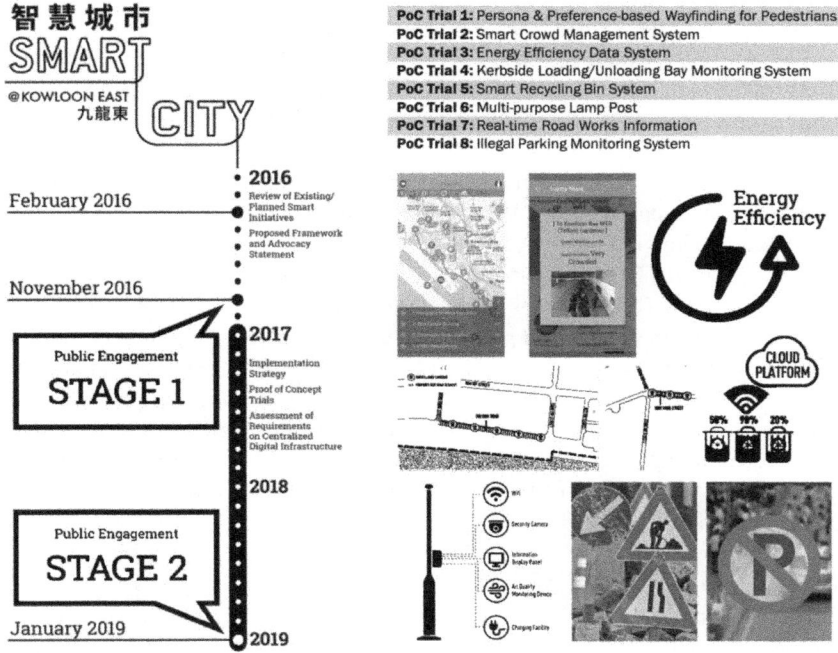

Source: Adapted from the Smart City @Kowloon East website (http://www.smartke.hk, viewed 15 May 2019).

Figure 3.4   PoC trials at the KE pilot project site

air-quality monitoring devices and information display panels for data collection and transmission, also can function as a wi-fi hotspot or an EV charging station. Utilisation of this facility is open to other areas, as it offers flexibility to operate in connection with various IoT devices and data transmission technologies.

Other institutional experiments and standalone development projects targeting Kowloon East are also under way outside the eight PoC trial sectors. Within Kowloon East, for example, ≥30 per cent of the entire project site will be reserved for open spaces, and stricter "green" standards are enforced as part of the land-lease conditions. New buildings in Kowloon East are required to pursue a BEAM Plus accreditation, and must meet at least the gold rating (Development Bureau 2017). With policy incentives, retrofitting existing public housing estates is also encouraged to strengthen green building features, such as grid-connected photovoltaic panels, and a smart meter system will be applied to all of the Kowloon East residential estates.

The Kai Tak Nullah upgrade project, whose completion is scheduled for 2021, is another standalone project worth mentioning. In terms of goals, the project can be positioned within smart environment and living domains, and a place-making approach has been taken to convert the 1.3 km long riverfront into an attractive green river corridor suitable for recreations and leisure (Harbourfront Commission 2013). A highlight of this project is two desilting compounds with vehicular access, which are under construction along the Kai Tak River (Hong Kong Green Building Council 2019). These two structures have acquired a Provisional Platinum rating under BEAM Plus, presenting some ideal features for future public construction projects, in terms of energy efficiency, water management and greening coverage. Water quality and odour along the Kai Tak Nullah and the Kowloon Bay are monitored 24/7 with the Environmental Air Quality Forecast and Water Quality Prediction System, and polluted discharge from the hinterlands into the Kai Tak Approach Channel and the Kai Tak Typhoon Shelter is treated with localised dredging and bio-remediation facilities.

CRITIQUES AND CHALLENGES AHEAD

Despite its comprehensive scope and ambitious goals, the *Hong Kong Smart City Blueprint* leaves room for further improvement and may meet with considerable challenges in implementation. First, an "institutional" lag has caused a substantial bottleneck in catching up with technological progress, and the smart-government initiative may be weak in the absence of strong civic engagement. Although the blueprint identifies an urgent need for reviewing "existing legislation and removing any outdated provisions which impede ICT development", it lacks concrete action plans. For example, popular ride-hailing services, such as Uber and Grab, are currently illegal in Hong Kong, and it is not certain whether and when the situation will change. This institutional barrier has been pointed to as a main interrupter of the transition toward a sharing economy and to local entrepreneurship seeking new business models (Lo & Yau 2017).[3] The lack of transparency and limited civic engagement in the public decision-making process represent another weak institutional dimension (Cheung 2011; EasyPark 2017). The smart-government domain aims to improve such weakness through ICT application, but the government may not have sufficient incentives to transform itself to fit such a model when external pressure is weak (Irvin and Stansbury 2004). It is not certain, either, whether civil society can create an effective push toward a smart government, given Hong Kong's weak tradition of society-led civic engagement (Lee et al. 2013).

Second, priority in public resource allocation needs some adjustment against investment redundancy and in favour of cyber security. The HKMA, for example, plans to introduce a new electronic payment system for the public transit system as part of its cashless payment programme, such as the FPS. It is not clear, however, whether developing an independent, brand-new payment system will be cost-effective, given the need for sizeable but inessential investment in dedicated hardware and infrastructure. Instead, a system upgrade in harmony with popular mobile e-payment systems coupled with near-field communication (NFC) or quick response (QR) code technologies may be a reasonable alternative. In addition, cyber security, which poses a main threat to IoT operations but tends to be low in the pecking order in public investment, needs to receive greater public attention (Joo & Tan 2018). Hong Kong's e-commerce system has already had alerts about potential cyber security risks, as exemplified by a temporary suspension of the electronic Direct Debit Authorisation (eDDA) service. The eDDA service is offered as part of the FPS, and its suspension occurred only 39 days after the initial operation of the FPS. Two fraudulent transactions were identified as the main culprit, and they caused an evaporation of HK$109,000.

Third, the public initiative for data sharing, a key ingredient for a smart-city operation, needs to be strengthened in terms of scope and speed. In the smart-mobility domain, for example, the government encourages public transport operators to open their data to the public, but franchised bus companies, such as Kowloon Motor Bus, Long Win Bus and New World First Bus Services, have shown strong reluctance. In Hong Kong, the private sector plays an important role in providing public transit services, except for government-owned rail systems (MTR), and it is cautious about sharing firm-level operation data with others. For the same reason, private car park operators also give only a half-hearted response to the government's voluntary call, making it hard to provide real-time parking availability information. At present, HKeMobility, a government-released mobile application for all-in-one transit information services, covers only 50 out of >3,000 car parks, and such limited coverage fails to satisfy the expectations of local drivers who need real-time parking availability information (Kao 2018; Transport Department 2019). The Parking.sg application, which covers >1,100 car parks or >70 per cent of Singapore's total parking spaces, may set a reasonable benchmark point for Hong Kong, speaking for the need for stronger public leadership in building a city-wide open data platform (Urban Redevelopment Authority (URA), Housing and Development Board (HDB) & Government Technology Agency of Singapore (GovTech) 2017).

Fourth, slow progress in developing 5G infrastructure may also cause a bottleneck in building a city-wide IoT network. In essence, a connected society works with big data and its real-time analytics and transmission, whose seamless processing exceeds the capacity of the contemporary 4G network (Yeung 2018). Quality 5G-class ICT infrastructure, which runs ≥10 times faster than the 4G network is thus not merely an option but a prerequisite for a smart city. In Hong Kong, the first-round high-frequency band allocation for 5G services (26 and 28 GHz) was completed in March 2019, opening the door for the potential launch of the city's first 5G services in 2020 (Office of the Communications Authority 2019). A series of follow-up auctions for mid-frequency bands (3.3, 3.5 and 4.9 GHz) are scheduled for mid and late 2019 (Chung 2018). The current progress and proposed schedule suggest that, in the global 5G race, Hong Kong is substantially behind first movers like South Korea, which completed all the frequency band allocation procedures by mid-2018 and officially began offering commercial 5G services in April 2019 (Park 2019). This delay in 5G transition contrasts with Hong Kong's earlier experience where it was among the world's earliest 4G technology adopters.

Finally, greater policy incentives seem necessary to promote smart-energy infrastructure and clean-energy vehicles. Hong Kong's progress toward a smart grid is slow and, at present, only a small existing building stock is equipped with smart meters. Policy incentives, such as non-peak-hour discounts or feed-in tariff, are also not strong enough to stimulate a significant system-level change toward a low-carbon economy. However, there are positive signs of change, such as an excess renewables resale programme. Starting on 1 October 2018, this programme allows Hong Kong households and firms to resell their excess electricity generated from rooftop solar panels or wind power installations (Chen 2018). Replacing the existing gasoline or diesel-powered vehicle fleets by clean-energy vehicles and promoting public investment in EV-supportive infrastructure is another area deserving increased policy attention. As of March 2019, 11,660 units of EVs were on the road, presenting an EV share of <1.4 per cent in the total motorised vehicle fleet (Environmental Protection Department 2019). This number shows a huge margin compared with global leaders, such as Norway, where EVs account for >30 per cent of the total active vehicle stock (International Energy Agency 2018). In the same month, 2,242 EV chargers were installed for public use in Hong Kong, suggesting a charger–vehicle ratio of <0.2 (Environmental Protection Department 2019).

CONCLUSION

Hong Kong has faced various emerging challenges in maintaining a decent quality of life for its citizens. Increased global competition and rapidly changing business environments require the city to further strengthen the core competencies and knowledge bases for its pillar industries and actively respond to new economic opportunities. Worsening congestion and ecological footprints also place serious constraints on the sustainable growth of the densely populated city with scanty natural endowments. A growing elderly population and declining housing affordability further extend the scope to a social dimension. The need for open governance and citizen participation also arises, as tackling such threats to sustainable development would require collective knowledge and increased transparency/accountability for effective public actions. In response to these challenges, Hong Kong has explored ICT-based solutions, proposing a grand smart-city vision.

During the last two decades, Hong Kong's smart-city development, while slow, has gradually evolved from a technology-centred, piecemeal approach to a people-oriented, comprehensive model. Similar to many other cases, earlier efforts in Hong Kong focused on ICT sector promotion and hardware/infrastructure development. Without a doubt, technologies and the physical infrastructure supporting their application and operation are essential for a smart city, but they are not the smart city itself. The smart city is not supposed to stand alone for its own function; instead, it is intended to integrate ICT into the existing urban system for synergy so that the city performs like a more efficient platform for various human activities. Accordingly, concerted actions and strategies toward grand visions are essential for a successful smart-city development. In their absence, however, Hong Kong's early leadership in certain ICT application areas, such as e-commerce and smart mobility, failed to introduce comprehensive system-level efficiency. The *Hong Kong Smart City Blueprint* announced in 2017 fills this gap and strives to better align various narrow-scoped projects toward greater smart-city visions and frameworks through improved integration and coordination.

Hong Kong is late in joining the global smart-city race but seems to have set the right direction for "Smart City 3.0". As declared in the blueprint, the ultimate goal of the city's smart-city initiatives is to improve public welfare and quality of life through ICT-based urban platform upgrading, going beyond its earlier techno-centric vision. The scope of the initiatives is extensive and covers all major domains of smart-technology application, such as smart mobility, environment, people, economy, government and living. Many smart-city projects in each domain are already underway, together with some pilot schemes, including the Kowloon East smart-city programme. Proposed vision and strategies may be subject to further improvement in certain respects and their implementation will likely meet with unforeseen challenges. But with a strong will, the Hong Kong Government plans to identify incomplete aspects and potential challenges with a living lab approach and address them through trial-and-error.

NOTES

1. Corresponding author (kmnam@hku.hk).
2. CBD2 also has the meaning of Hong Kong's second CBD, and is thus an acronym with dualistic concepts.
3. In this respect, Hong Kong needs to consider amending outdated law clauses with higher priority: (i) Section 52(3) and Regulation 37 of the Road Traffic Ordinance (Cap. 374); (ii) Regulation 37 of the Road Traffic (Construction and Maintenance of Vehicles) Regulations (Cap 374A); and (iii) Section 5(1) of the Hotel and Guesthouse Accommodation Ordinance (Cap. 349). The first two clauses curb car-sharing platforms by requiring taxi licences or restricting the installation of visual display units on motorised vehicles; likewise, the last clause forbids home-sharing platforms, requiring hotel or hostel licences for short-term rentals.

REFERENCES

BEAM Society 2019, 'BEAM Plus certified building', Beam Society, viewed on 15 May 2019, https://www.beamsociety.org.hk/en_beam_assessment_project_4.php.

Central Policy Unit 2015, *Research report on smart city*. Hong Kong: Government of the Hong Kong Special Administrative Region.

Chan, Y 1973, *The rise and growth of Kwun Tong: A study of planned urban development*. Hong Kong: Social Research Centre, Chinese University of Hong Kong.

Chen, L 2018, 'Lack of innovation and incentives holding back Hong Kong from becoming a smart city', *South China Morning Post*, 1 May.

Cheung, PTY 2011, 'Civic engagement in the policy process in Hong Kong: Change and continuity', *Public Administration and Development*, vol. 31, pp. 113–121.

Chung, K 2018, 'Hong Kong aims for cheaper, faster 5G roll-out with proposal to not charge telcos for spectrum', *South China Morning Post*, 27 July.

Cohen, B 2012, 'What exactly is a smart city?', Fast Company, viewed on 15 May 2019, http://www.fastcoexist.com/1680538/what-exactly-is-a-smart-city.

Development Bureau 2017, *Green government buildings (Development Bureau Technical Circular No. 2/2015 and Environment Bureau Circular Memorandum No. 3/2015)*. Hong Kong: Government of the Hong Kong Special Administrative Region.

Development Bureau and Planning Department 2016, *Hong Kong 2030+: Towards a planning vision and strategy transcending 2030*. Hong Kong: Government of the Hong Kong Special Administrative Region.

Dirks, S & Keeling, M 2009, *A vision of smarter cities: How cities can lead the way into a prosperous and sustainable future*. Somers, NY: IBM Institute for Business Value.

EasyPark 2017, *2017 Smart cities index*. Stockholm: EasyPark Group.

Environment Bureau 2013, *A clean air plan for Hong Kong*. Hong Kong: Government of the Hong Kong Special Administrative Region.

Environment Bureau 2017, *Hong Kong's climate action plan 2030+: Strengthening climate resilience*. Hong Kong: Government of the Hong Kong Special Administrative Region.

Environmental Protection Department 2019, *Promotion of electric vehicles in Hong Kong*, Environmental Protection Department, viewed on 15 May 2019, https://www.epd.gov.hk/epd/english/environmentinhk/air/prob_solutions/promotion_ev.html.

Fan, SC 1974, *The population of Hong Kong*. Paris: Committee for International Coordination of National Research in Demography.

Harbourfront Commission 2013, *An update on the planning and design of the Kai Tak development*. Hong Kong: Government of the Hong Kong Special Administrative Region.

Hong Kong Green Building Council 2018, *Hong Kong: Green building in action*. Hong Kong: Hong Kong Green Building Council.

Hong Kong Green Building Council 2019, *BEAM Plus online exhibition: Kai Tak development – Reconstruction and upgrading of Kai Tak Nullah (Desilting compounds 1 & 2)*, HK GBC BeamPlus, viewed 15 May 2019, http://greenbuilding .hkgbc.org.hk/projects/view/20.

Housing Department 2019, *Report on population and households in housing authority public rental housing (December 2018)*. Hong Kong: Government of the Hong Kong Special Administrative Region.

Hwang, JS 2020, 'The evolution of smart city in South Korea: The smart city winter and the city as a platform', in Joo, YM & Tan, TB (eds.), *Smart cities in Asia: Governing development in the era of hyper-connectivity*, pp. 78–92. Cheltenham, UK and Northampton, MA, USA: Edward Elgar Publishing.

IESE Business School 2018, *IESE Cities in Motion Index 2018*. Barcelona: IESE Business School.

International Energy Agency 2018, *Global EV outlook 2018: Towards cross-modal electrification*. Paris: International Energy Agency.

Irvin, RA & Stansbury, J 2004, 'Citizen participation in decision making: Is it worth the effort?', *Public Administration Review*, vol. 64, no. 1, pp. 55–65.

ITB 2017, *Hong Kong Smart City Blueprint*. Hong Kong: Government of the Hong Kong Special Administrative Region.

Joo, YM & Tan, TB 2018, 'Smart cities: A new age of digital insecurity', *Survival*, vol. 60, no. 2, pp. 91–106.

Kao, E 2018, 'Hong Kong government needs to improve access to public sector data to realise smart city ambition, academics say', *South China Morning Post*, 2 May, viewed 22 May 2019, https://www.scmp.com/news/hong-kong/health-environment/ article/2144201/hong-kong-government-needs-improve-access-public.

Komninos, N 2015, *The age of intelligent cities: Smart environments and innovation-for-all strategies*. Abingdon: Routledge.

Lam, J 2018, 'Time for Hong Kong government to get smart about modernising city, think tank says', *South China Morning Post*, 9 October.

Lee, EWY, Chan, EYM, Chan, JCW, Cheung, PTY, Lam, WF & Lam, WM 2013, *Public policymaking in Hong Kong: Civic engagement and state–society relations in a semi-democracy*. Abingdon: Routledge.

Legislative Council 2013, *Legislative Council panel on information technology and broadcasting public consultation on 2014 Digital 21 strategy*. Hong Kong: Legislative Council of the Hong Kong Special Administrative Region.

Legislative Council 2017, *Legislative Council panel on development – Initiatives of Development Bureau in the Chief Executive's 2017 policy address and policy agenda*. Hong Kong: Legislative Council of the Hong Kong Special Administrative Region.

Lo, C & Yau, C 2017, '22 Uber drivers arrested in undercover Hong Kong police operation', *South China Morning Post*, 23 May.

Mingtiandi 2018, 'DBS moves to Kwun Tong as SG bank cashes in on Hong Kong's red hot office market', *Mingtiandi*, viewed 22 May 2019, https://www.mingtiandi

.com/real-estate/finance-real-estate/dbs-moves-to-kwun-tong-and-cashes-in-on -hong-kongs-red-hot-office-market/.

Office of the Chief Executive 2015, *The 2015 Policy Address: Uphold the rule of law, seize the opportunities, make the right choices*. Hong Kong: Government of the Hong Kong Special Administrative Region.

Office of the Chief Executive 2017, *The 2017 Policy Address: We connect for hope and happiness*. Hong Kong: Government of the Hong Kong Special Administrative Region.

Office of the Communications Authority 2019, *Press releases: Offer of spectrum assignments in 26 GHz and 28 GHz bands for provision of 5G services*. Hong Kong: Government of the Hong Kong Special Administrative Region, viewed 22 May 2019, https://www.ofca.gov.hk/en/media_focus/press_releases/index_id_1891 .html.

Open Knowledge International 2017, *Global Open Data Index*, viewed 22 May 2019, https://index.okfn.org/place/hk/.

Park, JM 2019, 'S. Korea first to roll out 5G services, beating U.S. and China', *Reuters*, 3 April, viewed 25 May 2019, https://www.reuters.com/article/southkorea-5g/s -korea-first-to-roll-out-5g-services-beating-u-s-and-china-idUSL3N21K114.

Planning Department 2018, *Studies in progress, Planning Department*, viewed 22 May 2019, https://www.pland.gov.hk/pland_en/p_study/prog_s/index.html.

PwC Advisory Services 2017, *Report of consultancy study on smart city blueprint for Hong Kong*. Hong Kong: PwC.

Rating and Valuation Department 2018, *Hong Kong property review 2018*. Hong Kong: Government of the Hong Kong Special Administrative Region.

Schwab, K (ed.) 2016, *Global Competitiveness Report 2016–2017*. Geneva: World Economic Forum.

Söderström, O, Paasche, T & Klauser, F 2014, 'Smart cities as corporate storytelling'. *City*, vol. 18, no. 3, pp. 307–320.

Tang, W 2017, *Smart city 3.0*. Hong Kong: Smart City Consortium.

Townsend, AM 2013, *Smart cities: Big data, civic hackers, and the quest for a new utopia*. New York: W.W. Norton.

Transport Department 2019, *Transport in Hong Kong, Transport Department*, viewed 22 May 2019, https://www.td.gov.hk/en/transport_in_hong_kong/index.html.

URA, HDB & GovTech 2017, *Press releases: Parking.sg app available for motorcycles and heavy vehicles from 20 December*. Singapore: Government of Singapore, viewed 22 May 2019, http://www.nas.gov.sg/archivesonline/speeches/record -details/5e8ccd7f-e57d-11e7-be76-001a4a5ba61b.

WWF & GFN 2013, *Hong Kong ecological footprint report 2013*. Hong Kong: WWF-Hong Kong.

Yeung, R & Singh, H 2017, 'Why is Singapore much smarter than Hong Kong? City left trailing rival in Global Technology Index', *South China Morning Post*, 8 November.

Yeung, SC 2018, 'Is Hong Kong lagging behind in 5G development?', *EJ Insight*, viewed 22 May 2019, http://www.ejinsight.com/20180515-is-hong-kong-lagging -behind-in-5g- development/.

4. Smart Taipei City: understanding policy motivations, approaches and implementation

Shang-su Wu

INTRODUCTION

Taipei, officially known as Taipei City, is the capital of Taiwan and the political, economic and cultural centre of the island. The Taipei municipal government's establishment of a Taipei Smart City Project Management Office (TPMO) in 2016 demonstrates Taipei's smart city ambitions given the need to stay relevant and competitive in the digital age (Smart Taipei 2018a).

There is currently a paucity of literature, particularly of holistic studies, on Taipei as a smart city. The existing bibliography covers specific aspects such as the adoption of new modes of governance (Li et al. 2016, p. 5); the capability of its information and communications technology (ICT) ecosystem (Pan et al. 2011, p. 335; Alamsyah et al. 2016, pp. 114–116); and urban planning aligned with smart city notions of liveability (Wey & Hsu 2014). There are also sources in Mandarin which explain Taipei's smart city policies from a government perspective (Lin, C. 2017, pp. 82–91; Lin 2018, pp. 8–12).

Taipei presents a unique case of smart city development amidst international constraints stemming from the contested sovereignty of the island in which it is located. It should also be studied as one of a growing number of smart cities developing in Asia. This chapter seeks to contribute to a more comprehensive understanding of Taipei as a smart city by (1) presenting an overview of smart city policies in Taiwan at a national-level context; (2) examining smart city policies and the urban challenges they address in Taipei; (3) uncovering the motivation and rationale behind smart city policies; and (4) analysing Taipei's approach towards and key challenges in implementing them. In the process, it also pays attention to how institutional actors, arrangements and processes play a major role in Taipei's smart city push (Hollingsworth 2000, p. 601; Ostrom 2011, p. 8).

SMART CITY POLICIES: TAIWAN

The central government of Taiwan has been pursuing smart city development since the early 2010s. The executive branch of the central government, the Executive Yuan, aims to transform Taiwan into a 'Digital Nation and Smart Island', as articulated by President Tsai Ing-wen and other leaders (Ministry of Economic Affairs 2014).

One of the Executive Yuan's smart city policies is the 'Digital Nation and Innovative Economic Development Program 2017–25', or DIGI+, which is designed to enhance Taiwan's digital infrastructure, regulations, advanced research and development (R&D) capabilities and digital talents (Figure 4.1). Underscoring its role as a key policy component, stakeholders in the government's executive branch oversee DIGI. These stakeholders are part of an Executive Yuan Task Force supported by the Office of Science and Technology (Office of the President of the Republic of China (Taiwan) 2018).

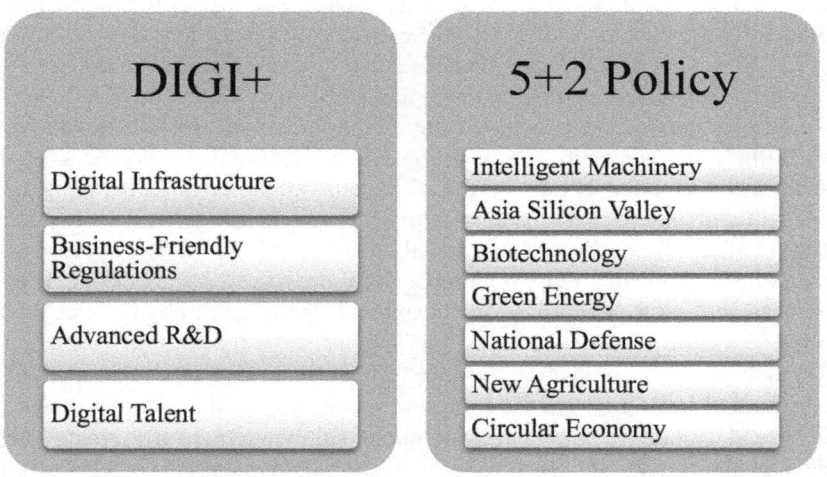

Source: Executive Yuan, Republic of China (Taiwan) 2018.

Figure 4.1   Overview of DIGI+ and the 5+2 policy

Complementing DIGI+ is a '5+2 Major Innovative Industries Policy' (5+2 Policy) meant to advance Taiwan's industrial growth from its traditional focus on contract manufacturing towards value-added business models (Figure 4.1). This policy started out with identifying '5 Pillar Industries' – intelligent machinery, Asia Silicon Valley, biotechnology, green energy and national defence – during Tsai's 2016 presidential campaign. It subsequently incorporated new agriculture and circular economy (Executive Yuan, Republic of China (Taiwan) 2018).

In implementing the 5+2 Policy, the central government has linked each of the targeted seven industries to specific localities in Taiwan. This is seen as an improvement from previous administrations, which had formulations similar to the 5+2 Policy but were insufficiently thorough in implementation at the local level. Geographical specificity would grant city- and county-level stakeholders such as mayors and local governments a sense of ownership of a central government policy. However, one downside to the design of the 5+2 Policy is the over-concentration of a certain industry and associated resources in one area to the detriment of opportunities available in other areas of Taiwan (Rickards 2017).

There are also smart city efforts at the regional level. Under a 'Smart Cities/Townships Regional Innovation Action Plan', the government envisions the development of urban and rural joint ecosystems to promote innovation and develop sustainable living environments. This entails concurrent top-down and bottom-up institutional approaches. From the top-down, central government agencies formulate mid- and long-term projects for smart urban and rural development, as well as coordinate funding from stakeholders in the private sector and other government agencies. From the bottom-up, local governments plan and prioritise smart city projects to meet local demands (DIGI+ 2018).

TAIPEI'S SMART CITY POLICIES

Smart City Framework

Taipei's municipal government adopts a '5+N' framework to guide its smart city policies. The '5' denotes five key areas closely related to the lives of its citizens: Smart Transportation, Smart Public Housing, Smart Healthcare, Smart Education and Smart Payment. Beyond Taipei's five official areas, an additional two – Participatory Governance and Environment Management – are worth highlighting. 'N' denotes innovation, which mirrors a key aspect of the central government's smart city policies. The municipal government adopts 'public–private partnership', 'citizen participation' and 'open government' as core visions in developing its smart ecosystem (Taipei 2018). It also envisions the city as a 'living lab', where the government provides non-financial resources for stakeholders to experiment with innovative solutions using the city as a testbed for new policies, projects and platforms (Figure 4.2) (Eden Strategy Institute & ONG & ONG Pte Ltd 2018).

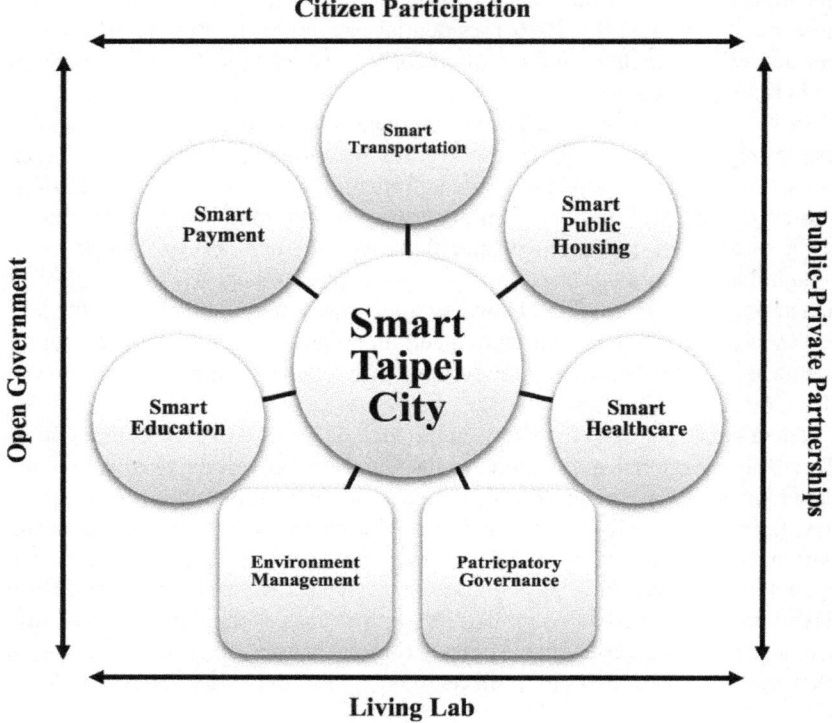

Source: Taipei 2018.

Figure 4.2   Taipei's smart city framework

These visions underpin the municipal government's efforts to create an inclusive and sustainable smart city ecosystem that encourages multiple institutions, including government agencies, startups and citizens, to take ownership as stakeholders. Examples of institutional arrangements and processes are public–private partnerships between government agencies and technology companies, startups' demonstration projects in certain neighbourhoods, Taipei residents' proposals for community projects and cloud-based platforms for open data access.

The TPMO, set up by the Taipei City Government Department of Information Technology (DoIT), functions as the primary city-level programme to facilitate stakeholder participation and the formation of institutional arrangements and processes needed for smart solutions within and beyond the '5+N' framework. Taipei's smart city policies are a function of urban challenges the city faces in the geographical, environmental, political, economic and societal dimensions. Of the seven areas identified (Figure 4.2), this section examines selected policies, projects and platforms in the areas of transportation, housing, healthcare, governance and the environment, where Taipei faces pertinent challenges.

Smart Transportation

Given Taipei's space constraints, the optimal use of land for transportation is essential. The municipal government has thus committed to shifting transport policies and infrastructure development away from motorised transport and towards systems of 'sustainable urban mobility' (Centre for Liveable Cities & Urban Land Institute 2017). Such a shift is exemplified through an approach to mobility known as Mobility-as-a-Service (MaaS), which encourages the consumption of mobility as a service rather than as an asset to be owned. In MaaS, the transportation services of stakeholders are integrated into a single platform such as Umaji, a journey-planner smartphone app that provides personalised travel advice to commuters. The app also incentivises commuters to use suggested public transportation options, thereby helping to manage transport congestion (ICLEI 2018).

To reduce private vehicle usage and promote alternatives, Taipei's municipal government, in partnership with private sector stakeholders, has implemented shared mobility options for bicycles, electric scooters and electric cars. Taiwan's well-known bike-sharing YouBike service, for instance, operates more than 13,000 bicycles across 400 rental stations in Taipei. The YouBike service has added new features due to public–private partnerships between the municipal government, technology vendors and research institutes. For example, a RunKeeper smartphone app developed by the National Taiwan University of Science and Technology and the Industrial Technology Research Institute (ITRI) works in tandem with Taiwan's EasyCard transportation card to help users track fitness information and businesses of interest based on location data (Fulco 2017).

Smart Public Housing

The municipal government uses smart public housing policies and projects as a means to provide affordable yet modern housing, attract young people back to the city and improve urban liveability. To meet these objectives, Taipei's mayor has plans to build 50,000 smart public housing units by 2022 through the public–private partnership model. As a case in point, Taiwan's first smart public housing project – the Dongming Public Housing complex – due for completion in 2019, leverages the expertise of stakeholders in the ICT industry such as Acer Group and Chunghwa Telecom to equip homes with smart features (Taiwan Today 2016; Department of Urban Development, Taipei City Government 2019).

In order to work around having a limited budget, the municipal government has instituted financial arrangements that secure alternative funding for smart public housing (Eden Strategy Institute & ONG & ONG Pte Ltd 2018). The city requires housing developers to allocate an additional 3 to 5 per cent of the construction budget for smart facilities and services. Other than improving Taipei residents' quality of life, smart public housing projects are also institutional arrangements that double as living labs for stakeholders in industries such as construction, ICT and energy to test Internet of Things (IoT) platforms, smart grids and other technologies (Taipei City Government 2018b).

Public housing is meant to be technologically progressive as well as socially inclusive. The municipal government offers subsidised rental rates, and units are reserved for different groups such as ethnic minorities, senior citizens, young professionals and entrepreneurs (Lee, I.C. 2018).

However, government development of smart housing alone would be insufficient, as public housing only comprises 0.8 per cent of total housing in Taipei, with the rest being privately owned (Teng 2018). Private housing projects are also adopting similar smart technologies, which is an essential requirement for smart housing to make a real impact on Taipei's liveability (Hsu 2018a).

Smart Healthcare

Taiwan is officially an aged society, with more than 14 per cent of the population aged 65 or older. Due to economic and societal trends, low fertility rates in Taipei, and Taiwan generally, compound the problems of an ageing population, which in turn constrains economic growth. An ageing population places competing policy demands between resources needed to support seniors and for economic development (The Straits Times 2018).

The TPMO promotes and facilitates projects in healthcare innovation, particularly meant for the elderly, as a way to address the challenges of an ageing populace. Given pressing healthcare needs, public–private partnerships are favoured for their ability to tap the resources and expertise of stakeholders from the private and non-profit sectors, and the community at large.

The homes and living environments of seniors are sites where such innovations and partnerships are applied. For instance, a joint public–private proposal between the municipal government's Department of Social Welfare and a number of small and medium-sized enterprises involves integrating various sensing devices in the homes of seniors to monitor their health-related data and other activities to create a safer living environment. It is also meant to alleviate the shortage of healthcare workers (Smart Taipei 2018c). Amenities in smart public housing complexes, another site of public–private partnerships, also include senior care centres and clinics (Lin, S. 2017).

For general healthcare, institutional arrangements such as a Healthcare IoT medical research alliance set up by stakeholders from the Taipei Medical University's College of Medical Science and Technology enable the development of innovations in fields such as acute care and chronic care (Lee 2017).

Participatory Governance

During Taiwan's authoritarian era, citizen participation in public affairs was extremely limited due to the centralised nature of decision making. With democratisation from the 1980s onwards, citizens began to demand reformed institutional arrangements that facilitate access to decision-making processes, which governments could not ignore with the advent of electoral competition (Li et al. 2016, p. 3).

E-government, which Taiwan implemented from as early as 1997, refers to the application of ICT to improve the quality and efficiency of government services (System Integration Promotion Alliance Project Office 2018). E-government has since evolved to smart governance in which technology is an enabler for greater public participation in governance (Alamsyah et al. 2016, pp. 20–21). A policy and process that operationalises smart governance is i-Voting, an online voting platform which enables citizen participation in the policymaking process (Netherlands Trade & Investment Office Taiwan 2017; Taipei City Government 2015).

Another platform to engage the public is the Citizen Participation Committee comprised of stakeholders from government agencies and non-governmental organisations (NGOs). The committee is organised into working groups that promote public participation in areas such as open data initiatives and 'participatory budgeting', which entails inviting citizens to submit budget proposals for community-related projects. The objectives and processes of these smart governance platforms are institutionally linked. For instance, i-Voting relies on successful open data initiatives that enable citizens to make informed decisions when voting, while budget proposals are voted on by citizens using i-Voting (Centre for Liveable Cities 2016).

Environment Management

Environmental pollution is a major challenge for Taipei. The congested nature of the city and a high population density (9,872 people per kilometre square), including the use of a sizeable number of motor vehicles such as scooters clogging up the city's road network, result in harmful air pollution (Department of Budget, Accounting and Statistics, Taipei City Government 2017; Day 2017).

Technology-enabled smart environment policies and platforms are used to address air pollution. In order to raise the public's awareness of pollution and instil the values of environmental responsibility among them, the municipal government, in collaboration with stakeholders such as the national-level Department of Environmental Protection, technology firms Edimax and Realtek, and the community launched an Air Box project in 2016 to monitor atmospheric particulate matter via sensors and IoT technology. Air Box devices, located in approximately 300 locations around Taipei, provide data on PM2.5 (particulate matter less than 2.5 micrometres in diameter), temperature and humidity, which is accessible to the public via a cloud-based platform (Smart Taipei 2018b). The Air Box project encapsulates the municipal government's smart city visions of public–private partnership, citizen participation and open government.

SMART TAIWAN AND TAIPEI: MOTIVATION AND RATIONALE

The political and corporate leadership in Taiwan strongly support and promote the smart city concept. At Taiwan's hosting of the Smart City Summit & Expo in March 2018, President Tsai Ing-wen and business leaders were reported to have hailed the smart city as a 'catalyst for the next stage in Taiwan's economic development' (Chung & Liu 2018, para. 1). Stan Shih, the founder of Acer Group, Taiwan's renowned ICT firm, viewed the smart city as an 'important platform for the transformation of Taiwan's information communication technology sector' (Chung & Liu 2018, para. 6). In Taipei, the incumbent mayor Ko Wen-je's election to a second term would ensure his continued push for smart city policies (Bodeen 2018).

Such rhetoric is underpinned by a unique motivation to develop a Digital Nation and Smart Island – a means to overcome challenges that arise from Taiwan's contested sovereignty and the resultant international constraints. Due to China's assertive claims over Taiwan, the latter's national government, in particular, faces hurdles in international economic and diplomatic activities such as pursuing free trade agreements (Liu 2016). Pressure from Taiwan's electorate for economic growth provides added motivation to do so (Bodeen 2018). Since Taiwan counts its ICT sector as a competitive advantage to leverage in overcoming such hurdles, the rationale for its smart city vision is to build on this industrial strength and move up the economic value chain from contract manufacturing to value-added innovation (Tan & Chia 2016).

The New Southbound Policy (NSP), a key foreign policy of the Tsai administration, aims to foster economic and people-to-people relations between Taiwan and 18 countries across South and Southeast Asia as well as Australia and New Zealand. Given the absence of formal diplomatic ties between

Taiwan and all the target countries under the NSP, there has been an emphasis on city diplomacy to help achieve the NSP's objectives. This approach under-scores the growing role of cities, such as Taipei, as nodes in global networks of economic, social and cultural influence and as facilitators of stronger transna-tional links between urban centres in different countries (Hsiao 2017).

The Taipei municipal government's smart city institutions are set to acquire a significant international dimension aligned with the central government's NSP and the city diplomacy approach. The municipal government formed an international smart city network – named the Global Organisation of Smart, or Go Smart – in March 2019. Go Smart is intended to be an international platform for Taipei to promote itself as a hub for smart city solutions and for stakeholders from Taiwan's other local governments and the private sector to internationally promote their smart city capabilities (Lee 2019).

Go Smart is set to comprise 27 cities or regions from around the world including the United States, United Kingdom, Malaysia and Australia, as well as Taiwan's five other special municipalities (Kaohsiung, New Taipei City, Taichung, Tainan and Taoyuan) and the private sector. Even companies and officials from cities in China are reportedly planned to be invited, underscoring the potential of Taipei's city diplomacy approach in overcoming constraints over contested national sovereignty (Li & Ke 2018).

Alluding to the crucial role that the smart city concept and cities such as Taipei play in helping Taiwan navigate its international constraints, a Taiwanese minister remarked, 'It is actually very difficult for [Taiwan] to put together a platform for international cooperation. This is huge for us' (Andrews 2018, para. 8).

While acknowledging the key role played by Taipei and a few other notable cities in the island, the smart city discourse in Taiwan also recognises the need to move beyond policies and initiatives for individual cities and explore solutions to be scaled up to regional and national levels. According to Taiwan Premier Lai Ching-te, in September 2017, the government gathered infor-mation pertaining to smart city efforts across the island to bridge a gap in development between Taiwan's six major municipalities – Taipei, New Taipei, Taoyuan, Taichung, Tainan and Kaohsiung – and other cities and counties, which lag behind the municipalities in terms of knowledge and resources. Policymakers intend to review and integrate the resources and expertise of advanced municipalities in order to create standardised models in domains, such as shared mobility schemes and electronic payment platforms, and transfer them to other parts of the island (Executive Yuan, Republic of China (Taiwan) 2017). The effective transfer of such policies, projects and platforms could then help create economies of scale across a wider domestic and even international market (New Southbound Policy Portal 2018).

ANALYSING SMART TAIPEI: APPROACH AND IMPLEMENTATION

According to the commissioner of Taipei's DoIT, the smart city is 'more process than name, more a method than a single objective' (New Southbound Policy Portal 2018, para. 7). It is thus imperative to understand how Taipei's smart city stakeholders and institutional arrangements work to realise a smart Taipei.

Like its smart city counterparts from Singapore, Bhubaneswar (India) and Dubai (United Arab Emirates), Taipei's municipal government set up a dedicated office, the TPMO, in order to promote and facilitate smart city policies, projects and platforms (Eden Strategy Institute & ONG & ONG Pte Ltd 2018). However, providing the required funding, expertise and other resources to deliver solutions on the ground would be beyond the remit of the municipal government alone.

To overcome this limitation, the city's practical approach, spearheaded by the TPMO, is to encourage and coordinate the supply, demand and funding of smart city initiatives. Rather than deliver everything itself, the municipal government provides access to the city's infrastructure, data and other relevant resources so that other stakeholders such as the private sector, NGOs and citizens can devise and deliver solutions. Such an institutional set-up and 'matchmaking' process is considered a best practice in overcoming resource limitations and still delivering policy outcomes (Anderson 2018; Smart Taipei 2018a). In the long term, to ensure the sustainability of this approach, the municipal government aims to convert the TPMO into a financially independent company that only receives partial government funding (Global City Teams Challenge 2018).

Smart city policy functions within Taipei's municipal government are mainly divided between the DoIT and TPMO. Institutional arrangements for smart solutions are formed through both top-down and bottom-up approaches.

Under the top-down approach, the DoIT and TPMO follow the lead of policies and directives set by the municipal government and assist various other municipal government agencies by matching them with relevant stakeholders able to offer solutions to their needs. The DoIT primarily manages the strategy and design of long-term smart city visions, while the TPMO focuses on practical planning and coordination (Taipei 2018). As an indication of its efficacy, this organisation of policy functions between the DoIT and TPMO could potentially be transplanted to the smart city apparatus of Taiwan's other local governments which have shown interest (Global City Teams Challenge 2018).

Under the bottom-up approach, one of the municipal government's institutional arrangements is the 'Taipei Smart City Industrial Field Pilot Program', which encourages the private sector to propose ideas to the TPMO and DoIT for evaluation and possible field-testing in Taipei (Taipei 2018). To encourage public participation in general, the TPMO website contains a section titled 'Something to say' for citizens to submit posts on problems they face. It also includes a section titled 'I have proposals' for companies and other stakeholders to propose solutions (Taipei City Government 2017). Such bottom-up initiatives that tap the resources and motivations of a wide range of stakeholders are a means to overcome existing policy and institutional inertia as Taipei, and Taiwan in general, negotiate a complex transition from a manufacturing economy to an innovative and entrepreneurial one (Funaiole 2018, p. 36).

This combination of top-down and bottom-up approaches has seen the TPMO engage more than 400 stakeholders including the private sector and research institutions, coordinate with more than 30 government agencies, and facilitate more than 130 projects in areas such as deploying Unmanned Aerial Vehicles (UAVs) for rescue efforts and shared mobility schemes (Taipei City Government 2017).

As part of its matchmaking function, the TPMO conducts an initial screening of a project proposal from a company or other stakeholders. It then discusses the feasibility of the project with the relevant municipal government agencies to devise a possible pilot project, which could subsequently be implemented if successful (Global City Teams Challenge 2018).

The concept of Taipei as a living lab enables solution providers to test their proposed solutions, develop use cases and build a brand name that can be exported to other parts of Taiwan and overseas. A living lab driven by non-government stakeholders is also a means to overcome the institutional inertia of government agencies that are unable to respond quickly enough to urban problems (Anderson 2018).

Policy Implementation Challenges

There is much optimism from stakeholders surrounding the potential economic and social benefits of the smart city concept for Taipei and Taiwan. In reality, achieving the desired effects of smart policies and initiatives is dependent upon the alignment of a complex set of variables. A number of policy implementation challenges are pertinent to Taipei: regulatory hurdles; tension between public participation and government oversight; and system integration.

Strict regulations and a rigid bureaucracy that restrict the nature and extent of activities that stakeholders can undertake are cited as a hurdle for Taiwan's smart city ecosystem. For instance, young companies in the digital economy that seek to be listed on local stock exchanges to increase their access to capital and grow their innovative capacities are required to have two years of profit, which is an unlikely scenario for new firms, although there are plans to ease such listing terms in certain situations (Fulco 2017; Hsu 2018b). Local governments also face regulatory challenges. Taipei's municipal government encountered challenges when implementing smart public housing projects due to the need to work with multiple stakeholders from the central government and private sector in order to enact the necessary regulatory and policy changes (Lum 2017).

In response to this issue, the central government has adopted a regulatory sandbox model to provide testbeds for experimentation and the potential application of solutions. An example is the Executive Yuan's approval of a draft bill for UAV innovation and experiments that reduce limits from traffic regulations on the testing of UAVs and autonomous vehicles (SCSE 2018). This directly benefits Taipei, which launched Taiwan's first autonomous vehicle test site in Match 2018 (Taiwan Today 2018).

While collaboration between different stakeholders is a vital part of Taipei's smart city institutional set-up, there are scenarios in which the interests of the different stakeholders involved may not be aligned. Under Taipei's smart governance policies, the implementation of the i-Voting platform has the potential to reach a large proportion of Taipei's citizenry due to the high level of Internet penetration in Taipei, which is an estimated 87.8 per cent (Open Signal 2018).

However, the i-Voting process presents a tension between facilitating public participation in policymaking and government oversight of the process. Under current i-Voting regulations, citizen proposals on an issue should be seconded by 1,500 people and should be subject to governmental reviews and discussions when formulating voting options (Taipei City Government 2018a). While citizens do have a stake in decision making, the process allows city officials to frame an issue and shape options in a manner expedient to them should there be competing demands for the government to balance. Even though tools such as i-Voting encourage smart governance in Taipei, excessive government interference, real or perceived, could discourage public participation and undermine the value of the process.

If a balance is attained between public participation and the extent of government involvement in the process, frequent and substantive voting exercises have the ability to be a more accurate and granular reflection of public feedback than elections every four years. They can also have a greater impact on public policies.

System integration is an implementation challenge for smart city policies, such as with autonomous vehicle development. If experimental projects on autonomous vehicles succeed, their actual integration into Taipei's road system may prove problematic. Narrow and congested road networks with numerous scooters and motorbikes present complicated traffic conditions that may overwhelm sensors and control systems, causing unmanned vehicles to malfunction (Day 2017). A first fatality involving a driverless Uber car in the

United States prompted calls in Taiwan for more rigorous testing to ensure the deployment of mature technology, fixed route operations to minimise confusion, and strong law enforcement to ensure the safety of drivers and pedestrians (Lee, H.Y. 2018). Furthermore, determining legal and insurance liabilities regarding autonomous vehicles remains a major unresolved challenge (Bogost 2018).

CONCLUSION

This chapter has sought to contribute to a more comprehensive understanding of Taipei's smart city push, demonstrated by the Taipei municipal government's establishment of a TPMO in 2016. Institutional actors, arrangements and processes play a major role in Taipei's smart city aim. Multiple institutional actors from government, industry, academia and the community are involved as stakeholders, while institutional arrangements and processes take shape through various policies, projects and platforms.

The central government's overall goal is to transform Taiwan into a Digital Nation and Smart Island (Ministry of Economic Affairs 2014). National-level policies such as DIGI+ and the 5+2 Policy are important initiatives aimed at enhancing Taiwan's digital infrastructure, regulations, advanced R&D capabilities and digital talent as well as driving its industrial growth, respectively. There are also efforts to coordinate smart city policymaking at a regional level under a Smart Cities/Townships Regional Innovation Action Plan.

This chapter, in particular, examined Taipei's smart policies in the pertinent areas of transportation, housing, healthcare, governance and the environment, which address the various challenges the city faces. The TPMO facilitates stakeholder participation and the formation of institutional arrangements and processes needed for smart solutions in these areas. Taipei adopts public–private partnerships, citizen participation, an open government and a living lab as core visions in developing its smart ecosystem (Smart Taipei 2018a).

In uncovering motivations and rationale, Taiwan has a unique motivation to push for smart city development – a means to overcome international constraints stemming from its contested sovereignty. The rationale for its smart city vision is to build on its industrial strengths in the ICT sector and move up the economic value chain from contract manufacturing to value-added innovation and services (Tan & Chia 2016).

An emphasis on city diplomacy to help overcome Taiwan's international constraints has seen the growing role of Taipei as a node in global networks of economic, social and cultural influence and as a facilitator of stronger transnational links between urban centres in different countries. To that end, the Taipei municipal government formed an international smart city network, Go Smart, which is aligned with the central government's foreign policy objectives.

It also seeks to build its credentials as an international hub for smart city solutions (Hsiao 2017).

Additionally, this chapter analysed Taipei's approach to smart city policy-making and the challenges in implementing policies. To overcome the limitations of providing the required resources and expertise to deliver solutions, the TPMO's practical approach is to play a matchmaking role by coordinating the supply, demand and funding of smart city initiatives among various stakeholders. Institutional arrangements for smart solutions are formed through both top-down and bottom-up approaches.

Achieving the successful implementation and desired effects of smart policies and initiatives is dependent upon the alignment of a complex set of variables. This chapter discussed a number of policy implementation challenges pertinent to Taipei: regulatory hurdles; tension between public participation and government oversight; and system integration.

REFERENCES

Alamsyah, N, Susanto, TD & Chou, TC 2016, 'A comparison study of smart city in Taipei and Surabaya', paper presented at the international conference on ICT for smart society, Surabaya, 20–21 July.

Anderson, CR 2018, 'Taipei's grand plan to create a citizen-centric smart city', *Disruptive. Asia*, viewed 20 March 2018, https://disruptive.asia/taipei-citizen-centric -smart-city/.

Andrews, J 2018, 'Taipei launches smart city network', *Cities Today*, 28 March, viewed 15 September 2018, https://cities-today.com/taipei-launches-smart-city-network/.

Bodeen, C 2018, 'Analysis: Strong opposition results challenge Taiwan's Tsai', *The Washington Post*, 26 November, viewed 26 November 2018, https://www .washingtonpost.com/world/asia_pacific/analysis-strong-opposition-results-challenge -taiwans-tsai/2018/11/26/4f21a1b6-f1dd-11e8-99c2-cfca6fcf610c_story.html?utm _term=.bf707ad9b5ee.

Bogost, I 2018, 'Can you sue a robocar?', *The Atlantic*, 20 March, viewed 9 April 2018, https://www.theatlantic.com/technology/archive/2018/03/can-you-sue-a-robocar/ 556007/.

Centre for Liveable Cities 2016, 'Fresh directions in urban governance', *Urban Solutions*, Issue 9 (July), viewed 15 September 2018, https://www.clc.gov.sg/docs/ default-source/urban-solutions/urb-sol-iss-9-pdfs/roundtable-basuki-purnama-len -wenje.pdf.

Centre for Liveable Cities & Urban Land Institute 2017, *Urban mobility: 10 cities leading the way in Asia-Pacific*, viewed 15 September 2018, https://asia.uli.org/wp-content/ uploads/sites/126/ULI-Documents/UrbanMobility-10CitiesLeadingtheWayinAsia -Pacific.pdf.

Chung, JF & Liu, KL 2018, '"Smart cities" to spark Taiwan's future economic development: officials', *Focus Taiwan*, 27 March, viewed 15 September 2018, http:// focustaiwan.tw/news/ast/201803270022.aspx.

Day, H 2017, 'You thought your commute was bad! Incredible footage shows "scooter waterfall" as thousands of bikers cascade down a bridge in Taiwan', *Daily Mail*, 18

April, viewed 22 February 2018, http://www.dailymail.co.uk/news/article-4421908/ Taiwanese-footage-shows-scooter-waterfall.html#ixzz57p5MyAM8.

Department of Budget, Accounting and Statistics, Taipei City Government 2017, *Short report on important statistics of Taipei City*, December, viewed 22 February 2018, http://w2.dbas.taipei.gov.tw/web/ShortReport.pdf.

Department of Urban Development, Taipei City Government 2019, *Nangangqu Dongming gonggong zhuzhai (Nankang district Dongming public housing)*, viewed 4 June 2019, https://www.hms.gov.taipei/#!/map.

DIGI+ 2018, *Intelligent cities*, viewed 15 September 2018, https://www.digi.ey.gov. tw/en/WaterFall.aspx?n=505F79F68C535012.

Eden Strategy Institute & ONG & ONG Pte Ltd 2018, *Top 50 smart city governments*, viewed 15 September 2018, https://static1.squarespace.com/static/ 5b3c517fec4eb767a04e73ff/t/5b513c57aa4a99f62d168e60/1532050650562/Eden -OXD_Top+50+Smart+City+Governments.pdf.

Executive Yuan, Republic of China (Taiwan) 2017, *Government aims to more evenly develop smart cities, country: premier*, 29 September, viewed 15 September 2018, https://english.ey.gov.tw/News_Content.aspx?n=3FA02B129BCA256C&sms= 925E4E62B451AB83&s=DA5A7E8E02B1C245.

Executive Yuan, Republic of China (Taiwan) 2018, *'Five plus two' innovative industries plan*, 30 March, viewed 15 September 2018, https://english.ey.gov.tw/ News_Content.aspx?n=0899B3FCC4B38357&sms=8BCD9CBBA95A001D&s= 9FDC09B0F3DB4F96.

Fulco, M 2017, 'Taiwan's cities smarten up', *Taiwan Business Topics*, 15 August, viewed 15 September 2018, https://topics.amcham.com.tw/2017/08/taiwans-cities -smarten/.

Funaiole, MP 2018, 'Building tomorrow's cities', in Glaser, BS & Funaiole, MP (eds.), *Perspectives on Taiwan: Insights from the 2017 Taiwan–U.S. policy program*, pp. 33–38. Lanham, MD: Center for Strategic & International Studies, Rowman & Littlefield.

Global City Teams Challenge 2018, *Taipei smart city living lab*, viewed 15 September 2018, https://gctc.opencommons.org/Taipei_Smart_City_Living_Lab.

Hollingsworth, RJ 2000, 'Doing institutional analysis: implications for the study of innovations', *Review of International Political Economy*, vol. 7, no. 4, pp. 595–644.

Hsiao, R 2017, 'Can city diplomacy promote the New Southbound Policy and Taiwan's international space?', *Global Taiwan Institute*, vol. 2, issue 29, viewed 15 September 2018, http://globaltaiwan.org/2017/07/26-gtb-2-29/.

Hsu, C 2018a, 'Farglory seeks 25% jump on "smart" apartment sales', *Taipei Times*, 13 January, viewed 15 September 2018, http://www.taipeitimes.com/News/biz/archives/ 2018/01/13/2003685659.

Hsu, C 2018b, 'TWSE to lower listing requirements', *Taipei Times*, 4 January, viewed 15 September 2018, http://www.taipeitimes.com/News/biz/archives/2018/01/04/ 2003685126.

ICLEI 2018, *Selected SEA cities and transport experts learn about urban mobility best practices in Taipei*, 30 April, viewed 15 September 2018, http://icleiseas.org/ index.php/2018/04/30/selected-sea-cities-and-transport-experts-learn-about-urban -mobility-best-practices-in-taipei/.

Lee, HY 2018, 'Uber crash prompts reflection on driverless car development in Taiwan', *Focus Taiwan*, 24 March, viewed 9 April 2018, http://focustaiwan.tw/ news/aeco/201803240009.aspx.

Lee, IC 2017, '"Smart" device used in ER rooms to prevent violence', *Taipei Times*, 28 May, viewed 29 February 2018, http://www.taipeitimes.com/News/front/archives/2017/05/28/2003671434.

Lee, IC 2018, 'Applications open for Taipei public housing units', *Taipei Times*, 1 August, viewed 15 September 2018, http://www.taipeitimes.com/News/taiwan/archives/2018/08/01/2003697769.

Lee, IC 2019, 'New smart cities group established at Taipei summit', *Taipei Times*, 28 March 2019, http://www.taipeitimes.com/News/taiwan/archives/2019/03/28/2003712327.

Li, J, Liu, X & Li, W 2016, 'City profile: Taipei', *Cities*, vol. 55, pp. 1–8.

Li, Y & Ke, W 2018, 'New smart city promotion body to be formed in Taipei in 2019', *Digi Times*, 3 August, viewed 15 September 2018, https://www.digitimes.com/news/a20180803PD205.html?chid=9.

Lin, C 2017, 'Chengshizhili yu zhihuichengshi chanyequdong Taibeicelue' (The strategy of city governance and smart city industries in Taipei), *Guotuyugonggongzhili* (*Public Governance Quarterly*), vol. 5, no. 4, pp. 82–91.

Lin, S 2017, 'Public-housing project breaks ground in Taipei', *Taipei Times*, 11 March, viewed 23 February 2018, http://www.taipeitimes.com/News/taiwan/archives/2017/03/11/2003666559.

Lin, W 2018, 'Tishenfuwuzhineng, gongsixieli dazao Taibeizhihuicheng' (Improving service for building up Taiwan smart city with public and private efforts), *Zhungfujiguan Zixuntongbao* (*Information Bulletin of Governmental Institutes*), no. 352 (April), pp. 8–12.

Liu, DN 2016, 'The trading relationship between Taiwan and the United States: Current trends and the outlook for the future', *Brookings*, November 2016, viewed 23 February 2018, https://www.brookings.edu/opinions/the-trading-relationship-between-taiwan-and-the-united-states-current-trends-and-the-outlook-for-the-future/.

Lum, KK 2017, 'Taipei City targets 20,000 rent-only public housing units by 2018', *Edgeprop*, 12 January, viewed 29 February 2018, https://www.edgeprop.my/content/1034124/taipei-city-targets-20000-rent-only-public-housing-units-2018.

Ministry of Economic Affairs 2014, *Chuangxin lehuo* (*Creativeness and Happiness*), no. 28, 8 May, viewed 9 November 2019, https://www.moea.gov.tw/Mns/populace/news/EpaperContent.aspx?kind=1&menu_id=5498.

Netherlands Trade & Investment Office Taiwan 2017, *Smart cities Taiwan: Opportunities for Dutch companies*, viewed 9 April 2018, https://www.rvo.nl/sites/default/files/2017/05/taiwan-ambition-and-development-of-smart-cities-v2.pdf.

New Southbound Policy Portal 2018, *An urban evolution – Smart cities, from Taiwan panorama*, viewed 15 September 2018, https://nspp.mofa.gov.tw/nsppe/news.php?post=138944&unit=410.

Office of the President of the Republic of China (Taiwan) 2018, *President Tsai attends opening ceremony for 2017 World Congress on Information Technology*, 11 September, viewed 14 September 2018, https://english.president.gov.tw/NEWS/5205.

Open Signal 2018, *The state of LTE*, viewed 19 June 2018, https://opensignal.com/reports/2018/02/state-of-lte.

Ostrom, E 2011, 'Background on the institutional analysis and development framework', *Policy Studies Journal*, vol. 39, no. 1, pp. 7–27.

Pan, JG, Lin YF, Chuang SY & Kao YC 2011, 'From governance to service – smart city evaluations in Taiwan', paper presented at the *International Joint Conference on Service Sciences*, Taipei, 25–27 May.

Rickards, J 2017, 'Will the "5+2 Industrial Innovation Program" catch on better than its name?', *The News Lens*, 4 December, viewed 15 September 2018, https://www .smartasiataiwan.com/en_US/industry/news/info.html?id=2A0F483D665E2592& totalCount=13&currentRow=9.

SCSE 2018, *Taiwan's executive Yuan approves bill promoting unmanned vehicle experimentation*, viewed 9 June 2018, https://en.smartcity.org.tw/index.php/en-us/ news/item/599-taiwan-s-executive-yuan-approves-bill-promoting-unmanned-vehicle -experimentation.

Smart Taipei 2018a, *About*, viewed 9 April 2018, https://smartcity.taipei/about.

Smart Taipei 2018b, *Air box makers education empirical project*, viewed 9 April 2018, https://smartcity.taipei/project/6.

Smart Taipei 2018c, *Smart service application in old houses of senior citizens*, viewed 15 September 2018, https://smartcity.taipei/project/39.

System Integration Promotion Alliance Project Office 2018, *E-government*, viewed 9 April 2018, https://www.sipa.org.tw/eg.aspx.

Taipei 2018, *Transformative Taipei*, viewed 15 September 2018, https://drive.google .com/file/d/1xdd8iCA-l78suZdhl8spwQpPhkLdjTgt/view.

Taipei City Government 2015, *Taipei i-Voting system in the works*, viewed 9 April 2018, https://english.gov.taipei/News_Content.aspx?n=A11F01CFC9F58C83&sms =DFFA119D1FD5602C&s=3BBDE32E253F5C32.

Taipei City Government 2017, *Smart city website goes alive*, viewed 15 September 2018, https://english.gov.taipei/News_Content.aspx?n=A11F01CFC9F58C83&sms =DFFA119D1FD5602C&s=4E5A58C673DEB72B.

Taipei City Government 2018a, *i-Voting*, viewed 19 June 2018, https://ivoting.taipei/.

Taipei City Government 2018b, *Taipei City smart communities of public housing implementation project*, viewed 23 February 2018, http://www.housing.taipei.gov .tw/ph-ae/smart-community.

Taiwan Today 2016, 'Nangang hosts Taiwan's 1st smart public housing project', *Taiwan Today*, 18 January, viewed 9 April 2018, https://taiwantoday.tw/news.php ?unit=10,23,10&post=21672.

Taiwan Today 2018, 'Taipei City Government unveils self-driving vehicle test site', *Taiwan Today*, 14 March, viewed 9 April 2018, https://taiwantoday.tw/news.php ?unit=2,6,10,15,18&post=130926.

Tan, J & Chia S 2016, 'A tiger that lost its roar, Taiwan pays price for not looking ahead', *Today*, 27 February, viewed 9 April 2018, https://www.todayonline.com/ world/asia/tiger-lost-its-roar-taiwan-pays-price-not-looking-ahead.

Teng, PJ 2018, 'Governors, experts gather in Taipei to talk about how to build a smart city', *Taiwan News*, 29 March, viewed 19 June 2018, https://www.taiwannews.com .tw/en/news/3393609.

The Straits Times 2018, 'State-run dating service is Taiwan's solution to ageing society', *The Straits Times*, 12 April, viewed 23 February 2018, https://www.straitstimes.com/ asia/east-asia/state-run-dating-service-is-taiwans-solution-to-ageing-society.

Wey, WM & Hsu J 2014, 'New urbanism and smart growth: Toward achieving a smart National Taipei University District', *Habitat International*, vol. 42, pp. 164–174.

5. The evolution of smart city in South Korea: the smart city winter and the city-as-a-platform

Jong-Sung Hwang

INTRODUCTION

A smart city is a basic building block for an intelligent society.[1] It provides common 'platforms': infrastructure and rules that facilitate transactions and service developments (Eisenmann et al. 2006; Bridgwater 2015) for smart technologies such as big data, artificial intelligence (AI) and robots. Without such common platforms, it becomes very difficult and costly to develop smart services, because each service then needs to build its own infrastructure. Similar to the way industrial societies and information societies were heavily dependent on physical social overhead capital (SOC) and the Internet, respectively, for their development, intelligent societies need the smart city as their foundation.

However, it should be noted that not all smart cities provide platforms for an intelligent society. In fact, it has been rare to find such smart cities. According to a study by ITU-T (2014), 'smart city' has been defined in many different ways; the study lists 116 definitions. Most definitions have little, if anything, to do with city-wide platforms. Instead, they define 'smart city' in terms of specific city services or solutions. Recently, the need to develop a 'smart city-as-a-platform' has arisen due to the advancement of smart technologies that require a new breed of city-wide platforms (Hwang 2016; Navigant Research 2018).

The experiences of South Korea (hereafter Korea) showcase why the 'smart city-as-a-platform' strategy is important. Korea began smart city projects in 2003, ahead of other countries, under the name 'ubiquitous city' or 'u-City'. u-City sparked great expectations and garnered support when it began, because it was thought to make the best of the advanced 'ubiquitous computing' (Weiser 1991) technologies across cities, and innovate city operations and services in a way that was not possible before. However, u-City took a technology-oriented vertical approach which focused more on applying technologies than innovating new processes for city operations and services and simply developed applications separately for each city service. With such a partial and fragmented approach, u-City was not able to meet the high expectations of the Korean public, who expected totally transformed city operations and citizen life. These disappointments pushed u-City development into a long-running slump or 'smart city winter', analogous to the idea of an 'AI winter' (Russell & Norvig 2003).

The smart city winter ended when the Korean Government adopted the Fourth Industrial Revolution (FIR) as a national development strategy in 2017. Soon after, Korea began to redesign its smart city strategy from a technology-oriented vertical approach to an innovation-oriented horizontal one which stressed continuous innovations, using technology as a tool, not an end, on the one hand and focusing on removing boundaries between city services through city-wide common infrastructure on the other. Korea's experiences can provide many practical lessons and insights, not because it has achieved much success, but because Korea might be the only country that has experienced the disillusionment of a smart city winter.

EVOLUTION OF THE SMART CITY

The term 'smart city' is an evolving concept encompassing several unique characteristics. Its connotations have changed over time, shifting from something akin to 'sustainable city' in the 1990s to 'digital-' or 'connected city' in the late 1990s and 2000s, and to 'citizen-centred city' in the mid-2010s (ITU-T 2014; Eremia et al. 2017). At the same time, 'smart city' is a 'boundary object' which bridges different disciplines to create common understanding and to facilitate cooperation (Opdam et al. 2015). It has 'different meanings in different social worlds but their structure is common enough to more than one world to make them recognisable, a means of translation' (Star & Griesemer 1989, p. 393). For instance, urban engineering and information and communication technology (ICT), two leading fields in smart city development, use the term with slightly different meanings. Urban engineering stresses the sustainability and development issues of a city, while ICT focuses on the digital side such as platform development and intelligent services. Many fields now have some sort of involvement in smart city development, so it is natural that 'smart city' is interpreted differently in different fields.

Due to the conceptual ambiguity, two opposite approaches have been applied to smart city development: a fragmented approach and a holistic approach. The fragmented approach is based on a reductionist view which understands a smart city through its components. A city becomes a 'smart city' when its core components become 'smart'.[2] For instance, Frost and Sullivan

(2013) use eight 'smart' categories to define this entity: smart building, smart energy, smart information technology, smart mobility, smart city planning, smart business, smart governance and smart citizenry. If at least five of these eight parts become smart, the city will be considered a smart city (Siemens 2015).

A holistic approach believes that a smart city has its own unique characteristics which are 'other than the sum of the parts' (Six Revisions 2010). In this approach, making a smart city is in essence different from merely making all parts of a city smart. The role of a smart city is to optimise the functions of the whole city, not to stop at finding solutions for individual problems. In this approach, a city is understood as a system that is 'a set of things ... interconnected in such a way that they produce their own pattern of behaviour over time' (Meadows & Wright 2008, p. 2). City-as-a-platform is a typical example of this holistic approach.

Looking back on the history of smart cities, we can find a shifting trend from a fragmented approach to a holistic one. Smart City 1.0, [3] or the first generation of smart city, began in the mid-1990s and was just one of many attempts (including 'sustainable city', 'eco-city', 'green city' and 'liveable city'),[4] to solve urban problems in an innovative way (Albayrak & Eryilmaz 2017). There were no clear distinctions between these strategies; many projects were regarded as 'smart city initiatives' just because they attempted to make a city better and more self-sustaining.

The next stage, Smart City 2.0, led by Korea's u-City projects in the early 2000s, adopted a new method that focused on the use of ICT to improve city efficiency and services. As ICT became a key driver for social change through digital transformation, 'smart city' began to take on a leading role in city innovation. Sensors and other information technologies were extensively applied to city operations such as security control and traffic management. Vertical integration, or integration within a field, of data and applications became one of the major methods for city innovation.

Smart City 3.0 began with the proliferation of big data after 2010. It attempted horizontal sharing and convergence of data beyond the boundary of each field in order to increase the accuracy of decision making and provide high-quality and customised services to citizens. Smart City 3.0 approached problems from a city-wide perspective, unlike Smart City 2.0, which had a field-specific perspective. More and more cities developed centralised city operation systems, such as Rio de Janeiro's City Operations Centre in 2010, and upgraded urban management and public services by combining data from different systems.

The latest development, Smart City 4.0, tries to embed advanced smart technologies into city services as well as the daily lives of its citizens. For instance, the city of Columbus in the United States is developing smart transportation systems which will deliver more diversified options to citizens by using data and well-connected networks (Torres 2018), and Singapore is building a city-wide platform called Virtual Singapore (National Research Foundation 2018). Smart City 4.0 aims to transform a city's structure itself, whereas preceding smart cities were focused on reforming city operations and services within the existing city structure. Advanced smart technologies need a city to be transformed into a city-as-a-platform that provides city-wide common platforms for data to flow through, new technologies to deploy and services to combine across individual fields.

The city-as-a-platform has two advantages; first, it allows for much easier use of new technologies and services. With only marginal investment and effort it is possible to introduce new technologies and services due to the common foundation of the city platform already in place. If a city does not have such a platform, each service provider has to develop and build the technological and institutional bases necessary for their services. The second advantage is the democratisation of technology; even individual citizens or small and medium enterprises (SMEs) can develop innovative services for themselves. This enables open innovation and bottom-up development in which anyone can actively engage and try their ideas freely. If these public platforms are not developed, a smart city may become a playground only for governments or global companies that are able to build their own service bases.

U-CITY AND THE 'SMART CITY WINTER'

The u-City project first appeared in Korea in February 2003, when Samsung announced its plan to construct a u-City in Hwaseong Dongtan, a new town which began construction in 2001 (Ahn 2003). The announcement had massive repercussions. Korea's three biggest cities, Seoul, Busan and Songdo in Incheon, followed its lead and declared plans to build their own u-City that same year. However, the appearance of u-City was not what was planned by the government. Even though the government's ICT strategy at the time, u-Korea, included long-term plans to realise a vision of 'ubiquitous commuting' (Weiser 1991) at the city level, it had not planned to develop a full-scale u-City that early in the technology's development.

The key problem was timing; the overall level of technology and smart city market maturity were not then developed enough to implement smart city initiatives efficiently and economically. For instance, while it was possible to use sensors and create real-time data from a single source for any specific purpose, it was very difficult to collect and analyse data from various sources, which is critical to making a city smart. The Korean Government thus had originally planned to develop technologies and infrastructure first before starting on smart city developments (Chin & Rim 2006).

Unlike the national government's plan, local governments and companies quickly joined the u-City trend. There were many shared interests among them. The heads of local governments, elected every four years, wanted big and challenging projects to show how innovative their cities were, regardless of feasibility. ICT companies expected u-City development to create new markets, especially since the ICT market in Korea had become saturated in the early 2000s and the construction industry was searching for a new city model that would have a competitive edge in markets both at home and abroad.

The u-City projects had a very good start. The media and the public initially showed great interest and had high expectations. They expected u-City to create a whole new society by both digitalising city facilities and processes and innovating new ways for cities to operate and people to live, leading to a better life at a personal level and economic development at a national level (Ahn 2003). Such enthusiastic responses in turn encouraged more and more cities to accept the u-City idea. A positive cycle was created. In three years, six out of the seven metropolitan cities in Korea had joined the u-City trend and by 2006 most of the new city development projects in progress adopted u-City as their main vision (Ministry of Information and Communication (MIC) 2006). Figure 5.1 shows the number of greenfield u-Cities which were built over newly developed areas.

Even though its appearance was considered premature by the government as explained earlier, u-City has contributed to solving a number of urban problems, like traffic congestion, crime and the low quality of public services.

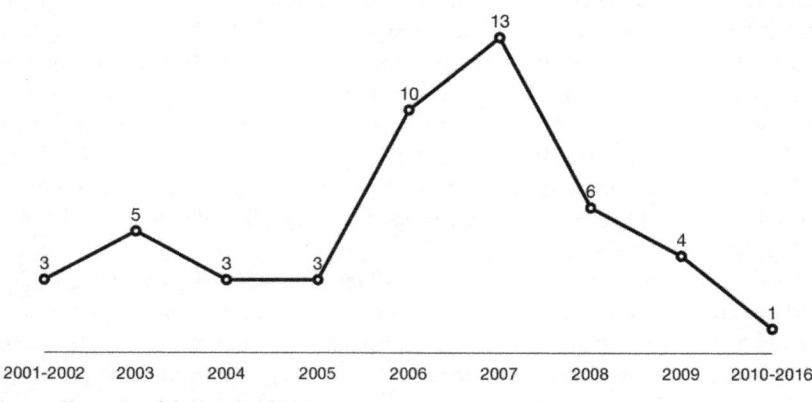

Smart City Winter in Korea

Source: Korea Land & Housing 2016.

Figure 5.1   Number of greenfield u-City projects based on start year

For instance, the Seoul transportation card system introduced in 2004 has revolutionised the payment process by developing a common payment platform for all forms of public transportation and charging, not based on the number of transfers but on the total distance a passenger travelled during a certain period of time. This new system greatly boosted citizens' preference for public transportation. Safety issues also have been substantially improved by integrating CCTV monitoring into districts. Location-based services that can track locations using mobile phone data have enhanced young students' safety going to and from school. Due to the solutions enabled by u-City development, Korea was recognised as a smart city leader and Seoul was selected by the International Telecommunications Union (ITU) as an exemplary case of a smart city in 2013 (ITU-T 2013).

However, u-City projects faced a major challenge when the first newly developed u-City, Hwaseong Dongtan, was unveiled in September 2008. The u-City fell far short of the expectations of Koreans, who wanted to see and experience a future city which was well advanced in terms of both digitalisation and transformation. Services in the u-City were hardly different from existing regular ones (Kim 2009), with the exception of some successful cases mentioned earlier such as the transportation card system, so that even many residents of the u-City did not see much evidence that they were living in one.[5] The number of services was not a problem: according to a government report, each u-City had introduced 19.3 new services on average, and as many as 33 new services, as in the city of Siheung (Ministry of Land, Infrastructure and Transport (MOLIT) 2013). The problem was the lack of innovation.

In order to deliver differentiated value, u-City services should have been developed based on a balanced combination of digitalisation and innovation, as in the cases of the Seoul transportation card and integrated CCTV (closed circuit TV) monitoring. However, most u-City services in practice relied heavily on digitalisation and made only marginal, if any, innovations in the way services were managed and delivered. As a result, many failed to differentiate themselves from existing services despite the large investment in technology. Moreover, u-City services worked independently of each other so that they could hardly create a synergic effect and provide meaningful value to residents. Behind this technology-oriented vertical approach was the fact that u-City was recognised as a product, and products usually stop developing once they are produced. Even though basic maintenance was provided, innovations and re-developments were stopped after a service was in place in most cases, resulting in u-Cities with many new technologies but insufficient innovation.

In some cases, the government's policies have worked as obstacles to innovation. The Korean Government developed guidelines for u-Cities and suggested a list of 228 services for u-Cities to develop (MOLIT 2009). However, these policies brought unexpected results, limiting choices rather than stimulating innovation. The guidelines were no more than examples to help local governments and developers to choose useful services. However, they were perceived as a sort of boundary that u-City should not cross, as criticism was raised that u-City caused excessive investment in urban development. In fact, houses in u-City were priced higher than others mainly due to the 'u-City' brand and maintenance costs increasing as more services were added. In the face of such criticism, it became increasingly difficult to experiment with services beyond the government's list.

Another point to note is that u-City development was concentrated in the capital city region. As of the end of 2012, 55 per cent of local governments in the capital region were engaged in u-City projects compared to 23 per cent in other regions. One of the u-City goals promoted by the Korean Government was to achieve balanced regional development, but, in reality, u-Cities were concentrated in the capital region, in contrary to the original goal (MOLIT 2013). Such imbalance was also seen between existing towns and new towns: 65 per cent of u-City projects were implemented in new towns, while run-down old towns benefited little (MOLIT 2013). This skewed distribution may be a result of a market mechanism, but it added to weakening social and political support for u-City developments. Because residents in new towns are on average wealthier than those in existing towns, u-City was scorned as a project for 'the haves'.

Disappointed by u-City development, Koreans began to re-think it. As shown in Figure 5.1, the number of greenfield u-City projects for new city development dropped sharply and only one u-City development project was added after 2010. The so-called 'smart city winter' had come. Many planned projects were cancelled, and smart city projects lost priority in government policies. Only the Korea Land & Housing Corporation, a state-run urban developer, proceeded with u-City services development as part of a city development package. The scope of u-City projects was reduced to focus only on the enhancement of citizens' safety from natural disasters and crime, and CCTV became central to smart city development while other innovative technologies were pushed aside. In 2010, even as the rest of the world began to pay attention to smart city technology, Korea closed the door to such trends and went into a long period of smart city hibernation.

To prevent a smart city winter from recurring in the future, it is important to understand why these problems occurred in the first place. It is obvious that timing was a key issue. Implementing u-City developments before the technologies and the market were mature enough to accept the new concepts was problematic. However, timing was not the only issue: despite the premature launch, u-City development has produced a number of successful cases, such as the transportation card and integrated CCTV monitoring.

Instead, we need to look at the mismatch between how u-Cities, and smart cities in general, were defined (the concept) and what was wanted from them (the goal). In most cases in Korea, u-City was considered a product, similar to smart cars and intelligent buildings. For instance, creating a u-City was thought of as a construction project: the title of the u-City law enacted in September 2008 was 'Act on the Construction, Etc. of Ubiquitous Cities' (The Korean Government 2009). In 2017, the title was changed to 'Act on the Development of Smart City', but by then the Korean Government had already understood u-City and 'smart city' in general from the point of view of a construction project for some time. This is a serious misconception of city innovation. Once products are manufactured, they stop evolving and innovating. 'Smart city' should be interpreted as a space where innovation continues unabated. The goal of a smart city is not to solve a single city problem but to achieve the ongoing sustainable development of a city. For this reason, 'smart city' should be understood not as a product, but as a platform.

There is a substantial difference between a product and a platform. A product achieves its best performance when completed. Adding new technologies and functions after production is as hard as making developing a new product (Bridgwater 2015). Therefore, a smart city as a product will gradually degrade after completion despite maintenance and require redevelopment at a certain point. In comparison, a platform does not have completion as a goal. A platform is not designed to complete any specific tasks; instead, it helps other products to provide services easily and in an improved manner. Likewise, a smart city as a platform will keep developing all the time and enable various contributors from the private as well as the public sector to innovate city operations and services. In some ways, a smart city can be thought of as both a product and a platform, but it is important to emphasise the platform aspect of a smart city in order for it to be sustainable.

Additionally, changes in Korea's national development strategies also accelerated the decline of the u-City. President Lee Myung-bak's administration beginning in 2008 tried to turn the national development strategy away from the digital industry toward the traditional manufacturing industry. The MIC in charge of ICT issues was dismantled. This shift accompanied fundamental changes to the governance of u-Cities, which was based on a two-tiered system consisting of MIC and MOLIT in order to promote the cooperation between the ICT and the city sectors. After MIC was dismantled, the governance was consolidated entirely under MOLIT, and more emphasis was placed on construction and physical development. As a result, the 'product' approach to the u-City became stronger than before and it became more difficult to take a city-as-a-platform approach in Korea.

SEARCHING FOR A NEW MODEL

It was not easy for Korea to end the smart city winter. Many Koreans had lost trust in smart city development from their experiences with u-City. Despite some successful cases, u-City was heavily product-oriented, fell short of expectations of digital transformation and was too fragmented to create substantial value. In fact, such disillusion was not restricted to u-City. Koreans had experienced ubiquitous computing services through IT839[6] and u-Korea projects since the early 2000s, and when they found the services were prematurely launched and not reliable enough to use in real life, their confidence in emerging technologies was shaken. According to the World Values Survey, the Korean respondents who approved of more emphasis on technology development declined from 72.2 per cent at its peak in 2005 to 66.6 per cent in 2010 (Lagos et al. 2014), after Koreans experienced ubiquitous computing technologies in person.

The smart city winter finally came to an end when President Moon Jae-In announced, contrary to his predecessor Lee Myung-bak, that smart city technology would return as a growth engine for the Korean economy in August 2017. Smart city was regarded as a critical cornerstone of the FIR, the new national development strategy led by the Moon administration.

The relationship between the real and cyber worlds, as shown in Figure 5.2, explains well why the FIR needs smart city for its platforms. Digitalisation since the mid-1990s could be thought of as an attempt to send the real world to the cyber, creating digital counterparts such as e-government, digital libraries and online banks. The Internet alone could cover most platform functions for digitalisation. In contrast, the FIR aims to move the cyber world (data and algorithms) to the real world; allowing AI to give instructions directly to machines, and autonomous cars and robots to then turn such instructions into actions without human intervention. For FIR technologies to work well, various kinds of platforms are needed beyond the Internet, such as location detectors, data hubs and digital twins. Smart city-as-a-platform could provide such infrastructures in a well-organised manner, so the Korean Government restarted smart city development for the FIR strategy.

In the second half of 2017, the Korean Government created the Korean Smart Cities Special Committee under the Presidential Committee on the FIR in order to link smart city to the FIR strategy. The Special Committee embraced the idea of city-as-a-platform and in January 2018 announced a new smart city strategy, which put emphasis on establishing platforms and supporting sustainable urban innovation. This was in contrast to the previous approach: building the components of a smart city in a fragmented way or on a one-off basis as though making a product (The Korean Government 2018).

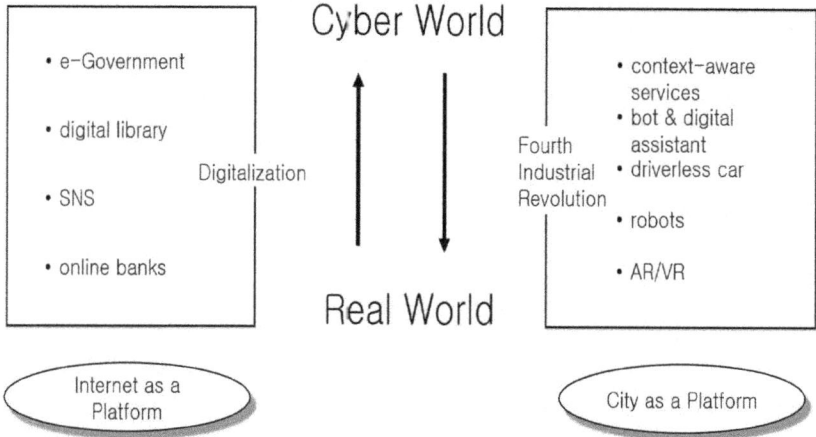

Note: SNS: social network service; AV/VR: augmented reality/virtual reality.

Source: Author.

Figure 5.2   Two ways to connect the real and cyber worlds

According to the new strategy, Korea's smart city will be created as an open city where citizens and companies can actively participate in city development and public policymaking and turn their ideas into reality.

Korea's new strategy has three characteristics. First, it takes a holistic approach. The biggest lesson Korea has learned from the smart city winter is that a fragmented approach is not effective in solving fundamental problems and in meeting people's high expectations. It should be noted that the optimisation of all parts does not necessarily lead to the optimisation of the whole. For instance, the traffic of a city does not improve simply by making all drivers use the fastest route suggested by a car navigation service, because it may create even more traffic by concentrating drivers on recommended routes. It is necessary to develop an optimised model at the city level and guide each driver based on it. In this sense, the Korean Smart Cities Special Committee seeks to place more emphasis on smart city platforms and architecture over individual services.

Second, Korea's approach is to associate smart cities with industrial development. In other words, Korea's smart cities aim not only to solve urban problems but also to boost the FIR-related industries by creating demands and promoting technology development. This is in line with Korea's 40-year-old industrial development strategies, like the Heavy and Chemical Industrialisation in the 1970s and the CDMA (code-division multiple access) digital mobile development strategy in the 1990s. Korea launched an intensive risk-taking strategy to accelerate economic development. In order to leapfrog to a more advanced economy, Korea often set very difficult goals like developing the world's first CDMA mobile service and achieved them by devoting various resources and policies to the plan and fostering close collaboration between the government, industries and academia (Lee 2013). Likewise, Korea's smart city initiative is an ambitious new project aiming not just at solving existing urban problems, but also at developing a future-city model which works as a platform; this is a difficult goal to achieve but if successful it will provide huge opportunities and benefits to the national and regional economy.

Third, the national government's role is greatly emphasised, another lesson learned from the smart city winter. In Korea, there are significant differences in each local government's capacity for smart city development. A handful of local governments, including Seoul's, can build smart cities on their own, while many local governments lack such capacity. Without national government support, many cities will be left behind in smart city development. In particular, cities experiencing notable population loss strongly need smart city solutions in order to maintain public services and quality of life, because intelligent technologies such as AI, robots and autonomous cars could enable those cities to provide similar levels of service to big cities, despite the decline of their population. As the percentage of shrinking cities nearly doubled to 28.5 per cent in 2015 from 14.6 per cent in 1985 (Koo et al. 2016), the national government's role is increasingly important.

CHALLENGES AHEAD

While Korea has emerged from the smart city winter, there is still the possibility of slipping into a slump at any time. Slow progress in meeting the expectations of citizens can be a significant problem. However, too-rapid progress might be an even larger problem: it may create public resistance if relevant institutions and policies fail to keep up with the technology advancement (ITU-D 2017; Hwang 2016), as in the case of the Luddite movement in the eighteenth century in which textile workers, displaced by machines developed during industrialisation, destroyed the new technology that was replacing them. To avoid a slump and ensure sustainable growth, smart technology and society need to develop simultaneously.

The most important challenge is how best to promote the use of big data while protecting the citizens' personal data and privacy. Smart cities provide citizens with intelligent services based on AI developed socially using data collected from citizens and applications across the city. Since AI is created by analysing experiences in society, smart cities need to secure as much data as possible about people's experiences and convert them into knowledge and services. The problem here is the fact that developing AI is in many cases at odds with protecting personal data. The conflict intensifies in a society that values privacy protection. In fact, Korea has very strong privacy protection regulations (Bauer et al. 2016). If no proper balance can be struck between privacy protection and data utilisation, it will be difficult to make use of socially created intelligence without running the risk of privacy infringement.

Another challenge is successfully implementing a holistic approach. There are two issues here. First is the issue of governance: integrating or reorganising the functions of a city is key to success under the holistic approach. However, city functions have long been divided into many parts and specialised organisations; it is hard to integrate or reorganise them again without a strong commitment and the support of the whole society. The other is the issue of institutional innovation. To take a holistic approach, the overall redesign of regulatory frameworks, which have been established in a fragmented way and sometimes work at cross-purposes, is required. The Korean Government has tried to address these issues by establishing the Korean Smart Cities Special Committee to coordinate public organisation and policies and by making laws to introduce a regulatory sandbox that is free from existing regulations. However, to properly execute the holistic approach, it is necessary to design a new governance model and a new, forward-thinking institutional framework.

Attracting investment is also essential to the success of smart city development. Investment for a greenfield u-City accounted for 1–3 per cent of overall development expenses, not an enormous share; in many cases, it actually costs less than landscaping. However, the size and period of investment will increase significantly for a smart city-as-a-platform. While u-City focused its investment on targeted services, 19.3 services on average per u-City, smart city-as-a-platform will develop various kinds of shared infrastructures across the city, in addition to individual smart city systems and services. Its investment is expected to be much greater than that of u-City. However, the city is essentially a public good where public interests need to be protected. This means that the investment of private capital in the city would need to have limitations and special measures taken to protect public interests. Finding a way to maintain public interests while receiving large-scale investments from the private sector will be the key to the success of smart cities.

Empowering citizens to play a leading role is also a challenge. There has not been a case where citizens have effectively led a u-City project in Korea (Hwang 2016). This is partially because many u-Cities were constructed as new city developments without human inhabitants, but the main reason is due to the erosion of regional communities in Korean cities during the era of rapid industrialisation (Kim et al. 2016). Unlike Europe's smart cities where citizens are leading solutions to urban problems in their neighbourhoods, Korean smart cities have been mostly led by developers or local governments. It is obvious that a truly citizen-centred smart city is not possible without citizen participation at all levels and all stages of smart city development. The future of Korea's new smart city strategy will depend on attracting citizens' interest, encouraging their participation and empowering them to take on active roles in building smart cities.

CONCLUSION

The concept of a smart city will continue to change. However, it is clear that a smart city will become the basis of an intelligent society. To ensure that smart technologies work properly and that there is continuous innovation across the city, the smart city must serve as a platform. The smart city winter Korea experienced for almost a decade demonstrates the type of results that follow when one approaches the smart city as a product instead of a platform.

However, it is not easy to take a smart city-as-a-platform approach. It requires incurring much higher costs and possibly more risks. Moreover, it would take a significantly longer time to show any progress, which may intensify conflicts among the stakeholders. It is highly probable that political leaders and bureaucrats will be lured into taking the easier and safer approach. However, the smart city winter shows that smart cities and intelligent societies cannot be established by making easy choices. Only those cities, countries and people who share a long-term vision and a willingness to take risks can transform their societies.

NOTES

1. 'Intelligent society' is not an established concept yet and has different meanings in different contexts. For instance, the International Telecommunications Union (ITU) stressed the role of digital technology and defined an intelligent society as 'one that successfully harnesses the potential of digital technology and connected devices and the use of digital networks to improve people's life' (ITU-D 2017, p.2). In this chapter, 'intelligent society' means one that automates city operation and services and enables citizens to do what they could not previously do through the use of intelligent technologies in order to transform society.
2. There is no shared definition of 'smartness'. Roughly it means a situation where machines or technology in general have the ability to acquire and apply knowledge and skills. In practice, many smart city projects define 'smartness' through their own goals and key performance indicators.
3. There are many different definitions of smart city development stages. The stage described here is based on Hwang (2016).
4. 'Sustainable city' is a city that does not compromise the ability of future genera-tions to meet their own needs, 'eco-city' means an ecologically healthy city, 'green city' aims at energy-efficient and low-carbon cities, and 'livable city' means a city with high quality of life. In practice, these city development model names are used interchangeably with each other or with 'smart city'.

5. According to data submitted to the National Assembly of Korea in 2017, only 6.8 per cent and 7.2 per cent of residents in u-Cities were aware of and satisfied with u-City services, respectively (Park 2017).
6. IT839 was an ICT development strategy initiated and managed by MIC from 2004 to 2008 aiming to develop eight new IT services, three key network infrastructures and nine new growing technologies.

REFERENCES

Ahn, KY 2003, 'samseong ⌈dongtan Usiti⌋ eotteohge sewojina?', *ZDNet Korea*, 4 June, viewed 19 May 2019, http://www.zdnet.co.kr/view/?no=00000010061962.

Albayrak, AN & Eryilmaz, Y 2017, 'Urban growth in sustainability perspective', *International Journal of Advances in Agriculture & Engineering*, vol. 4, no. 1, pp. 139–143.

Bauer, M, Ferracane, MF & van der Marel, E 2016, 'Tracing the economic impact of regulations on the free flow of data and data localization', viewed 23 December 2018, http://www.cigionline.org/sites/default/files/gcig_no30web_2.pdf.

Bridgwater, A 2015, 'What's the difference between a software product and a platform?', *Forbes*, 17 May, viewed 19 May 2019, https://www.forbes.com/sites/adrianbridgwater/2015/03/17/whats-the-difference-between-a-software-product-and-a-platform/#402d9e6856a6.

Chin, D & Rim, M 2006, 'IT839 strategy: the Korean challenge toward a ubiquitous world', *IEEE Communications Magazine*, vol. 44, no. 4, pp. 32–38.

Eisenmann, T, Parker, G & Alstyne, MW 2006, 'Strategies for two-sided markets', *Harvard Business Review*, vol. 84, no. 10, pp. 92–101.

Eremia, M, Toma, L & Sanculeac, M 2017, 'The smart city concept in the 21st century', *Procedia Engineering*, vol. 181, pp. 12–19.

Frost & Sullivan 2013, 'Global smart city market', viewed 19 May 2019, http://www.slideshare.net/FrostandSullivan/global-smart-city-market-a-15-trillion-market-opportunity-by-2020.

Hwang, JS 2016, 'Seumateusiti baljeonjeonmang-gwa hangug-ui gyeongjaenglyeog', *IT & Future Strategy*, NIA, vol. 6, pp. 1–42.

ITU-D 2017, *Creating the smart society: Social and economic development through ICT applications*, Study Group 2 Final Report. Geneva: ITU.

ITU-T 2013, *Smart cities Seoul: A case study*, ITU-T Technology Watch Report. Geneva: ITU.

ITU-T 2014, *Smart sustainable cities: An analysis of definitions*, Focus Group Technical Report. Geneva: ITU.

Kim, MJ 2009, 'U city nanmaegsang 'bulmandosi' ulyeo', *Digital Times*, 4 May, viewed 23 December 2018, http://www.dt.co.kr/contents.htm?article_no=2009050402010151611002.

Kim, SH, Cha, EH, Lim, KS, Cho, YK, Yoon, HW, Kim, MH, Chang, W & Chung, SH 2016, *Milae gugtobaljeon-eul wihan jisogganeunghan ma-eul yeongu*. Sejong: KRIHS.

Koo, HS, Kim, TH, Lee, SW, & Min, BS 2016, *Jeoseongjang sidaeui chugsodosi siltaewa jeongchaegbang-an yeongu*. Sejong: KRIHS.

Korea Land & Housing 2016, 'Chamgo: u-city juyo sa-eobhyeonhwang', Korea Land & Housing.

Lagos, M, Norris, P, Ponarin, E & Puranen, B 2014, 'World Values Survey: Round five & six – country-pooled datafile version', JD Systems Insitute, Madrid, viewed 19 May 2019, www.worldvaluessurvey.org/WVSDocumentationWV6.jsp.

Lee, K 2013, *Schumpeterian analysis of economic catch-up: Knowledge, path-creation, and the middle-income trap*. Cambridge: Cambridge University Press.

Meadows, DH & Wright, D 2008, *Thinking in systems: A primer*. White River Junction, VT: Chelsea Green Publishing.

MIC 2006, 'Huimanghangug silhyeon-eul wihan u-City guchughwalseonghwa gibongyehoeg', MIC.

MOLIT 2009, 'Ubikwoteoseudosi geonseolsa-eob eobmucheolijichim', MOLIT.

MOLIT 2013, 'Je2cha ubikwoteoseudosi jonghabgyehoeg: 2014–2018', MOLIT.

National Research Foundation 2018, *Virtual Singapore*, viewed 19 May 2019, http://www.nrf.gov.sg/programmes/virtual-singapore.

Navigant Research 2018, 'Smart city platforms', viewed 19 May 2019, https://www.navigantresearch.com/reports/smart-city-platforms.

Opdam, P, Westerink, J, Vos, C & de Vries, B 2015, 'The role and evolution of boundary concepts in transdisciplinary landscape planning', *Planning Theory & Practice*, vol. 16, no. 1, pp. 63–78.

Park, CS 2017, 'Leeheonseung, 1cheon 730eog tu-ib seumateusiti … manjogdo 7.2% bulgwa', *Yonhap News*, 10 October, viewed 23 December 2018, http://www.yonhapnews.co.kr/bulletin/2017/10/10/0200000000AKR20171010169900051.HTML.

Russell, SJ & Norvig, P 2003, *Artificial intelligence: A modern approach*. Upper Saddle River, NJ: Prentice Hall.

Siemens 2015, 'Smart cities: Pioneers with great business potential', Siemens Blog, viewed 23 December 2018, http://www.siemens.com/innovation/en/home/pictures-of-the-future/energy-and-efficiency/smart-grids-and-energy-storage-pioneers-with-great-business-potential.html.

Six Revisions 2010, 'Gestalt principles applied in design', viewed 23 December 2018, http://www.webpagefx.com/blog/web-design/gestalt-principles-applied-in-design/.

Star, SL & Griesemer, JR 1989, 'Institutional ecology, "translations" and boundary objects: Amateurs and professionals in Berkeley's Museum of Vertebrate Zoology, 1907–39', *Social Studies of Science*, vol. 19, no. 3, pp. 387–420.

The Korean Government 2009, 'Act on the construction, etc. of ubiquitous cities', viewed 19 May 2019, http://www.law.go.kr/lsInfoP.do?lsiSeq=94360&chrClsCd=010203&urlMode=engLsInfoR&viewCls=engLsInfoR#0000.

The Korean Government 2018, 'Dosihyeogsin mich milaeseongjangdonglyeog changchul-eul wihan seumateusiti chujinjeonlyag', Korean Government.

Torres, N 2018, 'How Columbus is accomplishing its smart city vision', Data-Smart City Solutions, viewed 23 December 2018, http://datasmart.ash.harvard.edu/news/article/how-columbus-accomplishing-its-smart-city-vision.

Weiser, M 1991, 'The computer for the 21st century', Scientific American Ubicomp Paper after Sci Am editing, viewed 23 December 2018, http://www.ics.uci.edu/~corps/phaseii/Weiser-Computer21stCentury-SciAm.pdf.

6. Facilitating innovation for smart cities: the role of public policies in the case of Japan

Masaru Yarime

INTRODUCTION

The eleventh goal of the United Nations Sustainable Development Goals (SDGs) is aimed at developing cities that are inclusive, safe, resilient and sustainable (United Nations 2015). In our efforts to achieve urban sustainability, smart cities are expected to play a critical role (Bibri & Krogstie 2017). Currently, many urban functions, including energy, transportation, and buildings, are currently undergoing a significant transformation following the advent of smart devices and equipment (Curley 2016). In the power sector, for example, a smart grid system can lower costs, integrate renewable energies, and balance loads, which contributes to improving energy efficiency and reducing carbon dioxide (CO_2) emissions. In the transportation sector, a dynamic congestion-charging system can adjust traffic flows and offer incentives to use park-and-ride schemes, depending upon real-time traffic levels and air quality, whereas car-to-car communication can manage traffic to minimise transit times and emissions and eliminate road deaths from collisions. These emerging innovations based on smart technologies are increasingly connecting various urban functions to create smart cities.

The idea of smart cities reflects different dimensions of complex technological assemblages. As such, there are significant differences in the nuances and emphases within the concept. These depend upon the specific contexts and conditions. In Europe, the focus is on creating an infrastructure that can use information collected and distributed among all connected users to ensure that various objectives of smart cities are achieved in more intelligent ways. In the United States, there is a specific emphasis on security, involving key features such as resilience against physical and cyber threats. Because there are numerous functionalities discussed as part of smart cities, there is also a wide range of benefits envisioned for societies when smart cities are implemented.

Given the potential benefits for societies that are facing the effects of air pollution and climate change and the high hurdles faced by such complex systemic technology areas, public policies are crucial in facilitating innovation for smart cities. With a variety of hardware as well as software for smart cities, the technologies, stakeholders, and institutions involved are diverse and influenced by economic, social, and environmental conditions. An in-depth examination of the processes of creating innovation will generate valuable lessons for public policies and institutional design. Utilising the experiences of industrialised countries to generate policies and institutional implications will be particularly important for many countries in the developing world, where urbanisation is proceeding rapidly in many major cities. This is generating difficult challenges for pursuing sustainability.

A systems approach is an effective way to examine the processes of creating innovation for smart cities. The framework of innovation systems is based upon the notion that the character of technological change is determined not only by the activities of researchers and companies but also by the broader societal and institutional structures (Soete et al. 2020). As there have not been any previous attempts to apply the systems of innovation approach to analysing the mechanisms of creating innovation in smart cities, the hope is that the discussion in this chapter will produce some useful findings with implications for public policy and institutional design. A sectoral innovation system consists of three main dimensions: knowledge, actors, and institutions (Malerba 2002, 2004). The knowledge dimension concerns the characteristics of knowledge and technological aspects that are specific to the sector. The actors involved in the innovation system exhibit heterogeneity, networks, and interactions among them. Institutions include formal ones such as policies, regulations, and standards as well as informal ones like norms, customs, and established practices. Therefore, innovation is a process that involves systematic interactions among a wide variety of actors for the generation and exchange of knowledge relevant to innovation, and its commercialisation is influenced by institutional conditions. In addition to the main actors in the innovation system – namely, universities, firms, and the government, other stakeholders including end-users also increasingly play an important role in creating innovation.

In this chapter, I examine the actors that are involved, the areas of knowledge and technological expertise, and the effects and impacts that have resulted from policy interventions in creating innovation for smart cities in Japan. To capture a broad understanding of the processes of introducing and implementing innovation, I discuss the major actors in academia, industry, and the public sector with specific knowledge and technological domains concerning smart cities and the institutional conditions and environments in which these stakeholders interact. I then discuss the role of policies in facilitating innovation for smart cities.

Based on this analysis of the experiences in Japan, I consider the lessons and implications for public policy as well as institutional design for creating innovation for smart cities in Asia and beyond.

KEY ACTORS INVOLVED IN THE DEVELOPMENT OF SMART CITIES

Japan's initiative on smart cities can be traced back to the government's promotion of renewable energy technologies, especially solar photovoltaics (PVs), since the 1970s. The Ministry of Economy, Trade, and Industry (METI), which was then called the Ministry of International Trade and Industry (MITI), has been in charge of policymaking for issues related to energy. Traditionally characterised by a strong relationship with the industry, METI's main mission has been to support industrial development through innovation (Yarime, 2015). Responding to the oil crisis of the 1970s, the government started R&D projects to stimulate the development and diffusion of new energy technologies (New Energy and Industrial Technology Development Organisation (NEDO) 2019). The Sunshine Project in 1974 was intended to develop alternative energy technologies, with a particular focus on PV technologies, whereas the Moonlight Project in 1978 encouraged energy-saving technologies. Based on these experiences, METI established NEDO in 1980. NEDO was reorganised in 2003 as an incorporated administrative agency under METI and has become the largest public R&D funding and management organisation focusing on the development of energy technologies. The organisation supported R&D projects aimed at grid-connecting technologies for renewable energy sources, including clustered PV generation, mega-solar generation, wind power stabilising and power quality management, and microgrids. Although these projects were not necessarily carried out under the name of smart cities, they addressed some of the functionalities associated with smart cities. Efforts to facilitate innovation specifically targeting smart cities started in 2010, when METI launched four large-scale smart city demonstration projects in Japan. These projects were called the Next-Generation Energy and Social Systems Demonstration Areas and have later been known as Smart Communities, supported by funding from the government. The objective of these projects was to create concrete cases of actually implementing smart energy technologies with relevant stakeholders, including the local authorities with a company coordinating with other participants (METI 2015b). Local governments that are allocated financial resources through the Smart Communities project and the Future Environment City project are particularly active in promoting smart city development. Some of the leading municipalities have partnered with private companies or universities. For example, the city of Kashiwa has been working closely with a real estate developer, Mitsui Fudosan, and the Fujisawa city has partnered with one of the largest electronic companies in Japan, Panasonic (METI 2016a).

Innovative efforts for smart cities were mainly conducted by large, well-established firms that have been long involved in working on electric power infrastructure (Yarime and Karlsson, 2018). With strong corporate networks and technological knowledge, these companies such as Hitachi, Toshiba, and Mitsubishi Electric have access to expertise in various aspects of smart city technologies, including renewable energy, distributed generation, electric vehicle charging infrastructure, and energy storage and security. Residential developers intend to provide end-users of electricity with smart housing technologies with significant potential for cost reductions. Consulting companies are also active in the development and provision of software components and services. Electric utilities, however, were not actively involved in developing smart cities, with their presence relatively invisible, at least initially.

Consortium and industry associations have played an important role in facilitating information exchange and standardisation among these stakeholders involved in smart city development. With a support from METI, the Japan Smart Community Alliance (JSCA) was formed in April 2010, following a recommendation from an internally produced roadmap that international standardisation efforts were needed. The secretariat has been hosted by NEDO. The organisation was initially led by Toshiba, and currently Hitachi serves as the chair of the organisation, which has more than 250 member companies as of February 2019 (JSCA 2019). The Energy Conservation and Homecare Network (ECHONET) Consortium was established in 1997 as a network of companies in smart housing, with an aim to develop open and universal standards of communication to support home networks connected to smart appliances (ECHONET Consortium 2019a). Currently the ECHONET Consortium is led by seven managing members representing the electronics industry in Japan, namely, Hitachi, Mitsubishi Electric, Nippon Telegraph and Telephone (NTT), Panasonic, Sharp, and Toshiba, as well as the largest power utility, Tokyo Electric Power Company (TEPCO) (ECHONET Consortium 2019b). Local associations have also been established for smart city development. For example, the Yokohama Smart Community Association is an association of local small- and medium-sized enterprises working in collaboration with smart city initiatives and acting as suppliers for the companies involved.

THE ROLE OF PUBLIC POLICIES IN FACILITATING INNOVATION FOR SMART CITIES

In the development of smart cities in Japan, public policies have played an important role in key aspects. They include a focus on resilience energy systems; liberalisation of energy markets for new entrants; financial support for promoting renewable energy; evolution of technological road mapping to participatory social experimentation; localisation of demonstration projects adjusted to economic environments, major actors, and technological orientation; and standard setting for smart meters and equipment.

Focus on Resilient Energy Systems

Energy has received a particular emphasis in the development of smart cities in Japan. Since the oil crisis in the 1970s, energy security has been widely understood as an area that needs to be maintained and strengthened (Yarime 2015). Hence, smart technologies that are considered to contribute to improving energy supply and efficiency are generally accepted without much concern about privacy or cyber security. After the Fukushima accident in 2011, the planned outages by the electric utilities led to strong criticism, as regions of less economic importance had to endure more blackouts. A priority is, thus, to strengthen the resilience of the energy supply against future disruptions and disasters such as earthquakes and typhoons by decentralising the energy supply (Kusunose 2019). Smart technologies utilising sophisticated information and communication technologies (ICTs) are expected to make such distributed energy systems feasible and reliable.

At the same time, there is a societal demand for reducing the environmental burden, including CO_2 emissions, by utilising renewable energy sources and improving energy efficiency. Solar photovoltaics, in particular, have been increasingly adopted to address the challenge of climate change (Yamada and Ikki 2017). As large-scale introduction of renewable energy sources makes the energy supply widely fluctuate and destabilise, it is critical to maintain a high quality of electricity in terms of voltage and frequency by flexibly adjusting energy use. Energy management systems (EMSs) based on demand–response systems enable management of the balance between energy supply and demand efficiently through collective participation of consumers (METI 2015b). It also becomes possible to reduce the capacities of thermal power generation prepared for peak energy consumption by establishing smart energy systems from a long-term perspective.

Liberalisation of the Energy Market

It is a crucial challenge to stimulate innovation for smart cities by encouraging new entrants to the market. The energy market in Japan was traditionally heavily regulated, with ten regional monopolies of vertically integrated electric power companies (EPCOs) in charge of providing electricity. While large engineering firms maintained long-term relationships with the EPCOs in supplying technologies and services, the monopolistic structure of the electricity market and the uncertainty about future policies and regulations have discouraged innovative activities by entrepreneurs and startups. Lack of competition in the electricity market continued until the middle of the 1990s.

Then, the Japanese Government started to introduce market reform in the energy sector with the aims of securing a stable supply of electricity, reducing electricity rates, and expanding consumer choices and business opportunities (METI 2018). In 1995, independent power producers (IPP) were allowed to enter the electricity market for the first time. Competition was introduced to the retail market for capacities of more than 2,000 kW, which accounted for 26 per cent of the market, in 2000 and for those of over 500 kW, representing 40 per cent of the market, in 2004. At the same time, the third-party access to grid lines was also regulated. In 2005, competition in the retail market was expanded to the segment of capacities of more than 50 kW, which covered 62 per cent of the market. The wholesale power exchange (JEPX) was established together with a supporting body for transmission in wider areas, and accounting separation was introduced for the transmission/distribution sector. The rule of wheeling rates was revised in 2008.

The liberalisation of the energy market by the Japanese Government has been accelerated recently (METI 2015a). The Amended Electricity Business Act enacted in November 2013 established the Organisation for Cross-regional Coordination of Transmission Operators (OCCTO) in 2015 to promote wide-area electrical grid operation. The retail market of electricity has been fully liberalised since April 2016, and furthermore separation of power generation and power transmission will be completed in 2020 (METI 2018). As these policy measures for the liberalisation of the energy market are in place, many companies have entered the sector bringing in entrepreneurial initiatives. The share of new entrants to the retail electricity market reached approximately 12 per cent in 2017, with the total number of retail companies being more than 400 (METI 2018).

Support for the Introduction of Renewable Energy Sources

The policy instruments to promote the introduction of renewable energy sources are also important in creating demands for technological development and adoption. The shutdown of nuclear power plants following the Fukushima accident in March 2011 has effectively accelerated the expansion of renewable energy as a strategy to make up for lost power generation and to reduce Japan's dependence on imported oil and natural gas. The government announced in June 2011 a target of putting PV systems on the top of 10 million roofs by 2030 (METI 2015b). The feed-in-tariff (FIT) program has been implemented since 2012 to encourage the installation of renewable energy, particularly solar PVs. The Strategic Energy Plan, enacted in April 2014, has further accelerated the introduction of renewable energy sources so that the energy produced by solar PVs will be increased to 53 GW and by wind to 10 GW by 2030. Revised FIT for PVs is expected to account for more than 80 per cent of newly installed capacity in the 2020s. As FIT has induced a significant amount of investment in PVs, however, there are many facilities that have not yet started to produce solar energy. Out of the capacities of 53 GW of solar power approved for FIT by the end of the fiscal year 2015, only those of 18 GW had been actually installed (METI 2019). Installations of PVs have slowed as some utilities denied additional grid access to new solar farms because the existing grid infrastructure was not set up in consideration of incorporating large-scale adoption of renewables such as solar and wind power, and further deployment would disrupt the operations of the grid. In the strategic plan, a particular emphasis is placed on the importance of R&D and demonstration of transmission and distribution equipment, and it is specified that regional or interregional grids for renewables will be established.

Evolution of Technological Road Mapping to Societal Experimentation

METI and NEDO have traditionally played an important role in influencing the direction of R&D and supporting the development of technologies in energy. Initially, NEDO focused on supporting the development of specific technologies related to energy saving and renewable energy sources, particularly solar PVs. Experts in academia and industry were invited to discuss key opportunities and challenges in strategically important areas and to agree upon future roadmaps for technological development from a long-term perspective (Yasunaga, Watanabe, Korenaga 2009). A process of developing roadmaps was particularly implemented for solar PV technologies (NEDO 2014). The NEDO PV 2030 roadmap was initially published in 2004 and was subsequently revised in 2009 as NEDO PV 2030+ and again in 2014 as NEDO PV Challenges. The main focus was placed on technical issues concerning PVs from the perspective of experts working in this field (RTS Corporation 2009).

As PVs were increasingly integrated into smart cities, technological road mapping by experts has evolved to facilitate societal experimentation with relevant stakeholders. After large-scale projects were initiated for testing technologies to connect renewable energy sources to the electricity grid in the field in 2000, smart community projects started to actually implement new technologies together with users in actual communities (Morozumi 2012). The evolution of technology-oriented road mapping to more participatory societal experimentation is important in expanding the scope of the knowledge by getting various types of stakeholders involved so that wider societal needs and expectations concerning smart cities are incorporated into the design and process of the projects.

Localisation of Demonstration Projects on Smart Cities

Smart city projects were implemented in the four cities of Yokohama, Toyota, Keihanna and Kitakyushu in the period from 2011 to 2014 (METI 2015b). They were mainly aimed at verifying emerging advanced technologies concerning smart cities, including cogeneration, renewable energy, energy storage, electric vehicles, and EMSs. There was also another objective to contribute to developing robust business models with the active participation of relevant stakeholders, including local communities and residents, as well as technology providers in the private sector.

These projects were locally adjusted, considering the specificities of the economic and social conditions and contexts (METI, 2015b). The smart city project in the metropolitan city Yokohama intended to promote the large-scale introduction of renewable energy and electric vehicles with 4,000 households equipped with home energy management systems (HEMS), ten large-scale buildings, and multiple storage batteries. The project in Toyota local production of energy for consumption locally was the target, involving 67 households equipped with solar panels, household fuel cells, storage batteries, and advanced transportation systems including electric vehicles and plug-in hybrid vehicles. In the Keihanna Science City project, the visualisation of energy for control and management was introduced to a housing complex of 700 households equipped with HEMS, and the feasibility of consulting business on energy saving was also examined. As Kitakyushu has a specially designated area for energy supply, the smart city project aimed to optimise the use of various sources of energy supplied by large steel and metal companies, and a dynamic pricing system was also tested in 180 households.

These demonstration projects contributed to implementing technological integration, reliability, and learning by facilitating collaboration with relevant stakeholders in the specific local conditions (METI 2016b). Some of the emerging technologies, such as demand response (DR) and vehicle to X (V2X), showed promising results. The peak electricity demand in factories was actually reduced with the use of EV storage batteries. Large-scale adoption and intensive learning with trials and errors become possible by conducting the demonstration projects, effectively inducing a decline in the prices of component technologies and the costs of operating energy systems. Smart city projects were especially important in providing collaborative platforms in which novel technological functionalities were tried out. The tightly knit groups involved in the smart city projects helped to produce valuable knowledge and to encourage the sharing of that knowledge among the stakeholders involved.

Standard Setting for Emerging Technologies

Standard setting for component technologies played an important role in supporting the development of smart cities. Initially the level of activities at the ECHONET Consortium for smart housing systems remained relatively low (ECHONET Consortium 2018b). After the Fukushima accident in 2011, the government started to promote standardisation and to provide financial support to consumers for purchasing HEMS (National Policy Unit 2012). Technologies have been developed subsequently in the area of home appliances and equipment for ECHONET Lite Specifications, a standard that can work with standard protocols in Japan and abroad (ECHONET Consortium 2018a). HEMS for managing the various types of energy used in the home in smart ways has made the data on the amounts of electric power generation and utility usage – as well as gas and water used – visible on monitors and other screens, facilitating smart control of HEMS-compatible home appliances and household devices. The ECHONET Lite Specifications have also provided the function of shared communication protocols that are necessary to achieve two-way communication between HEMS controllers and various home appliances and household devices. With HEMS controllers adopting ECHONET Lite and devices compatible with HEMS, it has become possible for different manufacturers' products to be connected together for use. ECHONET Lite has been recommended as an open standard interface on HEMS by the International Standardisation Working Group of JSCA, which has promoted an increase in interest and membership in the ECHONET Consortium (ECHONET Consortium 2018a).

Proprietary standards among competing providers initially slowed down the take-off of the market. Then the Open Automated Demand Response (OpenADR) 2.0 technology standard was adopted – following feasibility, interoperability and connectivity testing – in the summer of 2013 (Ishii 2015). With an application programming interface (API), the efficient development of applications became possible, including HEMS and building energy management systems (BEMS). The adoption of HEMS had a significant impact on driving Japan's smart household appliance industry, with LED lights, smart thermostats, plug-in electric vehicles, rooftop solar, demand-flexible water heaters, battery energy storage, and other appliances now integrated within the information technology (IT) network.

Recently, different standards have been emerging in various sectors, particularly in fields related to what is called the Internet of Things (IoT). Through the IoT, virtually everything will be connected for information exchange and communication so that many activities that were formerly conducted separately can now be coordinated with each other efficiently. New standards – such as ZigBee and Bluetooth Low Energy – are currently under rapid development, leading to an urgent need to consider cooperation and coordination among various stakeholders in different sectors (Ray 2015). Furthermore, the electric vehicle has become an innovation area of critical importance. While currently at too early a stage to contribute significantly, the development and diffusion of electric vehicles will benefit considerably from smart city innovations, particularly integrated with renewable energy development, notably solar PVs. Currently, the electric utilities allow only limited amounts of electricity to be connected to the grids, due to concerns about the grid capacity to absorb the fluctuations and interruptions in the electricity produced from renewable energy sources. Smart energy systems integrating solar PVs and electric vehicles have a significant potential for reducing CO_2 emissions while securing energy supplies in cases of disasters and disruptions (Kobashi & Yarime 2019).

CONCLUSION

The analysis of Japan's experience with developing smart cities reveals that large companies in the electric and electronics industries played an important role in facilitating innovation with technical knowledge and expertise on renewable energy, energy storage, community energy management, and applications for home appliances and electric vehicles. The policies and regulations introduced by the public authorities supported the process of creating innovation with economic incentives to promote renewable energy technologies, liberalisation of energy markets for new entrants, evolution of technological road mapping to participatory societal experimentation, localisation of demonstration projects reflecting specificities, and standard setting for smart technologies.

At the same time, several challenges of implementing system transformation for urban sustainability still remain. There needs to be a clear vision of what kinds of smart cities should be established that can be matched to feasible plans for implementation. Strong leadership for projects and transparency in the process of decision making and implementation are also important. In this respect, universities, in particular, are expected to play a key role in generating concrete solutions and strategies to tackle the dynamic, complex challenges of smart cities (Yarime et al. 2012; Trencher et al. 2014a, 2014b). To this end, utilising universities as a platform for social experimentation through collaboration and networking among academia, industry, and the public sector will enable contributions to learning and innovation for urban sustainability.

The experience and expertise of the private sector is also a particularly crucial ingredient when implementing multi-stakeholder collaborations with the public sector aimed at triggering institutional reforms favouring innovation for smart cities. Robust business models are currently missing, which has the effect of discouraging private companies from taking over the demonstration projects that have mainly been financed by the public sector. It is also critical to nurture human resources with skills and capacities necessary to understand and integrate technical and societal dimensions of smart cities. While large established companies tend to have advanced technological expertise and capabilities concerning various instruments and facilities in smart cities, local governments and communities do not necessarily possess sufficient knowledge of or experience with technical measures (Yarime 2017). Under the existence of the significant degree of asymmetry of knowledge and expertise between large technology companies on the one side, and local government and communities on the other side, we need to consider how it would be possible to secure serious and active participation of end-users in an equal and equitable manner for jointly facilitating innovation. As smart cities consist of various types of hardware and software, coordination among different standards is also indispensable for facilitating the development and adoption of technologies for smart cities.

Policy measures and instruments require careful coordination in an institutional landscape at the macro level, while the same needs to be done for specific technologies at the micro level. Liberalisation of energy markets would have an effect of encouraging new entrants and entrepreneurship and consequently creating innovation and competition. Technological road mapping mainly conducted by experts in specific fields can be extended to implement societal experimentation through collaboration with stakeholders in various sectors. Standard setting plays a crucial role in securing interconnectivity among various devices and equipment in different sectors while innovative technologies continue to emerge.

Located at the complex intersection of economic development and environmental change, cities play a central role in our efforts to move towards sustainability. Currently, over half of the world's population lives in urban areas, and more than two-thirds of the world's population is expected to be urbanised by 2050 (United Nations Department of Economic and Social Affairs 2018). While cities are the engines of economic growth, accounting for a significant part of global energy consumption, they are at the same time producing harmful pollutants and greenhouse gas emissions and are vulnerable to natural disasters. These issues are interconnected with one another, with their dynamic interactions changing in highly complex and unpredictable manners. Reducing air and water pollution and improving energy efficiency while securing energy supply and maintaining resilience to disruptions and disturbances would be a formidable task. The hope is that Japan's experience with facilitating innovation for smart cities will provide useful lessons and implications for addressing the sustainability challenge in Asia and beyond.

ACKNOWLEDGMENTS

This chapter has significantly benefited from collaboration with Martin Karlsson.

REFERENCES

Bibri, SE & Krogstie, J 2017, 'Smart sustainable cities of the future: An extensive inter-disciplinary literature review', *Sustainable Cities and Society*, vol. 31, Supplement C, pp. 183–212.
Curley, M 2016, 'Twelve principles for open innovation 2.0', *Nature*, vol. 533 (19 May), 314–316.
ECHONET Consortium 2018a, 'The ECHONET Lite Specification, Version 1.13', ECHONET Consortium, July 6.
ECHONET Consortium 2018b, 'The ECHONET Lite: This is the Future of Home Life', ECHONET Consortium, Tokyo.
ECHONET Consortium 2019a, 'Greetings', viewed 7 June 2019, https://echonet.jp/organization_en/#organization-01.
ECHONET Consortium 2019b, 'List of members', viewed 7 June 2019, https://echonet.jp/kaiin_kigyo_en/.
Ishii, H 2015, 'Initiative for establishment of smart society by industry–government–academia collaboration', *China–Japan Science and Technology Innovation Interdisciplinary Salon 2015*, Beijing, China, November 25.
JSCA 2019, 'Members of Japan Smart Community Alliance', viewed 6 June 2019, https://www.smart-japan.org/english/memberslist/index.html.
Kobashi, T & Yarime, M 2019, 'Techno-economic assessment of the residential photovoltaic systems integrated with electric vehicles: A case study of Japanese households towards 2030', *Energy Procedia*, vol. 158, pp. 3802–3807.
Kusunose, N 2019, 'NEDO's Approach for Improvement of Resiliency', Smart Community Department, New Energy and Industrial Technology Development Organization, Smart Community Summit 2019, June 4.
Malerba, F 2002, 'Sectoral systems of innovation and production', *Research Policy*, vol. 31, no. 2, pp. 247–264.
Malerba, F (ed.) 2004, *Sectoral systems of innovation: Concepts, issues and analyses of six major sectors in Europe.* Cambridge: Cambridge University Press.
METI 2015a, 'Bill for the act for partial revision of the Electricity Business Act and other related acts (outline)', Ministry of Economy, Trade and Industry, Tokyo, March.
METI 2015b, 'Japan's policy on smart community', Smart Community Policy Office, Agency for Natural Resources and Energy, Ministry of Economy, Trade and Industry, Tokyo, April 15.
METI 2016a, 'Current Situation of Smart Community Projects in Japan: Examples of Implemented Projects', Division of Energy Saving and New Energy, Agency for Natural Resources and Energy, Ministry of Economy, Trade and Industry, Tokyo, June 7.
METI 2016b, 'Next Generation Energy and Social System Demonstration Projects: Summary and Future Activities', Division of Energy Saving and New Energy,

Agency for Natural Resources and Energy, Ministry of Economy, Trade and Industry, Tokyo, June 7.

METI 2018, 'Electricity system and market in Japan', Electricity and Gas Market Surveillance Commission, Ministry of Economy, Trade and Industry, Tokyo, January 22.

METI 2019, 'Other Issues to be Examined about the Act on Special Measures Concerning Renewable Energy', Agency for Natural Resources and Energy, Ministry of Economy, Trade and Industry, Tokyo, November 18.

Morozumi, S 2012, 'Strategy of smart community in NEDO and New Mexico Japan–US demonstration project', New Energy and Industrial Technology Development Organization, Kawasaki.

National Policy Unit 2012, 'Green Policy Outline – From the Birth of Green Energy Revolution to Its Growth', Cabinet Secretariat, November.

NEDO 2014, 'NEDO PV Challenges', New Energy and Industrial Technology Development Organization, Kawasaki.

NEDO 2019, 'History of NEDO', viewed 16 January 2019, https://www.nedo.go.jp/.

Ray, B 2015, 'A Bluetooth & ZigBee Comparison For IoT Applications', LinkLabs, October 28.

RTS Corporation 2009, 'Research on the Review of PV Roadmap Toward 2030 (PV2030)', New Energy and Industrial Technology Development Organization, Kawasaki, March.

Soete, L, Verspagen, B & ter Weel, B 2010, 'Systems of innovation', in Hall, BH & Rosenberg, N (eds.), *Handbook of the economics of innovation*, pp. 1159–1180. Amsterdam: Elsevier.

Trencher, G, Bai, X, Evans, J, McCormick, K & Yarime, M 2014a, 'University partnerships for co-designing and co-producing urban sustainability', *Global Environmental Change*, vol. 28, pp. 153–165.

Trencher, G, Yarime, M, McCormick, KB, Doll, CNH & Kraines, SB 2014b, 'Beyond the third mission: Exploring the emerging university function of co-creation for sustainability', *Science and Public Policy*, vol. 41, no. 2, pp. 151–179.

United Nations 2015, 'Transforming our world: The 2030 agenda for sustainable development', Resolution adopted by the General Assembly on 25 September 2015, A/RES/70/1.

United Nations Department of Economic and Social Affairs 2018, 'World urbanization prospects: The 2018 revision', Population Division, UN DESA, New York.

Yamada, H, Ikki, O 2017, 'National Survey Report of PV Power Applications in Japan 2016', Photovoltaic Power Systems Programme, International Energy Agency, Paris.

Yarime, M 2015, 'Integrated Solutions to Complex Problems: Transforming Japanese Science and Technology,' in Baldwin, F and Allison, A (eds.), *Japan: The Precarious Future*, New York: New York University Press and United States Social Science Research Council (SSRC), pp. 213–235.

Yarime, M 2017, 'Facilitating data-intensive approaches to innovation for sustainability: Opportunities and challenges in building smart cities', *Sustainability Science*, vol. 12, no. 6, pp. 881–885.

Yarime, M, Karlsson, M 2018, 'Examining Technological Innovation Systems of Smart Cities: The Case of Japan and Implications for Public Policy and Institutional Design,' in Jorge Niosi (ed.), Innovation Systems, Policy and Management, Cambridge, UK: Cambridge University Press, pp. 394–417. (2018).

Yarime, M, Trencher, G, Mino, T, Scholz, RW, Olsson, L, Ness, B, Frantzeskaki, N & Rotmans, J 2012, 'Establishing sustainability science in higher education institutions: Towards an integration of academic development, institutionalization, and stakeholder collaborations', *Sustainability Science*, vol. 7, Supplement 1, pp. 101–113.

Yasunaga, Y, Watanabe, M, Korenaga, M 2009, 'Application of technology roadmaps to governmental innovation policy for promoting technology convergence,' *Technological Forecasting and Social Change*, vol. 76, pp. 61–79.

PART II

Smart city initiatives of two Asian giants

7. The smart city policy of India and its governance implications

Souvanic Roy and Tathagata Chatterji

INTRODUCTION

This chapter analyses the Smart Cities Mission launched by India's national government in 2014, with the ambitious goal of transforming Indian cities into drivers of the country's economic aspirations. This is done in an effort to understand the challenges of implementing big data and a knowledge-intensive urban agenda in a developing country with a large, mostly poor, digitally divided population that is undergoing a rapid urban transformation.

The Smart Cities Mission (which combines the features of a policy document and a financial package) is one of the most significant milestones in India's urban development trajectory. While cities started to gain greater attention in the national policy arena with initial economic reforms in the early 1990s, the Jawaharlal Nehru Urban Renewal Mission (JNNURM) (2005–2014) has been the watershed for urban India's development trajectory in terms of outlay, coverage, sectoral interventions and a slew of administrative, fiscal, managerial and legislative reforms (Shivaramakrishnan 2011). The Smart Cities Mission is an ambitious successor to JNNURM and intends to accelerate urban sector reforms and bring about transformative change in city management practices. It also makes an emphatic political statement by acknowledging the potential of cities to drive the twenty-first-century Indian economy, overcoming decades of rural bias in the policy arena. It is not a standalone policy, but rather part of an overarching nation-building agenda, centred on information and communication technology (ICT) applications such as Digital India, Skilling India, the National Digital Literacy Mission and a cashless digital economy.

The scope of the Smart Cities Mission covers 100 cities and has a budgetary outlay of Indian rupee (INR) 48,000 crore (US$750 billion) by the national government, with matching contributions by state governments and local urban governments (Ministry of Urban Development 2017). The national policy framework allows sufficient leeway to select specific projects based on precise local contexts. It has enabled the introduction of several innovative and unique approaches, which sets it apart from its Asian counterparts.

The mission has created a regime of sub-national competition among cities for access to federal funds, which incentivises better-performing cities. Moreover, it has introduced an area-based approach towards urban development. Almost four-fifths of fund allocations are targeted at the urban renewal of specific precincts in the selected cities. These 'smartified' areas are expected to act as 'lighthouses' and trigger development in other areas of the cities. Most importantly, the mission has introduced a new institutional mechanism, in the form of special purpose vehicles (SPVs), in each of the selected cities as single-point nodal agencies responsible for the implementation of projects. The SPVs operate as 'companies' and have been entrusted with wide-ranging governance responsibilities.

In this chapter, we examine India's smart city policy through the lens of the good governance framework. The term 'governance' has come to be defined in various ways, as it had gained increased traction in academic as well as popular discourse on public policy issues in recent decades. Here, following Chhotray and Stoker (2009), we note, 'Governance is about the rules of collective decision making in settings where there are a plurality of actors or organisations and where no formal control system can dictate the terms of the relationship between these actors and organisations' (p. 3).

As we try to understand and examine the operationalisation of an urban policy – including its drivers, agencies and its institutional mechanism – in a developing country context, the good governance framework is an appropriate analytic lens. The framework has been widely applied in practice to analyse the governance process and policy outcomes and to drive political and institutional reforms in developing countries.

The good governance framework, as defined by the United Nations, includes a set of eight characteristics – participatory, consensus oriented, accountable, transparent, responsive, effective and efficient, equitable and inclusive and follows the rule of law (United Nations Development Program (UNDP) 1997) – which we apply to test the smart cities policy agenda. Our research revolves around two research questions: What are the drivers of the smart cities initiative in India? And, what challenges and concerns have emerged during the process of its implementation? We then analyse the smart cities initiative through the good governance framework.

The chapter is organised as follows. The first section discusses the drivers and agencies of the new economy and enclave urbanism in the post-liberalisation period and their influence on the smart cities initiative in the country. The next section examines the salient features of the initiative and its implementation process. This is followed by an analysis of the initiative and its operationalisation using the good governance framework. The last section concludes with the emerging challenges and concerns and their governance implications for Indian cities.

NEW ECONOMY AND ENCLAVE URBANISM: ACTORS AND DRIVERS

The Indian government's approach towards cities had undergone a radical shift, following the nation's gradual integration with the global economy from the 1990s onwards. During the earlier period of state-led developmentalism, government policies sought to encourage balanced regional development by channelling industrial investments towards backward regions and away from big cities. However, economic policies lacked spatial focus and urban development and city building were essentially seen as by-products of industrial location policy. In contrast, urban development is now increasingly seen as a crucial growth driver itself and the smart city its latest leitmotif.

The genesis of smart cities in India is not an isolated phenomenon, but an intrinsic part of the socio-political changes in the post-1991 era, including the rise of a new middle class and the nation-building aspirations of policy elites who sought to capitalise on the success of the Indian IT services industry in the global business process outsourcing market.

Impressive economic growth since the 1990s has significantly improved living standards and fuelled middle-class consumerism. The proportion of the middle class increased from 28.9 per cent in 1998–1999 to 50.3 per cent in 2011–2012[1] (Krishnan & Hatekar 2017). This growth story was largely spearheaded by the ICT sector. Starting at the bottom, within two decades, India has become the dominant player in the global IT services market. Consequently, the ICT sector has become the main source of employment in the organised sector in cities and symbolic of middle-class career aspirations (Upadhya 2007).

However, it is also important to note here that integration with the global economy and market-led policies have increased socio-economic inequalities in India, as in several other emerging economies (Datt et al. 2016). A high gross domestic product (GDP) growth rate did not translate into a proportionate increase in organised sector jobs. The urban employment scenario in India is characterised by high informality, a low wage structure, adverse service conditions and exclusion from the social safety net (Roy 2016). Consequently, the socio-economic character of Indian cities is predominantly informal, whether in relation to employment or other aspects of urban life.

Nevertheless, the IT sector's international image, its export earnings and the ability of Indian tech entrepreneurs to penetrate the global knowledge economy have created a particularly favourable impression among policymakers (Gilbertson 2017; Upadhya 2007). Successive government policies over the past two decades have encouraged the ICT sector through various fiscal and non-fiscal benefits. Support for the ICT sector received further momentum following the election of the National Democratic Alliance (NDA) government under Prime Minister Modi in 2014. Taking into account the pivotal role played by the aspirational youth population during the election, the NDA government launched several programmes oriented towards a more ICT-enabled future (e.g., the Digital India Mission, the National Digital Literacy Mission) along with the Smart Cities Mission.

Apart from incentivising the ICT sector, the Smart Cities Mission also carries forward several neoliberal city building ideas, which began to circulate in the early 2000s with the launch of the JNNURM programme (2005–2014), the passage of the Special Economic Zone Act (2005) and the commissioning of the Delhi Mumbai Industrial Corridor (DMIC) project in 2007.

First, the JNNURM sought to improve the functioning of Indian cities by strengthening urban management processes through an infusion of information technology and the construction of new civic infrastructure by leveraging market-friendly financial instruments. The programme encouraged urban local bodies (ULBs) to adopt management information systems tools to improve inter-organisational coordination; plug leakages in the delivery of civic infrastructure through service-level benchmarking; augment the collection of municipal revenue and user charges; and improve the citizen interface through e-governance applications. The JNNURM also encouraged ULBs to access market-based instruments such as municipal bonds, land value capture and public–private partnerships (PPPs) to finance high-value infrastructure projects.

Second, from 2005, national fiscal policies sought to direct foreign investments to the development of greenfield industrial estates and private cities, such as Special Economic Zones (SEZ) and Special Investment Regions (SIR) outside the municipal governance system. These new economic spaces enjoy 'tax-free status' and are exempt from several labour laws and municipal zoning regulations. The ICT industry, which had earlier shown a preference for locating in high-tech, high-security, suburban industrial parks as opposed to inner city areas, began to cluster in SEZ areas to leverage additional benefits (Chatterji 2017).

The DMIC project has also expanded its scope for the development of industrial clusters and smart cities through market capital. The flagship national industrial corridor involves the construction of eight new cities in Phase 1, as manufacturing and logistical hubs, each with population of 2 million, along with a high-speed road and rail network (DMIC 2018). Cities along the 1,500 km corridor are expected to be developed as smart cities, funded by Japanese and other international investors and are to be administered by SPV.

Complementing the moves of the national government, entrepreneurial state governments have also started to promote greenfield smart cities to attract investment. For example, the Gujarat government initiated the development of the Dholera SIR, Dahej Port and Petro-chemical Investment Region and Gujarat International Finance Tec (GIFT) City as the building blocks of the state's economic strategy (Datta 2015). Similarly, Andhra Pradesh has started to build Amaravati, its smart technology-enabled state capital, which can compete with older and larger urban agglomerations to attract new investments.

Development strategies for these new-age cities share certain commonalities in terms of their ideational orientation, institutional frameworks of governance, funding patterns and infrastructure support systems, which subsequently informed the conceptualisation of the Smart Cities Mission.

First, these new-age cities are oriented towards the external world. They draw inspiration from other Asian smart cities (e.g., Songdo, Tianjin) and are being planned and designed by foreign consultants. Further, they are actively seeking foreign investment as part of the development process (especially from Japan, Singapore, UK, Germany, France and the United States). For example, Singapore's government is providing techno-economic support for the development of Amaravati in Andhra Pradesh (Livemint 2016).

Second, ICT companies are becoming increasingly involved in the planning process. IBM formulated the ICT Master Plan for Dighi Port Industrial Area in the DMIC area and CISCO prepared the ICT Master Plan for four other smart cities in the DMIC project.

Third, the projects are being routed through the PPP model to build urban infrastructure and are being financed through market-friendly financial instruments such as infrastructure bonds, real estate-oriented mutual funds and debt bonds.

Fourth, to facilitate development of such exclusive new economic spaces, new institutional arrangements are being rolled out (e.g., SEZ, SIR). Unique legal features allow these areas to be governed by SPVs, bypassing the prevailing institutional architecture of democratically elected local urban bodies.

It is conceivable to see the Smart Cities Mission as Stage II of post-liberalisation urban policy, taking off from JNNURM-era neoliberal funding arrangements. The mission seeks to facilitate a more technology-enabled urban future to improve the delivery mechanism of civic infrastructure by leveraging India's strength as a software powerhouse. However, using such a straightforward narrative to analyse the policy would be rather problematic. Unlike the greenfield projects (e.g., Dholera, Amaravathi) discussed above, the Smart Cities Mission (Ministry of Urban Development 2015) is targeted at improving the functioning of existing cities with hugely diverse socio-economic contexts, high incidences of poverty and informality and complicated multi-scalar governance arrangements.

India's urban scenario presents a paradoxical picture of economic vibrancy and small-scale entrepreneurial energy (often in the unorganised sector), fuelled by a youthful population and thriving in a messy, chaotic physical setting. Almost every Indian city had grown in an unplanned fashion with a high degree of informality. More than 1 million cities in the country are home to 40 per cent of the slum population. Large sections of the urban population remain deprived of decent shelter, basic services, livelihoods, affordable means of mobility and a voice in development. The process of urban development has resulted in inequities and sharp social divisions in a contested urban landscape where affluent gated communities lie in contrast to impoverished slums and informal settlements.

With ICT-led 'bypass urbanism' unfolding in the peri-urban areas of large cities, a sharp digitally divided and splintered landscape is emerging, where technology-enabled 'smart' enclaves, vernacular, rural settlements and informal urban neighbourhoods coexist in close proximity (Graham 2002). The lifestyle aspirations of the upwardly mobile, educated, urban middle class, which are tied to the global economy, and the survival needs of the urban poor, which are stuck to the local market economy, have turned urban spaces into arenas of conflict (Chattopadhyay 2017; Jaikumar & Sarin 2015). A crucial challenge confronting India's urban governance system is how to address the growing disconnection between these two sections of urban society.

The proactive role of the state in emphasising cities as engines of economic growth through the development of a knowledge-intensive service sector, as well as the emergence of a neo-rich middle class as a driver of smart cities and the institutional bypass used for their realisation, have created apprehension with regard to their validity in the Indian context from the angle of social sustainability and inclusivity (Roy 2016). Burte (2014) argues that the fundamental problem of Indian cities is not that they are 'un-smart' but that they are dysfunctional in terms of governance capacity and the delivery of basic civic infrastructure. Improving their functional ability is more to do with the control and management of land, the universal provision of basic infrastructure and the ability to address issues of urban informality (Bhide & Burte 2015).

Elected municipal governments continue to remain weak in India's federal governance structure. A major effort was made by the nationwide government in 1992 to strengthen local urban governance systems and encourage participatory urban planning processes by demarcating the administrative and financial powers of municipal governments vis-à-vis state governments. However, the reluctance of state governments to shed power got in the way and the implementation of such reforms has been partial and uneven across the country (Shivaramakrishnan 2011).

Physical and economic planning activities continue to be guided by parastatal development authorities or State Town Planning Directorates as many state governments are unwilling to devolve urban planning responsibilities to elected municipalities (Mohanty 2014). Similarly, financial policies of municipal bodies by and large continue to be guided by state governments. Consequently, municipal governments remain weak and bereft of technical, managerial and financial expertise.

As the most ambitious city rejuvenation initiative in post-independent India, the Smart Cities Mission attempts to address critical urban infrastructure problems by bridging the efficiency gap in the delivery of civic services through technology retrofitting (smart solutions) and transforming parts of cities as world-class enclaves of the highest standard of living to attract the aspiring middle class. The chapter recognises the contradictions between the current problems of cities the mission seeks to address and the content and processes, or modes, used for its realisation. The lens of good governance seems appropriate to analyse the emerging contradictions and issues and inform course correction for the mission, which we discuss in further detail in the following section.

THE SMART CITIES MISSION AND ITS IMPLEMENTATION

Salient Features

The Smart Cities Mission was launched in June 2015 as a centrally sponsored scheme covering 100 cities for a period of five years. However, slow progress led to the extension of the project for an additional three years.

The mission document underscores the need to drive economic growth and improve people's quality of life in cities through urban development and harnessing technology that could lead to smart outcomes (Ministry of Urban Development 2015). The overarching aim is to increase competitiveness and improve the quality of life in cities through the provision of core infrastructure, as well as to attract investment and improve operational efficiency in service delivery through institutional, physical, social and economic infrastructure, mediated by real-time data monitoring and the application of ICT and digital technologies. The mission states its 'objective is to promote cities that provide core infrastructure and give a decent quality of life to its citizens, a clean and sustainable environment and application of smart solutions' (Government of India 2015).

The mission intends to create a replicable model that would act as a lighthouse for other aspiring cities. It has adopted a combination of a Pan-City approach with smart solutions and Area-based Development. The Pan-City component envisages overhauling, upgrading or installing an infrastructure component embedded with smart elements to benefit the greater city. The

Table 7.1   Smart solutions envisaged under the Smart Cities Mission

e-Governance and Citizen Services	Public Information, Grievance Redressal, Electronic Service Delivery, Citizen Engagement, Video Crime Monitoring
Waste Management	Conversion of Waste to Energy and Fuel, Conversion of Waste to Compost, Treatment of Wastewater, Recycling and Reduction of Construction and Demolition Waste
Water Management	Smart Meters and Management, Leakage Identification, Preventive Maintenance, Water Quality Monitoring
Energy Management	Smart Meters and Management, Renewable Sources of Energy, Energy Efficient and Green Buildings
Urban Mobility	Smart Parking, Intelligent Traffic Management, Integrated Multimodal Transport
Others	Telemedicine and Tele-education, Incubation and Trade Facilitation Centres, Skill Development Centres

Source: Prepared by the authors based on data from the Ministry of Urban Development 2015.

Area-based Development approach involves urban renewal of a specific zone or neighbourhood through retrofitting existing infrastructure with smart solutions, constructing energy efficient new buildings and engaging in comprehensive place making in a phased manner.

Each city formulates its own smart city proposal, including the vision for the city, a plan for resource mobilisation and intended outcomes in terms of infrastructure upgrades through the features indicated in Table 7.1.

Selection Process of Cities

A total of 99 smart cities are distributed among states based on a formula of equal weightage (50:50) to the urban population of the state and the number of statutory towns in the state. Each state will have a certain number (at least one) of potential smart cities. Here, the objective of the mission is to provide equal opportunities for all states in a cooperative federal environment (Ministry of Urban Development 2015). The mission adopted the unprecedented two-stage 'City Challenge' competition to shortlist cities within each state (Stage I) and, subsequently, an intra-city competition was held to select the winning cities in different stages (Stage II).

For Stage I, 13 criteria were proposed for use by the state governments to shortlist cities for Stage II. Among the 13 criteria, a higher weight (60 per cent) was assigned to attributes of institutional systems/capacities, ability to self-finance and past track record on implementing the JNNURM programme. The parameters include: trend of revenue (taxes, fees, charges) collection; contribution of tax revenue, fees and user charges, and rents to the city budget receipts; share of operation and maintenance cost of the water supply met through user charges; contribution of internal revenue sources used for capital works; previous track record in completion of JNNURM projects; and achievement of envisaged reforms. The highest-scoring potential smart cities were shortlisted and recommended for participation in Stage II.

In Stage I, shortlisted cities were given a grant of INR200 million to prepare a detailed smart city proposal for Stage II, with the support of external empanelled consultants. This is crucial as each proposal should contain the model chosen, with an emphasis on retrofitting, redevelopment, or greenfield development, along with Pan-City elements embedded with smart solutions. The evaluation criteria emphasise: a city vision and strategy; the extent of consultation with citizens and other stakeholders, including investors groups; the use of social media and mobile communication technologies for consultation; whether citizen aspirations match the civic vision; the use of smart solutions in Pan-City and Area-based Development implementation plans; and proposed financing, including a revenue model to attract private participation.

Cities covered under the Smart Cities Mission include: 24 state capitals; 24 business and industrial centres; 18 cultural and tourism centres; five port cities; and three educational and healthcare hubs. In terms of city population, 70 are classified as 'Class I' cities (population over 100,000) by the definition of the Census of India. Nine cities have a population below 100,000, while 20 cities are million-plus urban agglomerations. On the whole, the mission covers 99.5 million of India's 450 million urban population in 2018 (Housing and Land Rights Network 2018).

Implementation and Financing

Under the Smart City Mission, the national government has allocated financial support of INR480 billion over five years, for an average INR1 billion per city per year. An equal amount, on a matching basis, must be contributed by state governments and municipalities. The project cost for each smart city varies based on the nature of the development model and the execution capacity of the city. In view of the institutional deficiencies of municipalities and fragmented jurisdictions of governance among different state agencies, the Smart Cities Mission has proposed the creation of an SPV for the smooth and efficient implementation of its activities, including planning, appraisal, approval, fund disbursement, implementation, management, operation, monitoring and evaluation of the projects in each selected city.

The SPV structure is created to ensure efficiency in planning and implementation of projects and to overcome problems of interagency coordination in urban development and multiplicity of jurisdictions. It is a single-point decision-making entity expected to mobilise funds for urban infrastructure augmentation through market-based instruments (Taraporevala 2017). The SPV is a limited company under the Companies Act of 2013 and is managed by independent directors, nominees of the union and state governments, as well as the concerned municipalities. The state government and municipalities are the joint promoters, with 50:50 equity shareholding. The private sector or financial institutions can be considered for an equity stake, provided that the state government and municipality together have majority shareholding and control of the SPV (State Government: Urban Local Body: Private Sector shareholding may be 40:40:20 or 30:30:40) (Ministry of Urban Development 2015).

The shareholders need to ensure a substantial revenue stream for the SPV to make itself sustainable by raising additional resources from the market. The SPV can explore the possibility of joint ventures, subsidiaries, PPPs, turnkey contracts and other arrangements for this purpose. Several state governments (e.g., Tamil Nadu, Gujrat and Punjab) have established financial intermediaries that can be tapped for support. The rest of the fund may be mobilised from conventional and non-conventional sources: state governments' and municipalities' own resources (user charges, impact fees, land monetisation, debt, loans), additional resources linked to the acceptance of recommendations of the Fourteenth Finance Commission, municipal bonds, pooled finances and tax increment financing among others.

From the nationwide pool of 99 shortlisted cities, the first 20 cities, in order of ranking, are: Bhubaneswar, Pune, Jaipur, Surat, Kochi, Ahmedabad, Jabalpur, Visakhapatnam, Solapur, Davangere, Indore, New Delhi Municipal Council, Coimbatore, Kakinada, Belagavi, Udaipur, Guwahati, Chennai, Ludhiana and Bhopal. Some of the key success factors of the first 20 winning proposals are (DFID–MoUD 2016):

1. A business plan approach that creates a win–win situation for the government, citizens and private sector.
2. Doing more with less by making existing infrastructure and services smart using technology.
3. Proposing a financially viable model using grants as returnable surplus.
4. Packaging various parts of the proposal into an integrated solution.
5. Prioritising high-impact and low-cost solutions.
6. Developing a mechanism to continuously engage citizens in plan formulation and implementation.
7. Impactful consultations with stakeholders (strong partnership with local/state/national government), expert inputs from city planners and collaboration with the private sector.

A study on the financial allocations made by the top 90 cities found that the Area-based Development components, which had been allocated 80.8 per cent of the funds, were focused on improving small portions (a little over 4 per cent) of the total area of the cities. Only residual portions are directed towards the Pan-City component, which has impacts on larger areas of cities (Taraporevala 2017).

Analysis of the Pan-City proposals of 20 cities reveals that traffic control measures top the list of priorities. Most of the cities opted for features such as a centralised command and control centre to streamline real-time interagency coordination and management of city-level data and an intelligent traffic management system, including synchronised traffic signalling, real-time tracking of bus movements through mobile apps, smart parking and smart bus stops with a public information system (Ministry of Urban Development 2017).

Other prioritised projects observed in more than four cities are: a common card system; smart metering for water; CCTV surveillance; an emergency response system; wi-fi connectivity; a platform for citizen engagement and citizen services; a city dashboard; grievance redressal through the web; an app and mobile phone; LED street lighting; GPS tracking and optimisation of routes of garbage trucks; the management of solid waste collection and disposal; e-healthcare; and water flow monitoring.

Within the Area-based Development component, the predominant projects are: slum redevelopment and affordable housing; rejuvenation of parks and public spaces; lake and riverfront development; restructuring of the Central Business District and market redevelopment; retrofitting the historic urban fabric; development of multimodal transit hubs and transit-oriented development. Incubation/startup centres were also proposed by six cities. Smart solutions proposed to be integrated with retrofitting and redevelopment projects are: rainwater harvesting; LED/solar streetlights; renewable energy initiatives; smart signalling; smart parking; wi-fi connectivity; e-rickshaws; public bike sharing; and barrier-free walkways.

SMART CITY PARADIGM THROUGH THE LENS OF THE GOOD GOVERNANCE FRAMEWORK

Considering the present and future needs of society and the need to incorporate the voices of the vulnerable and the marginalised, the UNDP developed a framework of good governance to facilitate inclusive decision making for development. The tool provides an analytical lens comprising eight attributes of Participation, Rule of Law, Transparency, Responsiveness, Consensus, Equity and Inclusiveness, Effectiveness and Efficiency, and Accountability for examining the intent and process of decision making, the extent of participation of stakeholders and compliance with the existing legal provisions of the society (UNDP 1997). A summary of these governance implications is listed in Table 7.2.

This section discusses the salient features and the implementation process of smart cities with respect to the eight attributes of good governance:

1. **Participation:** A few cities, like Bhubaneswar, which had a top ranking in the competition due to its robust public consultation process, engaged in several rounds of consultation with various sections of society, including slum dwellers and the urban poor. By and large, however, citizen engagement in smart city proposal formulation was through online platforms, which has a greater reach to the digitally empowered urban middle class than the poor. This resulted in community protests in some places. Another particularly worrying aspect is long-term institutional arrangements for public participation. The nodal agency for rolling out of the Smart City Mission – the SPV – is a corporate entity through public–private equity participation and is managed by techno-managerial elites. There is apprehension that enabling SPVs as the key agency may take power away from elected municipal counsellors and thereby reduce the scope for participatory democracy.

2. **Rule of law:** Neither the planning process, nor the authority for city management through SPVs has statutory backing under the existing legal provisions of municipal planning in India. The planning and institutional short cuts are projected as obligatory passage points to bypass the contested terrain of the urban political landscape and truncate the role of elected municipalities. Without statutory legal backing, the future of the projects could become uncertain with change in the government and such uncertainty could hinder new investment flows.

3. **Transparency:** Presently, all major documents pertaining to the mission – its guidelines and processes – are available in the public domain. Similarly, the composition of the SPVs and projects initiated by them are available online at the website of the Ministry of Urban Affairs. However, there are concerns about how the situation will unfold in the future. As noted earlier, the SPV model has scope for the private sector to hold up to a 40 per cent equity stake. It may be possible that private investors would seek to dilute transparency to further their business interests by awarding contracts and the right to use proprietary technology.

4. **Responsiveness:** The overarching emphasis of the initiative on technology-enabled service delivery and governance modelled in line with cities of the developed world (e.g., intelligent command and control centres, traffic management systems, smart water metering) tends to ignore the structural inadequacies of Indian cities, leading to inefficiencies in resource allocation and utilisation. The leapfrogging approach to city mod-

Table 7.2   *Summary of governance implications for the Smart Cities Mission*

Attributes of good governance	Salient features of the initiative	Weaknesses/limitations
Participation	Citizen engagement in smart city proposal (SCP) formulation predominantly through online platforms.	The voice of the urban poor and marginalised are not well represented.
Rule of law	The planning process for SCPs and city management through SPV do not have statutory backing under the existing legal provisions for municipal planning.	Planning and institutional bypass eliminating the role of elected municipal councillors.
Transparency	As of now, documents pertaining to the Smart Cities Mission and its various implementation processes are transparent and available in the public domain.	Privatised mode of project execution through the SPV brings into question issues of transparency, especially regarding awarding of sub-contracts and outsourcing.
Responsiveness	The leapfrogging approach to city modernisation through technology-enabled service delivery and governance is modelled in line with cities of the developed world.	Lack of customisation of proprietary technologies hinders their adaptability to Indian cities.
Consensus	Fast-track mode of consultant-driven SCP formulation and a top-down planning approach.	Inadequate scope for consensus building due to fast-track planning. A mechanistic approach to consensus building may lead to community resistance and tardy implementation of smart city projects.
Equity and inclusiveness	Financial allocation for the Area-based Development component, which serves only a small part of the city and a small portion of the population, is much higher compared to Pan-City initiatives, further perpetuating intra-city inequalities.	The basic issues of informality, migration and life and livelihood challenges of the urban poor in Indian cities are being bypassed. It is also doubtful whether land monetisation-oriented Area-based Development is appropriate for small towns.

Attributes of good governance	Salient features of the initiative	Weaknesses/limitations
Effectiveness and efficiency	The efficiency of smart cities is envisaged to rely on technology-enabled service delivery. The delivery pattern has been centralised in the form of SPVs to fast-track decision making.	The role of the SPV in the institutional architecture of urban governance is not clearly defined, which may cause tensions with elected municipal officers and other parastatal bodies.
Accountability	SPVs constituted as public limited companies are likely to be accountable to their shareholders instead of citizens.	The SPVs are directly accountable neither to citizens nor to elected municipal councillors. The efficacy of the indirect modus of accountability is questionable.

Source: Authors.

ernisation does not take into consideration the customisation of proprietary technologies for adaptation to the multi-dimensional problems and hetero-geneous character of Indian cities. How such smart technologies play out in cities with large incidences of poverty and informality and a broad digital divide remains to be seen.

5. **Consensus:** In their urge to fast-track the submission of smart city pro-posals within the stipulated time window, most cities attempted to reach a consensus of diverse stakeholders in a hurried manner and identified a list of projects under the Pan-City initiative and Area-based Development. Subsequently, the number of cases of reported community protests and resistance has raised questions about the legitimacy of the fast-track mode of city development, and the illusion of consensus around the contested issues of land acquisition, eviction notices to communities for project execution and conservation of urban commons continues to be exposed. Moreover, collection of user charges could be politically problematic in several cities. Presently, efficiency of the collection of user charges for municipal services is abysmally low. Consequently, Indian cities are caught in the 'low income–low quality services–low level of expectations' equilibrium trap. Efforts to raise fees without political consensus at the local level cause tension and delay project implementation.

6. **Equity and inclusiveness:** Though the mission speaks about inclusion and equity as one of its guiding principles, the disproportionately higher funding allocated for Area-based Development raises concern. As dis-cussed previously, the scope of Area-based Development covers only a small part of any city. Another area of concern is that the competitive selection process compelled cities to pay more attention to revenue gener-ation and beautification to entice investors. In this process, basic structural

issues in these cities regarding informality, migration and livelihood challenges of the urban poor did not receive due attention.

7. **Effectiveness and efficiency:** The Smart Cities Mission has placed major
emphasis on increasing operational efficiency in urban services delivery.
However, the focus is on the application of high-end technology by SPVs
and the outsourcing of its management to external consultants. First, the
need to augment the managerial and technological capabilities of municipalities has not been adequately addressed. Thus, who will monitor the
efficiency parameters for the functioning of smart city projects is questionable. Second, the financing strategies for smart cities are overwhelmingly
oriented towards market-based instruments, including land monetisation.
It is doubtful whether such strategies would be effective in small towns,
where land markets are less vibrant compared to bigger cities. Third,
projects identified for funding were hurriedly identified to meet the tight
three-month time frame of the national government, which may cause difficulties in implementation. Projects under the Pan-City component, such as
water flow metering or tracking of vehicles involved in garbage collection,
require a high degree of coordination with line agencies undertaking the
operations. There is a lack of clarity in regard to how such interfaces will
be managed. Moreover, water supply and sewage systems of several cities
are presently managed in a piecemeal and ad-hoc manner. Retrofitting
them with advanced technologies could be problematic and difficulties in
resolving such operational details could delay project implementation.

8. **Accountability:** Management of smart cities has been entrusted to SPVs,
which are constituted as public limited companies and are likely to be
accountable to their shareholders instead of citizens. The Smart Cities
Mission guidelines are silent about the future institutional architecture
of urban management. First, the working relationship between the SPV
and elected municipal officials is a matter of serious concern. Municipal
bodies are elected by people and are accountable to citizens. Although the
municipalities have equal stake to participate in the SPV board, how such
power will be exercised is questionable. Second, the relationship between
the SPV and the planning agency – the concerned urban development
authority – is another open question. It is unclear who should coordinate
the interfaces between the various public and private urban network service
providers. For the time being, smart city SPVs have constituted the executive head of the Urban Development Authority or the Commissioner of the
Municipality as the Chairman of the Board of the SPV. However, this is an
ad-hoc arrangement and must be institutionalised. A formal consultative
platform for the SPV and other stakeholders is yet to take shape.

CONCLUSION

This chapter sought to explore the driving force behind the Smart Cities Mission and the challenges and concerns emerging in the process of implementing big data and a knowledge-intensive urban agenda in a developing country with a large, mostly poor, digitally divided population across diverse geographies and varied institutional landscapes. Subsequently, we analysed the smart cities policy through the lens of the good governance framework – a widely accepted instrument used to analyse governance characteristics in developing countries.

Projects funded under the Smart Cities Mission are divided into Pan-City and Area-based Development components. The Pan-City projects are targeted at injecting smart technology and advanced monitoring mechanisms for selected urban infrastructure on a city-wide scale but account for only one-fifth of the project budget. In contrast, Area-based Development, which accounts for four-fifths of the project budget, is targeted to improve a small part of a city through comprehensive place making and act as a 'lighthouse' for the rest of the city.

Our analysis shows that the formulation of a smart city plan is a top-down, consultant-driven, modularised, business-plan-centric, outcome-oriented, project management approach towards cities. In taking this form, it further reinforces the tradition of centralised technocratic master planning, rather than attempting to reinvent planning as a continuous process.

The Smart Cities Mission has introduced the SPV system of urban governance. As a single-point nodal agency, the SPV framework is expected to ensure efficiency in planning and implementation of projects, smooth interagency coordination and overcome hindrances posed by the multiplicity of jurisdictions. The corporation-like structure of the SPV is also expected to make it easier to access capital markets for fund mobilisation.

While SPV has been previously used in building urban infrastructure (i.e., airports, metro railways, SEZ and even greenfield industrial city projects) in a more controlled environment with a limited number of governance actors in India, this is the first time that such an institutional arrangement has been entrusted with managing a fully functional urban settlement with hugely complex socio-economic contexts, including high incidences of poverty and informality.

Looking through the lens of the good governance framework, we raise the following doubts and concerns regarding the long-term sustainability of the SPV-driven smart city governance model:

1. Could the centralised and techno-managerial decision-making process of the SPV reduce the role of elected municipal counsellors in the governance process? Would this further shrink the space for participatory planning?
2. The absence of statutory backing has made long-term prospects of the project uncertain, especially in the case of changes in the government. Could such uncertainties have impacts on private sector investment flows into SPVs?
3. How does the corporate governance mode of SPVs influence transparency in awarding sub-contracts and sharing information in the public domain?
4. How responsive are SPVs in providing needs of the urban poor?
5. How would fast-track planning play out at the time of project implementation? Would inadequacies in public consultation cause community resistance and slow down the projects?
6. Would the business plan-oriented approach of SPVs turn cities more exclusionary and increase the digital divide?
7. How would SPVs manage interfaces with other institutional actors to bring more efficient service delivery in a multi-layered governance framework?
8. There are apprehensions about the long-term accountability of the SPV framework. To whom would they remain answerable: to their shareholders or to the public?

India's ambitious Smart Cities Mission is still at an early stage of implementation. How the project unfolds will considerably depend on how its SPV-centric governance framework negotiates the issues and concerns noted in this chapter.

NOTE

1.   The bulk of the increase was in the lower-middle category of USD 2.0 to 4.0 per person per day. For details, refer to Krishnan and Hatekar (2017).

REFERENCES

Bhide, A & Burte, H 2015, *Smarter urbanisation, not (just) smart cities*, NCAS Discussion Paper Series. Pune: National Centre for Advocacy Studies.
Burte, H 2014, 'The smart city card', *Economic and Political Weekly*, vol. 49, no. 46, pp. 22–25.
Chatterji, T 2017, 'Modes of governance and local economic development: An integrated framework for comparative analysis of the globalizing cities of India', *Urban Affairs Review*, vol. 53, no. 6, pp. 955–989.

Chattopadhyay, S 2017, 'Neoliberal urban transformations in Indian cities: Paradoxes and predicaments', *Progress in Development Studies*, vol. 17, no. 4, pp. 307–321.

Chhotray, V & Stoker, G 2009, *Governance theory and practice*. London: Palgrave Macmillan.

Datt, G, Ravallion, M & Murgai, R 2016, *Growth, urbanization and poverty reduction in India, NBER working paper no. 21983*. Cambridge, MA: National Bureau of Economic Research.

Datta, A 2015, 'New urban utopias of postcolonial India: "Entrepreneurial urbanization" in Dholera smart city, Gujarat', *Dialogues in Human Geography*, vol. 5, no. 1, pp. 3–22.

DFID–MoUD 2016, 'Workshop on proposal upgradation of second round smart cities challenge', New Delhi, 6–10 June.

DMIC 2018, 'About DMICDC', Delhi–Mumbai Industrial Corridor Development Corporation Limited, viewed 20 April 2018, http://dmicdc.com/about-DMICDC/implementation-framework.

Gilbertson, A 2017, 'Aspiration as capacity and compulsion: The futures of urban middle-class youth in India', in Stambach, A & Hall, KD (eds.), *Anthropological perspectives on student futures: Youth and the politics of possibility*, pp. 19–32. New York: Palgrave Macmillan.

Government of India 2015, 'What is a smart city?', viewed 19 May 2019, https://www.india.gov.in/spotlight/smart-cities-mission-step-towards-smart-india.

Graham, S 2002, 'Bridging urban digital divides: Urban polarisation and information and communication technologies (ICT)', *Urban Studies*, vol. 39, pp. 33–56.

Housing and Land Rights Network 2018, 'India's Smart Cities Mission: Smart for whom? Cities for whom?', viewed 19 May 2019, https://www.hlrn.org.in/documents/Smart_Cities_Report_2018.pdf.

Jaikumar, S & Sarin, A 2015, 'Conspicuous consumption and income inequality in an emerging economy: Evidence from India', *Marketing Letters*, vol. 26, no. 3, pp. 279–292.

Krishnan, S & Hatekar, N 2017, 'Rise of the new middle class in India and its changing structure', *Economic and Political Weekly*, vol. 52, no. 22, pp. 40–48.

Livemint 2016, *Amaravati: A capital idea, but how feasible?*, viewed 28 March 2018, https://www.livemint.com/Politics/fpmYw43wSuaTHwfTgTlpgK/Amaravati-A-capital-idea.html.

Ministry of Urban Development 2015, 'Smart cities mission', Government of India, viewed 4 July 2015, http://smartcities.gov.in/.

Ministry of Urban Development 2017, 'Smart cities mission', Government of India, viewed 2 December 2017, http://smartcities.gov.in/.

Mohanty, PK 2014, *Cities and public policy: An urban agenda for India*. New Delhi: Sage.

Roy, S 2016, 'The smart city paradigm in India: Issues and challenges of sustainability and inclusiveness', *Social Scientist*, vol. 44, no. 5–6, pp. 29–48.

Shivaramakrishnan, KC 2011, *Re-visioning Indian cities: The urban renewal mission*. New Delhi: Sage.

Taraporevala, P 2017, 'India's municipalities can learn from the funding pattern of smart cities', *Hindustan Times*, viewed 6 September 2017, https://www.hindustantimes.com/opinion/india-s-municipalities-can-learn-from-the-funding-pattern-of-smart-cities/story-GEERJBC368ENdFcUl9jvyM.html.

UNDP 1997, *Governance for sustainable human development*. New York: UNDP.
Upadhya, C 2007, 'Employment, exclusion and "merit" in the Indian IT industry',
    Economic and Political Weekly, vol. 42, no. 20, pp. 1863–1868.

8. Smart cities in China: development background, policy measures and implementations

Xinhui Yang and Lin Ye

INTRODUCTION

With the rapid development of urbanisation in China, a series of problems – including overpopulation, traffic congestion, environmental pollution and resource shortages – have emerged in many cities, especially in megalopolises such as Beijing, Shanghai, Guangzhou, Shenzhen and others. In order to tackle these urgent urbanisation-related problems, the Chinese national and local governments have dedicated their efforts to finding new modes of urban governance. In recent years, the smart city concept has emerged as an effective solution. At present, China has published three batches of national smart city pilot programmes that cover more than 290 cities. By June 2016, 95 per cent of sub-provincial cities and 76 per cent of prefecture-level cities in China – over 500 cities – explicitly mentioned that they would build smart cities in their Government Work Reports or their 13th Five-Year Plans (Wulian Zhongguo 2016). In this context, we ask the following important questions: What were the specific contexts for developing existing smart cities in China? What key policy measures have the Chinese national and local governments taken in developing smart cities? What main development fields were chosen by the local governments when they decided to build smart cities? And what development outcomes have been achieved in these fields?

This chapter begins by introducing the development background of China's smart city programmes. The second section summarises the course of their development by exploring the supporting policies in chronological order as well as the evaluation index systems of China's smart cities. It also lists the smart city pilot programmes and analyses their respective characteristics. Using a case study of the Sino–Singapore Suzhou Industrial Park (SSSIP), the subsequent section focuses on the introduction of some development practices carried out by the SSSIP government and their development outcomes.

Finally, the chapter concludes by highlighting the implications for those individuals seeking to understand the development background, supporting policy measures, implementation practices and outcomes of China's smart city programmes. We believe that by building scientific and well-developed smart cities, China can better deal with the problems stemming from urbanisation.

BACKGROUND OF SMART CITY DEVELOPMENT IN CHINA

According to the data from the National Bureau of Statistics of the People's Republic of China (NBSPRC) (2018), the population of permanent urban residents in China reached 813.47 million by the end of 2017 – an increase of 20.49 million from the end of 2016. The proportion of the urban population in the total population (urbanisation rate) was 58.52 per cent, representing an increase of 1.17 per cent over the end of the previous year. In the last 40 years, the urbanisation rate of China has increased rapidly from 17.92 per cent in 1978 to 58.52 per cent in 2017, with an annual average growth rate of 3.163 per cent. The experience of urbanisation in developed countries suggests that the urbanisation rate of a country reaching to 40–60 per cent not only marks the acceleration of urbanisation but also potentially the onset of 'urban disease' outbreaks resulting from urbanisation (Liu 2014). Urban disease refers to a series of economic, social and ecological problems arising from the imbalance between the carrying capacity of urban resources and environment, and the speed and scale of urbanisation (Jiao 2015). Notably, it includes social diseases (such as overpopulation and traffic jams), economic diseases (such as urban poverty and unemployment) and ecological diseases (such as a short-age of resources and environmental pollution). Although the current level of urbanisation in China is lower than that of developed countries, the rapid urbanisation of the metropolises in China may suffer a concentrated outbreak of one or more of these urban diseases.

Chinese metropolises such as Shanghai, Beijing, Guangzhou and Shenzhen are already encountering the issue of overpopulation. According to data from the NBSPRC, the urban population of China is increasing by 20 million per year, thereby producing unprecedented congestion in Chinese cities. For instance, according to data published by *Shanghai Pudong New Area Statistical Yearbook* (Li & Zhang 2017), the population density of Lujiazui Sub-district in the Shanghai Pudong New Area reached 24,863 people per square kilometre. Meanwhile, the population density in Dongcheng and Xicheng Districts in Beijing are 20,330 and 24,144 people per square kilometre, respectively, according to statistics released from the Beijing Municipal Bureau of Statistics (2018). The population density of Guangzhou is 13,000 people per square kilometre (Guangzhou statistics bureau 2018), which is already much higher than in other major cities worldwide. The rapid increase in urban population and density have brought tremendous challenges to urban management, resulting in an insufficient supply of public services, such as public transportation, water and electricity, which is creating difficulties for daily work and the life of urban residents. In order to control the urban population, the Beijing government clearly specified in the Beijing Urban Master Plan (2016–2035) – issued by the Beijing Municipal Planning and Land Resources Management Committee (2017) – that the maximum resident population in Beijing will be 23 million by 2020.

Accompanying the issue of overpopulation is the increasing number of private vehicles, which has led to deteriorating traffic conditions in many Chinese cities. Figure 8.1 suggests that the number of total private automobiles in Guangzhou and Shenzhen exceeded 2 million in 2011, while that of Beijing exceeded five million in 2013. This large increase in urban private vehicles, in addition to the faultiness of urban road planning, has resulted in serious urban traffic congestion. Many local governments have taken a series of measures to alleviate the pressure of their urban traffic conditions, such as car plate lotteries, car plate limiting, and the most recent policy from Guangzhou named 'Drive Four Days, Then Stop Four Days'. On 30 June 2012, the Guangzhou

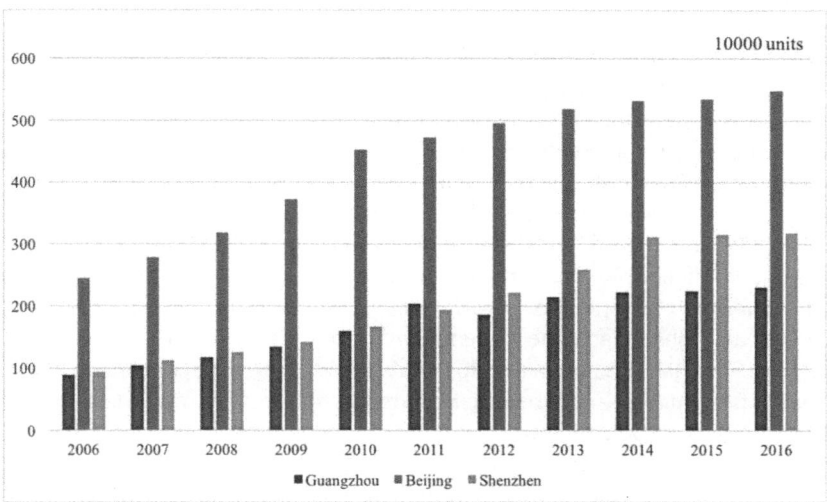

Source: The statistical yearbooks of Guangzhou (Guanghzoushi tongjiju 2006–2016), Beijing (Beijingshi tongjiju 2006–2016) and Shenzhen (Shenzhenshi tongjiju 2006–2016).

Figure 8.1   *Total civil automobiles of Guangzhou, Beijing and Shenzhen in China, 2006–2016*

Municipal Traffic Management Committee held a press conference titled 'Regarding the Systematic Improvement of Urban Traffic and the Control of the Total Number of the Small and Medium-sized Civil Automobiles in Guangzhou'. A policy for the improvement of urban traffic conditions in Guangzhou was issued alongside a notice on the control and management of the total number of small and medium-sized private vehicles in Guangzhou, which stated that the increment of private automobiles in Guangzhou would be controlled to a certain extent by the policy of the car plate lottery. After implementing this policy, the number of private automobiles in Guangzhou was reduced from 2.04 million in 2011 to 1.86 million in 2012. However, the number of cars continued to grow from 2013, thereby exacerbating the transportation pressure in the city.

According to the report 'Analysis of the Traffic Conditions of Major Chinese Cities' released by Amap and Alibaba Cloud Computing Co. Ltd. (2017), Guangzhou ranked sixth in the ranking of traffic congestion index (TCI) in China; however, its TCI ranked first in the afternoon rush hours, with a high index value of 2.158. This represents an urgent challenge for the Guangzhou government in terms of improving traffic conditions. On 15 May 2017, the Guangzhou Traffic Commission (GTC) signed a cooperation agreement with Amap that would fully utilise the big data platform to improve traffic management and decision-making analysis. The cooperation involved the fields of traffic information mining and analysis, public transport services and building a 'smart transportation' system for the era of big data. With the help of this system, the GTC can obtain various types of information, including city traffic congestion points, traffic incident points and traffic anomaly points, among others. More importantly, the GTC can publish information such as route guidance to the public in order to effectively alleviate traffic congestion and provide more intelligent and custom services for citizens.

China's urbanisation has not only brought a series of urban diseases to its cities, such as overpopulation, traffic congestion and environmental pollution but has also challenged the government's ability to provide urban public services. This has resulted in a serious imbalance between supply and demand in education, medical treatment, employment, housing and other key public services. These problems force city managers to seek improved management models and find new ways to improve the efficiency of their cities. By developing smart cities that include the provision of smart transportation, smart medical systems, smart education, smart environmental protection, and other smart services, they can solve these otherwise intractable problems.

INTRODUCTION TO SMART CITY DEVELOPMENT POLICIES

Since 2011, many Chinese cities have proposed or promoted the development of smart cities. In order to maintain the healthy development of smart cities, the Chinese government not only released many supporting policies and evaluation index systems but also conducted three batches of smart city pilot projects that aimed to explore the development model and accumulate development experiences.

Supporting Government Policies

As previously mentioned, many Chinese cities announced that they would establish smart cities from 2011 onwards. As a result, a number of departments in the Chinese central and local governments aimed to regulate and lead the development of the smart city concept in China. To date, the development of the Chinese smart city has successively undergone three stages: spontaneous exploration at the local government level; multi-ministry co-management at the central ministries and commissions level; and top-level coordination and resource integration at the national level (Ye et al. 2018).

In the first stage, the local governments formulated and promulgated their own smart city planning, which was suitable for their own development according to their actual respective conditions. However, from the beginning of the second stage, the central governments, including many ministries and commissions, began to promulgate programmatic regulations and policies to ensure the healthy development of Chinese smart cities. For instance, in May 2012, the Ministry of Culture and Tourism of the People's Republic of China (MCTPRC) announced 18 smart tourism pilot cities including, among others, Beijing and Chengdu. In November 2012, the Ministry of Science and Technology of the PRC (MSTPRC) issued the 'Notice on Carrying out the Development of the Pilot of Smart Cities' and officially announced approximately 20 smart city pilots, including Nanjing and Wuxi, in October of the following year. In 2013, the National Administration of Surveying, Mapping and Geoinformation of China (NASMGC) published the 'Pilot Technology Guide for the Development of the Space–time Information Cloud Platform for Smart Cities' and set nine cities – including Taiyuan and Guangzhou – as development pilots for the platform. In November 2013, the Ministry of Industry and Information Technology of the PRC (MIITPRC) (2013) launched the China–EU cooperation project for developing smart cities and selected 15 cities for the project. In August 2014, the National Development and Reform Commission of PRC (NDRCPRC) (2014) and MIITPRC published the 'Guidance on Promoting the Healthy Development of the Smart City' that announced a specific goal: 'by the end of 2020, a number of smart cities with distinctive features will be built to ensure and improve citizens' livelihood services, and to provide the government with innovative approaches to social management'. Table 8.1 summarises a series of important policies and regulations issued by Chinese national ministries to promote smart city development.

Smart City Pilot Programmes in China

Apart from promulgating a series of policies and regulations (see Table 8.1), the Chinese government also released three batches of smart city pilot projects that were spread throughout the country and across many industries. These pilot projects aimed to strengthen the comprehensive application of modern science and technology in the urban planning, construction, management and operation of existing cities, districts and towns. According to the Tentative Management Measures for the Pilots of the National Smart City (Version 2012) and the Evaluation Index System of the National Smart City Pilots (including district and town) (Version 2012), there are 90 smart city programmes in the first batch of pilots, including 37 at the city level, 50 at the district and county level and three at the township level. In May 2013, the MOHURDPRC (2013) released the 'Notice on Carrying out the Work of the Declaration of National Smart City Pilots' in 2013, which included 103 smart city pilots in the second batch of pilots, including 83 at the city and district level and 20 at the county and town level. The third batch of smart city pilots, a total of 84, was released in April 2015 by MOHURDPRC. On 24–25 October 2015, guided by the Economic Daily, a forum titled 'Global Wisdom City (Beijing) Summit Forum' with the topic 'The New Future of the City, the Finding and Release of the New Development Impetus' took place as part of the 'Economic and Technological Cooperation Between 100 Cities and 100 Enterprises' conference held by the Economic Journal in Beijing. A total of 2,800 smart city projects were showcased at the conference, which would considerably influence the development process of Chinese smart cities. On 26 January 2016, the 'Workshop of the Smart City Standards and Evaluation (Year 2016)' was organised by the National Smart City Standardisation General Group. In the workshop, the NDRCPRC and OCCAC, together with the Inter-Ministerial Coordination Working Groups on Smart Cities, issued 100 pilot projects for new smart cities.

To date, there have been a total of 290 smart city pilots released by the MOHURDPRC in China. In terms of the final total number of smart city pilots, there are 14 provinces whose number of smart city pilots is greater than ten. Shandong and Jiangsu provinces have the largest number of pilots at 26 and 24, respectively. Figure 8.2 presents the number of these pilots throughout

Table 8.1   *The main supporting policies and regulations for developing a smart city, 2012–2017*

Date issued	Name of issuing organisation	Policies and regulations	Key points
November 2012	Ministry of Housing and Urban–Rural Development of the People's Republic of China (MOHURDPRC)	Tentative management measures for the pilot of the national smart city	Guiding and managing the application and implementation of national smart city.
August 2013	The State Council of the People's Republic of China (TSCPRC)	Opinions of the State Council on promoting information consumption and expanding domestic demand	Clearly proposing to speed up smart city development, and declaring that the cities with appropriate conditions can start building the smart city, and encouraging all kinds of market entities to participate in the development of smart cities together.
March 2014	TSCPRC	National plan for new urbanisation (2014–2020)	Clearly promoting smart city development, and pointing out the direction in which it should develop.
August 2014	NDRCPRC, MIITPRC, MSTPRC, Ministry of Public Security of the People's Republic of China, Ministry of Finance of the People's Republic of China, Ministry of Natural Resources of the People's Republic of China, MOHURDPRC, Ministry of Transport of the People's Republic of China (MTPRC)	Guidance on promoting healthy development of smart cities	Announcing the development of a batch of smart cities with distinctive features by the end of 2020, largely enhancing the aggregation and radiation function of smart cities, clearly improving the advantage of comprehensive competitiveness, ensuring and improving citizens' livelihood services, and so on.

Date issued	Name of issuing organisation	Policies and regulations	Key points
January 2015	MCTPRC	Guidance on promoting the development of smart tourism	Announcing, by the end of 2020, to greatly improve smart tourism services, to strengthen smart management in tourism, to enhance the ability of smart tourism to mine and analyse big data, and to form a systematic network of smart tourism value chains.
May 2015	NASMGC	Notice about the work related to the transformation and upgrading of a digital city to a smart city	Proposing some suggestions for how NASMGC can play a fundamental and leading role in the development of smart cities and how to promote the healthy development of smart cities.
October 2015	Standardisation Administration of the People's Republic of China (SAPRC), Office of the Central Cyberspace Affairs Commission (OCCAC), NDRCPRC	Guidance on the development and application of the standard system and evaluation index system of a smart city	Announcing that, by the end of 2020, a total of about 50 items of standard formulation work in the field of smart city will be completed to promote the standardisation of existing relevant technology and application standards, and to realise the comprehensive application of the evaluation index system of smart city.
August 2016	NDRCPRC, OCCAC	Task division of the inter-ministerial coordination working group on new smart city development (2016–2018)	Specifying the tasks and responsibilities – a total of 26 items – of the 25 departments which are the members of the Inter-ministerial Coordination Working Group.
November 2016	NDRCPRC, OCCAC, SAPRC	Notice on organising the evaluation work of new smart cities to promote the healthy and rapid development of new smart cities	Researching and laying down the evaluation indicators, evaluation work requirements and evaluation organisation methods of new smart cities.

Date issued	Name of issuing organisation	Policies and regulations	Key points
January 2017	MTPRC	Action plan for promoting smart transportation development (2017–2020)	Announcing that, by the end of 2020, the goals of intelligent infrastructures, intelligent production organisations, intelligent transportation services and intelligent decision making and supervision will gradually be realised.
September 2017	NASMGC	The outline of development of smart city time–space big data and cloud platform	On the basis of the original digital city geo-spatial structure, relying on the urban cloud support environment to build the spatial and temporal data of the smart city, the space–time large data and the space–time information cloud platform, to build the urban space–time infrastructure and to develop the intelligent special application system, to accumulate experience for the comprehensive application of the time–space infrastructures in the smart city, to promote the transformation of digital city to smart city.

Source: Data compiled by authors.

China. China's smart city pilots are primarily concentrated in China's eastern coastal cities and second- and third-tier cities in eastern areas, since the economic base of these cities is better than those of other cities.

However, Chinese local governments have encountered a number of problems in the process of developing smart cities. For example, they were often trapped by the previous models of urban construction and repeatedly built smart cities as vanity projects of dubious quality, leading to mass wastage of financial and human resources. Gao and Wang (2015) assert that China can only standardise and promote the efficient, rapid and healthy development of smart cities after standardising and unifying the technical requirements, project implementation requirements, testing and certification methods and evaluation index system of smart city development.

Evaluation Index Systems of Chinese Smart Cities

Despite the relatively recent development of the smart city, China has actively explored a smart city evaluation index system. The evaluation index system for smart cities (EISSC) consists of a series of scientific indicators involving

Note: Taiwan, Hong Kong and Macau, are excluded.

Source: Map drawn by authors with data from the MOHURDPRC (Shujuguan, 2016).

Figure 8.2   Three batches of smart city pilot programs in China

both 'hard' and 'soft' indices that are able to quantify the outcome of smart city development. This is to be an action guide for building smart cities and plays an important role in guidance, supervision, evaluation and feedback in the development process.

Notably, more than ten influential EISSCs have been released by Chinese governments or scholars since 2010. In July 2012, the China Institute of Communications (2012) promulgated their EISSC, which is composed of four dimensions that include smart infrastructure, smart application, a support system and a value realisation, with 19 second-level indicators and 57 third-level indicators. In December 2012, the Shanghai Pudong Smart City Research Institute promulgated its EISSC (Version 2.0), which comprises six dimensions that include the following aspects of a smart city: infrastructure, public management and services, information services for economic development, humanistic quality, citizens' subjective perceptions, and soft environmental development. These dimensions are further divided into 18 second-level indicators and 37 third-level indicators. In 2013, the China Software Test Centre (2013) announced its EISSC, which includes three first-level indicators, eight second-level indicators, 36 third-level indicators and 53 fourth-level indicators. In December 2016, according to the needs of the national urban development strategy, the OCCAC and NDRCPRC – along with other national ministries – formulated the New Smart City Evaluation Index (Version 2016). This new EISSC emphasises citizen-centricity in its service to urban residents. Namely, it comprises eight first-level dimensions, including improving public services, precise governance, ecological liveability, smart facilities, information resources, network security, innovation and reform, and public perception. This has become the main standard for the development and evaluation of Chinese smart cities and also an important means for the orderly development of smart cities throughout the country, since it provides a strong guarantee and guidance for new urbanisation construction in China. Table 8.2 presents other EISSCs in China and how their indicators are mostly divided into three levels.

However, upon further study, we discovered several shortcomings in the EISSCs. First, most of them unduly emphasise economic development and infrastructure construction. Second, most existing EISSCs are relatively complex. Hence, it is difficult to measure and evaluate them in the developing stages of smart cities. Third, the existing indicator system can be divided into three types, prophase indicators, metaphase indicators and anaphase indicators; however, the anaphase indicators have a poor correlation with smart cities development at early stages (Zhu 2013). Finally, differences in smart cities' development in different regions and their industrial characteristics were not considered. Notably, there should be differences in the EISSCs for different smart city projects because they require different development emphases.

DEVELOPMENT PRACTICES OF SMART CITIES IN CHINA

The main reason why Chinese national and local governments value the development of smart cities and have zealously released a series of supporting policies is that the emergence of the smart city provides a potential solution to the problem of urban diseases while also helping to improve urban management efficiency. As previously mentioned, the sustainable development of metropolises in China is limited by urban diseases, including overpopulation, traffic jams, environmental pollution, a shortage of resources and an insufficient supply of services. Under these circumstances, Chinese national and local governments have turned to smart cities to solve their problems. In other words, smart transportation, smart education, smart government services, smart medical systems and other smart initiatives are the main domains of smart city development in China. For instance, in smart transportation systems, local governments can reduce the pressure of traffic congestion, which makes cities operate more efficiently. With the smart government service system, local governments can improve the efficiency of the government's provision of services while urban residents can truly achieve the goal of doing things

Table 8.2   *The brief introduction of main EISSCs in China*

Date issued	Name of issuing organisation	Name/source of EISSC	1st-level indicators	2nd-level indicators	3rd-level indicators	4th-level indicators
Aug. 2011	China Wisdom Engineering Association	The Chinese smart city (town) development index	3	23	86	362
July 2012	China Institute of Communications	White paper on smart city technology	4	19	57	None
July 2012	Smart City Research Institute of Ningbo City	Research on the development of the evaluation index system for smart cities in China (Gu & Qiao, 2012)	7	21	48	None
Dec. 2012	Shanghai Pudong Smart City Research Institute	The evaluation index system for smart cities (Version 2.0)	6	18	37	None
Nov. 2012	MHURDPRC	The evaluation index system for the national smart city pilots (including district and town)	4	11	56	None
Jan. 2013	China Software Test Centre (CSTC)	Research report on the evaluation index system for smart cities	3	8	36	53
June 2013	MIITPRC	The evaluation index system for smart cities (draft)	3	9	45	None
Aug. 2014	CCID Consulting Co Ltd	Research on Chinese smart city evaluation and empirical strategy	5	15	57	None
Dec. 2016	National Development and Reform Commission of the People's Republic of China	New smart city evaluation index in 2016	8	21	54	None

Source: Data compiled by authors.

without leaving their homes. In the case study of smart transportation in the SSSIP, the following section discusses how smart city technology was implemented to solve the traffic jams and discusses its outcomes.

A Case Study of the Smart Transportation System in the SSSIP

In 1994, the SSSIP was proposed and developed to share country development experiences from Singapore to help the development in China. It now covers an administrative area with a total of 278 square kilometres (Suzhoushi tong-jiju 2018) and is located in the municipal district of the Suzhou city, Jiangsu province. In 2016, a Government Services Cloud Platform was built in SSSIP and it began the era of the smart city in China. There are various reasons why the SSSIP was chosen as the case study region. Singapore is a country with ample experience in smart city development and SSSIP is a flagship cooperation project between Singapore and Suzhou city that involves a wide range of aspects, such as urban planning, technological development and infrastructure construction. It is well known that Singapore announced its third information technology (IT) master plan in 1992, named IT2000: The Intelligent Island, which was the first concept for developing an intelligent city in the world. To date, Singapore remains a country with advanced urban planning and an efficient urban operation system, which means that Suzhou city can learn much from Singapore's experiences. In the context of collaborating with Singapore, the development level of the smart city in SSSIP can reflect the most advanced smart city development in China to some extent, while providing valuable lessons for future improvement.

There are many smart city development domains in the SSSIP, such as smart education, smart medical services, smart government services, smart tourism and smart transportation. We chose the smart transportation domain in the SSSIP as a case study because traffic congestion is one of the most severe and representative urban diseases facing local Chinese governments (Zhang 2014). Of the 655 cities in China, 66.67 per cent have traffic jams during rush hour, with traffic congestion in first-tier cities being very serious. The traffic congestion time in Beijing has increased from an average of 3.5 hours a day in 2008 to 5 hours in 2014, with an average vehicle speed of only 15 kilometres per hour (Xiang 2014). Traffic congestion causes urban residents to travel more expensively and inefficiently, which causes psychological anxiety and irritability in people and more traffic accidents, resulting in increasing economic losses. At the same time, the frequent change of speed in the process of traffic congestion aggravates air pollution in the city, which seriously threatens the health of urban residents. Therefore, it is typical and representative to choose smart transportation as a case study in this section.

The *Suzhou Statistical Yearbook* (Suzhoushi tongjiju, 2017) showed that the permanent population of the SSSIP was 0.8078 million in 2016, with a high population density of 2,906 persons per kilometre and a high urbanisation rate of 99.55 per cent. Figure 8.3 presents the total number of automobiles in the municipal district of Suzhou city and that of Suzhou city in its entirety from 2006 to 2016. The total number of automobiles in Suzhou city exceeded 1 million in 2009 and 2 million in 2013, while the total number of automobiles in the municipal district exceeded 1 million in 2013. The rapid urbanisation and increasing number of automobiles in Suzhou city has brought some severe urban diseases such as traffic congestion, with the SSSIP being no exception. In order to relieve traffic pressure in the SSSIP, the Suzhou Industrial Park Administrative Committee (SIPAC) aimed to develop its smart transportation project as part of the smart city programme in the SSSIP.

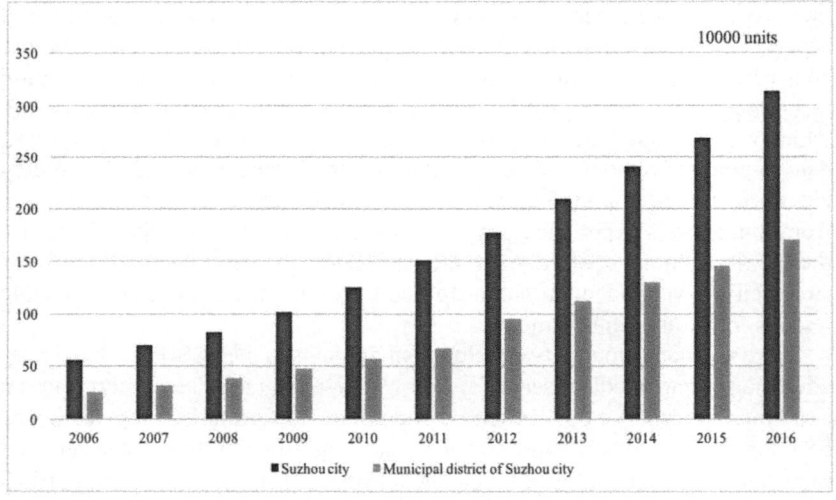

Source: Suzhou Statistical Yearbook (Suzhoushi tongjiju 2006–2016).

Figure 8.3   *Total civil automobiles of Suzhou city and municipal district of Suzhou city*

On 30 September 2015, the State Council of the People's Republic of China issued the 'Approval of the General Plan for the Comprehensive Opening-up and Innovation Test in Suzhou Industrial Park'. Under the framework of this approval, the SIPAC issued the 'Planning of Suzhou Industrial Park Smart Big Transportation (2015–30)' in 2016, which borrowed advanced experiences from Singapore's smart urban transportation development. The plan was completed by the Planning and Construction Committee of the SSSIP. Tongji University and Shanghai Urban Construction Design & Research Institute provided intelligent top-level planning and design of the urban integrated transportation system in the SSSIP by integrating and converging the big traffic data. Figure 8.4 shows that, overall, planning considers the international frontier technologies and real demands of the SSSIP and builds a comprehensive traffic information centre with the framework of one Integrated Transportation Information Centre and eight Smart Application Subsystems (1C+8S). With the help of this framework, the SIPAC can manage the bus, control, parking, logistics, maintenance and other various demands of the SSSIP. According to the data released by SIPAC (2016), by 4 August 2016, the smart transportation system of the SSSIP had covered the 'two vertical avenue and two horizontal avenues' (including Modern Avenue, Jinjihu Avenue, Xinghu Street and Xinggang Street), covering a total of 250 kilometres of roads and 137 intersections with 112 SCOOT signal machines by Siemens, 27 large signal machines, 263 electronic police systems, 16 traffic guidance screens, one changeable lane indicator and 1,958 sets of geomagnetic and coil detectors with an accuracy of 95 per cent. With the help of these devices, a daily average of more than 500,000 high-definition pictures with 5 million pixels of cards and 700,000 traffic flow data points were collected at intersections. Additionally, the smart transportation system was equipped with a powerful video surveillance system with 425 card-port cameras, 78 intersection ball machines and 49 high-altitude 'hawkeye' surveillance cameras over the entire area that produced nearly 30 terabytes of video files daily, which were saved for 30 days.

The overall planning involves a complex, scientific and highly efficient system. Figure 8.5 vividly shows the operation mechanism of one application file named Smart Decision, and how this smart transportation system makes decisions based on traffic flow data collected by smart perception devices. The system works as follows: the traffic flow data of each intersection and road direction are dynamically collected in real time by magnetic induction coils and monitoring devices (smart perception); these collected data are saved to the Cloud Computing Centre through network transmission (Data Integration and Management Platform). Then, the Intelligent Processing Centre accurately calculates the timing of traffic lights for each intersection and direction according to the traffic volume of each intersection and direction (Decision Support Platform). Finally, the Cloud Computing Centre automatically sends its decision to each traffic light and its referential routes to the users' APPs (Public Services Platforms).

The smart transportation system of the SSSIP integrates information from different parts of the SIPAC, such as the traffic police system, network operation, municipal property, urban management as well as bus companies and other industries. With the help of this system, travel routes can be analysed,

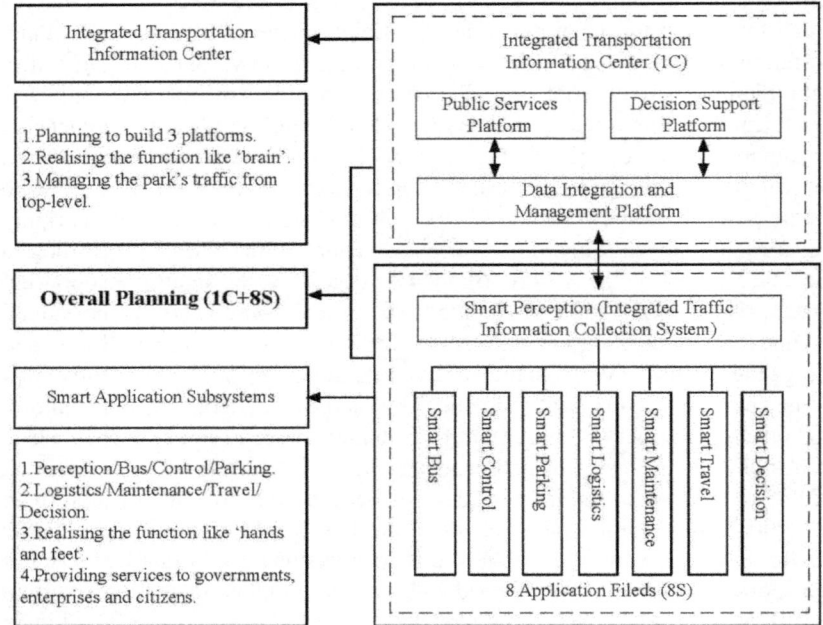

Source: SIPAC (Guihua jianshe weiyuanhui 2016).

Figure 8.4   *The overall planning of smart big transportation in SSSIP*

making travel more convenient by obtaining real-time and accurate information for vehicle operation before and during passengers' trips through the
information publishing system, which will improve the attractiveness of public
transport and play an important role in promoting the healthy development of
transportation.

For example, on 22 April 2015, an ambulance raced to Kowloon Hospital
on Modern Avenue in the SSSIP with an unconscious child. Whenever the
ambulance approached an intersection, the signal lights instantly turned green
based on the control of the smart transportation system and its staff remotely
controlling the signal lights on the road. Thanks to this smooth operation, the
ambulance crew saved that child's life.

Smart transportation systems can also reduce the pressure on public transport and improve the overall speed of bus operation, thereby making buses the
first choice for people when they travel. According to the data released by the
Suzhou Industrial Park Planning Exhibit Hall, the average number of automo-

Source: Photos by authors, taken at the Suzhou Industrial Park Planning Exhibit Hall.

Figure 8.5   Smart transportation system in SSSIP

biles stopped on the road in the SSSIP has reduced by 17.13 per cent, while traffic delays on main roads has reduced by 15.36 per cent and average car speed has increased by 15.44 per cent. Furthermore, the capacity of roads in the SSSIP has increased by 16 per cent and instances of car accidents on main roads has decreased by 5 per cent. Additionally, the system can help to reduce operational costs and increase the efficiency of public transport enterprises while also improving their management level.

With the help of the smart transportation systems, public transport enterprises can realise scientific scheduling and reasonable dispatching for public transportation drivers, which can improve the operational efficiency of transportation infrastructure while reducing the probability of traffic congestion, traffic accidents, noise and exhaust emissions from public transport vehicles, fuel consumption costs for public transport vehicles and the overall cost of the public transport enterprises.

CONCLUSION

With the rapid development of China's economy, increasing urbanisation has brought serious urban diseases to Chinese metropolises. Since 2006, the Chinese national and local governments have begun to explore the development of smart cities to solve this growing problem. The national government issued a series of supporting policies, regulations and evaluation index systems to promote the healthy development of smart cities. Moreover, the MOHURDPRC released three batches of smart city pilots from 2013 to 2015 and issued 100 pilots for new smart city programmes in 2016 that aimed to accumulate experience in smart city development. Apart from the action of issuing policies and smart city pilots, Chinese governments formulated a series of evaluation index systems for smart city development to ensure the healthy and rapid development of smart cities across China.

The emergence of the smart city suggests a new opportunity that can have a significant impact on Chinese urban management, especially in dealing with urban diseases, such as overpopulation, traffic jams, environmental pollution, resource shortages and the insufficient supply of services. There exists a potential opportunity for Chinese national and local governments to transform traditional urban management into modern urban governance. From the case study of the smart transportation system of the SSSIP, it can be concluded that, by developing its smart transportation system, the SIPAC in Suzhou city is striving to solve its traffic dilemma, with the system achieving some excellent outcomes. For example, both the average number of automobiles stopping and traffic delays on the main road were reduced, while the average car speed increased.

Based on the case study of the smart transportation system in the SSSIP, we realise that the development of smart cities in China remains a complex project that requires many participants from the national government, local governments, enterprises, experts, scholars and citizens, with each participant playing an important role. For example, enterprises and scholars can take part in the development of the evaluation index system, while citizens play an important role as both customer and supervisor in the process of smart city development. Therefore, this chapter proposes that the development of Chinese smart cities should include defining the leading role of governments while maintaining the tenet of 'people-oriented' to account for citizens and make full use of market forces.

ACKNOWLEDGEMENTS

This research is jointly supported by Ministry of Education Social Science Key Research Centre Major Project 16JJD630013, National Natural Science Project Fund No. 71673111, Major Project of Sichuan Provincial Social Science Key Research Base of Regional Public Management Information Research Centre QGXH19-01.

REFERENCES

Amap, Alibaba Cloud Computing Co. Ltd 2017, 'Zhongguo zhuyao chengshi jiaotong baogao', viewed 17 July 2018, http://www.199it.com/archives/678630.html.

Beijing Municipal Bureau of Statistics 2018, Beijingshi tongji nianjian, viewed 26 July 2019, http://tjj.beijing.gov.cn/nj/main/2018-tjnj/zk/indexch.htm.

Beijing Municipal Planning and Land Resources Management Committee 2017, 'Beijing chengshi zongti guihua (2016–2035)', viewed 16 July 2018, http://zhengwu .beijing.gov.cn/gh/dt/t1494703.htm.

Beijingshi tongjiju 2006–2016, Beijing Statistical Yearbook, viewed 26 July 2019, http://tongji.cnki.net/kns55/navi/HomePage.aspx?id=N2019010235&name=YOFGE &floor=1.

China Institute of Communications 2012, 'Zhihui chengshi jishu baipishu jianjie', viewed 17 July 2018, https://wenku.baidu.com/view/b740f2cf9ec3d5bbfd0a744b .html.

China Software Test Centre 2013, 'Zhihui chengshi pinggu zhibiao tixi yanjiu baogao', viewed 17 July 2018, https://wenku.baidu.com/view/e7edef43c850ad02de804153 .html.

Gao, L & Wang, C 2015, 'Zhihui chengshi de guojia biaozhun yu pingjia tixi', Frontiers, no. 17, pp. 16–26.

Gu, D & Qiao, W 2012, 'Woguo zhihui chengshi pingjia zhibiao tixi de goujian yanjiu', Future and Development, vol. 35, no. 10, pp. 79–83.

Guangzhou statistics bureau 2018, 'Guangzhoushi gequ tudi mianji he renkou midu', viewed 26 July 2019, http://www.gzstats.gov.cn/gzstats/gzsq/201701/1036f733 bf244636b9b62af6f9a504f1.shtml.

Guangzhoushi tongjiju 2006–2016, Guangzhou statistical yearbook, viewed 26 July 2019, http://tongji.cnki.net/kns55/navi/HomePage.aspx?id=N2014050070&name= YKDNA.

Guihua jianshe weiyuanhui 2016, 'Suzhou gongye yuanqu zhihui dajiaotong guihua (2015–2030) bianzhi wancheng', viewed 26 July 2019, http://news.sipac.gov.cn/ sipnews/jwhg/2016yqdt/12/201612/t20161205_508431.htm.

Jiao, X 2015, 'Chengzhenhua Jincheng Zhong "chengshibing" wenti yanjiu: hanyi, leixing ji zhili jizhi', On Economic Problems, no. 7, pp. 7–12.

Li, L & Zhang, J 2017, Shanghai Pudong New Area statistical yearbook, viewed 26 July 2019, http://tongji.cnki.net/kns55/navi/YearBook.aspx?id=N2017120307& floor=1.

Liu, Y 2014, 'Chengshibing fangzhi yu zhihui chengshi jianshe', Journal of Henan Science and Technology, no. 5, pp. 251–252.

MIITPRC 2013, 'Zhihui chengshi pinggu zhibiao tixi (zhenqiu yijiangao)', viewed 17 July 2018, http://www.tranbbs.com/Smart_City/d_p_city/Smart_City_115044 _2.shtml.

MOHURDPRC 2013, 'Guojia zhihui chengshi (qu/zhen) shidian zhibiao tixi (shixing)', viewed 17 July 2018, http://www.iotworld.com.cn/html/News/201212/ 273a70851010a260.shtml.

NBSPRC 2018, 'Zhonghua renmin gongheguo 2017 nian guomin jingji he shehui fazhan Tongji gongbao', viewed 15 July 2018, http://www.stats.gov.cn/tjsj/zxfb/ 201802/t20180228_1585631.html.

NDRCPRC 2014, 'Guanyu cujing zhihui chengshi jiankang fazhan de zhidao yijian', viewed 16 July 2018, http://www.ndrc.gov.cn/gzdt/201408/ W020140829409970397055.pdf.

Shenzhenshi tongjiju 2006–2016, *Shenzhen statistical yearbook*, viewed 26 July 2019, http://tongji.cnki.net/kns55/navi/HomePage.aspx?id=N2017060065&name= YZEEE&floor=1.

Shujuguan 2016, 'Zhihui chengshi shidian mingdan daquan', viewed 26 July 2019, http://www.cbdio.com/BigData/2016-02/19/content_4632137.htm.

SIPAC 2016, 'Yuanqu daolu jiaotong guanli zhinenghua zhilu, zhineng jiaotong "linghang" zhihui chengshi', viewed 26 July 2019, http://www.sipac.gov.cn/dept/ ghjswyh/gzdt/201608/t20160804_449611.htm.

Suzhoushi tongjiju 2006–2016, *Suzhou statistical yearbook*, viewed 26 July 2019, http://data.cnki.net/area/Yearbook/Single/N2009030135?z=D10.

Suzhoushi tongjiju 2017, *Suzhou statistical yearbook*, viewed 26 July 2019, http://www .sztjj.gov.cn/SztjjGzw/tjnj/2017/zk/indexch.htm.

Suzhoushi tongjiju 2018, *Suzhou statistical yearbook*, viewed 26 July 2019, http://www .sztjj.gov.cn/SztjjGzw/tjnj/2018/zk/indexch.htm.

Wulian Zhongguo 2016, '2017 nian quanguo zhihui chengshi jiang chaoguo 500 ge, nijia shangbang le ma?', viewed 15 July 2018, http://www.50cnnet.com/show-89 -128993-1.html.

Xiang C 2014, 'Zhongguo chengzhenhua jinxing zhong de "chenshibing" jiqi zhili', *Journal of Xinjiang Normal University (Edition of Philosophy and Social Sciences)*, vol. 35, no. 2, pp. 45–53.

Ye L, Fan Z & Yang X 2018, 'Woguo zhihui chengshi jianshe jingyan tansuo ji tuijin celue', *Journal of University of Electronic Science and Technology of China (Social Sciences Edition)*, vol. 20, no. 5, pp. 97–102.

Zhang G 2014, 'Zhongguo chengzhenhua Jincheng zhong "nongcunbing" he "cheng- shibing" jiqi zhili', *Journal of Liaoning University (Philosophy and Social Sciences Edition)*, vol. 42, no. 3, pp. 18–24.

Zhu M 2013, 'Zhihui chengshi pingjia zhibiao tixi huigu ji youhua jianyi', *Mobile Communications*, no. 3–4, pp. 23–25.

PART III

Second-tier cities and smart city development

9. Bureaucratic readiness for smart city initiatives: a mini study in Yogyakarta City, Indonesia

Arif Budy Pratama and Satria Aji Imawan

INTRODUCTION

According to a 2016 report by the United Nations, 65 per cent of the world's population will be living in cities by 2030. Asia is no exception; the urban population is expected to double, and more areas will be covered by cities. In Southeast Asia, rapid development and economic growth have caused a decrease in the quality of life in urban areas (Thuzar 2011). Such economic activities cause serious urban problems, including housing issues, low sanitation standards, congestion, crime, environmental damage and pollution. Statistics Indonesia has predicted that 82.37 per cent of Indonesia's population will live in the country's cities by 2045. Consequently, Indonesian cities will likely encounter various urban problems as they grow.

Recently, smart city initiatives have been implemented to overcome the negative effects of urban development. However, implementing such initiatives is no easy task. Indeed, it is dilemmatic, as smart city development brings not only benefits but also potential problems (Skou & Echsner-Rasmussen 2015). In a similar vein, Anthopoulos (2017) has used the term 'utopia' to highlight the gap between what *should be* and what *is*. He points out that these 'smart utopias' can still experience various community problems, even with their 'smart' designation. This pessimistic view indicates that smart city programmes are of unclear benefits, and open to criticism as over-valued or overly futuristic (Hollands 2015). The main criticism of smart city initiatives emphasises their concepts and methods, corporate visions, overlooking of citizens, 'splintering urbanism', and mismatch between sustainable visions and local needs (Angelidou 2017).

While literature on smart cities and urban governance has mushroomed, researchers frequently stress the stakeholders, vendors and technological aspects of such initiatives while minimising the role of public bureaucracy.

In an Indonesian context, studies of smart cities have also grown significantly in the past five years. Many have concentrated on technological discourses (Mayangsari & Novani 2015; Sasono et al. 2016; Wiseli et al. 2017). This chapter, however, reaches beyond the technological by focusing on the readiness of the bureaucracy as one of the most important institutions in smart city implementation. Given the fact that most of Indonesia's smart city programmes have been initiated by local governments, we argue that the connection between government bureaucracy and smart city initiatives requires investigation. As such, this chapter will focus on the readiness of local governments in implementing smart city initiatives, drawing on the perceptions of civil servants tasked with the implementation of smart city policies. Following Pratama (2017) and Wihantoro et al. (2015), we use the term *bureaucracy* instead of *government agency* or *government organisation* to denote the characteristics of Indonesian public institutions as hierarchical, law-based bodies with professional civil servants and specialised technical knowledge of rules and procedures.

This chapter is structured as follows. First, it presents the smart city concept as defined by national Indonesian policy, thereby highlighting how governments at the local and national level perceive smart cities as concepts and as policy agendas. Second, using Yogyakarta City as our locus of study, we identify the dimensions that shape the perceived bureaucratic readiness for smart city initiatives. Last, we identify policy sets as well as their administrative implications before providing concluding remarks.

SMART CITIES IN INDONESIA: POLICIES AND ACTORS

The Indonesian smart city policy is explicitly delineated in the National Medium-Term Development Plan (RPJMN) for 2015–2019. In the national urban development policy, the smart city concept is positioned as an umbrella agenda. Policies and strategies are directed towards the realisation of competent and technologically based urban services and grounded in local cultures. Such goals may be realised by: (a) developing the economy through city branding which supports national branding; (b) providing Information and Communication Technology (ICT)-based public infrastructure and services; and (c) building innovative, creative and productive social capital and community capacity. As such, the smart city concept at the national level is still abstract, as the substance and dimensions of smart cities – necessary for bureaucracies to implement smart city programmes – have yet to be elaborated. As a result, there are no clear policy guidelines along which Indonesia's cities can implement smart city programmes using the policy language and concepts. There is a lack of conceptualisation, which may produce multiple interpretations and vague policy agendas during policy formulation.

In practical venues, the multi-level governance model which has been implemented in Indonesia since regional autonomy was granted in 1999 has allowed local governments to design their own development agendas, including smart city initiatives. To support such initiatives across the municipalities and regencies of the country, the central government enacted Regulation of the Minister of Communication and Informatics, Republic of Indonesia, Number 14/2016 about Guidelines for the Use of Communication and Informatics Software by Government Ministries and Agencies. This regulation states that smart cities should be managed by all government sectors in Indonesia. However, it only defines smart cities narrowly, as digital or electronic cities, thereby emphasising the use of information and technology in the delivery of urban services; such a definition does not conform with the current debate on smart cities. The absence of clear definitions, thus, may produce misperceptions of 'what is meant by smart' in various stakeholders' urban development activities (Pratama 2018).

There are also several non-governmental organisations that deal with smart city initiatives. For example, the Smart City Framework, Version 1, which was formulated by the Association of Indonesian Intelligence Initiatives (APIC), defines smart city initiatives as being intended to improve quality of life by managing resources in an effective, efficient, innovative and integrated way. These measures fall into three categories: economic, social and environmental. Each domain has clusters and services, including smart mobility, smart energy and smart health. This framework also offers a distinction between e-government and smart cities and underscores the importance of formulating and defining indicators for smart cities.

Indicators for smart cities have also been formulated by Smart City Rating Indonesia (RKCI), an initiative of the Bandung Institute of Technology (ITB) in conjunction with the Smart City and Community Innovation Centre (SCCIC) and Association of Municipal Governments throughout Indonesia (APEKSI). RKCI identifies five types of smart city indicators: (1) smart people; (2) smart energy; (3) smart economy; (4) smart infrastructure; and (5) smart services. Their scope implies that these indicators have a more expansive definition than that of Regulation of the Minister of Communication and Informatics, Republic of Indonesia, No. 14/2016, which only includes smart people and smart services as smart city indicators.

The Ministry of Public Works and Housing identifies seven indicators of smart cities: (1) smart development planning; (2) smart green open spaces; (3) smart transportation; (4) smart waste management; (5) smart water management; (6) smart building; and (7) smart energy. Meanwhile, the Ministry of Internal Affairs of Indonesia identifies 13 indicators: (1) smart economy; (2) smart society; (3) smart environment; (4) smart health; (5) smart mobility; (6) smart, safe and secure city; (7) smart way; (8) digital government readiness;

(9) integration readiness; (10) infrastructure readiness; (11) competitive eco-system; (12) innovative ecosystem; and (13) financial technology ecosystem.

These indicators and measurements show that government organisations in Indonesia have different understandings of smart cities. Despite the divergent definitions of various organisations, there is similarity in their dimensions. Most of the above conceptual frameworks underscore the importance of sustainable development values such as society, economy, environment, and good governance in the smart city concept. A comprehensive approach to defining smart cities must not focus only on technical approaches, such as building ICT support, but also on social approaches such as smart government, smart living and smart society.

This is reflected in the National Medium-Term Development Plan 2015–2019, which states that smart cities should be oriented towards brand-ings, such as the best products, human resources, social character and cultures, e-business, innovation and entrepreneurship development. The Ministry of National Development Planning has already communicated this definition to all of Indonesia's local governments. This shows that, although no consensus exists, there is nevertheless an effort to define the core concepts of smart cities, as well as indicators and measurements, in Indonesia.

From a policy direction perspective, the Indonesian public sector is the main actor in realising Indonesia's smart cities. Special attention should be given to local governments at the municipality or regency level, as they act as the spearheads of urban development. National policy requires all Indonesian government entities, in all regions and municipalities, to provide professional public services using technology. This can be observed, for instance, in the implementation of e-public services, development of ICT-based public services, application of service standards, preparation of Standard Operational Procedures (SOP) for various types of services, implementation of integrated services, implementation of One-Stop Integrated Services, and the establishment of ICT-based complaint units.

Given their scope, smart city programmes in Indonesia are mostly initiated by the local governments with the authority to govern local urban development. Admittedly, various stakeholders are also involved in smart city implementation, including state-owned enterprises, private technology-based corporations and academia (Mayangsari & Novani 2015; Yunianto 2016). However, the bureaucracies of local governments – as the executors of smart city policy – are likely dominant, as signified by the emergence of smart city initiatives in many Indonesian cities. Smart city projects have mushroomed in the last three years, with local governments as the main initiators.

Endorsed by the Ministry of Communication and Informatics, Ministry of Interior, Ministry of Public Work and Housing, Ministry of National Development Planning and the Presidential Office, a 100 Smart Cities movement has been implemented since 2017 (Kemkominfo 2018). In 2017, 50 cities were involved in this programme; another 50 cities became involved in early 2018. This movement is regarded as providing the momentum through which the Indonesian government can commit to developing smart cities as part of its urban development agenda.

YOGYAKARTA SMART CITY: IDENTIFYING PERCEIVED BUREAUCRATIC READINESS

The implementation of smart city programmes can be considered a transformational process through which changes are applied to grant citizens a better quality of life (Ibrahim et al. 2018). This means that smart city development must involve a process of change, the successful management of which is related to organisations' readiness to face it (Napier et al. 2017). Many empirical studies have demonstrated that organisational readiness matters in the processes of change and organisational reform (Todnem 2007; Oreg et al. 2011; Vakola 2013). Lack of readiness and organisational inertia, therefore, is frequently the cause of organisations' failure to change (Armenakis et al. 1993).

It is thus important to assess organisations' level of readiness when planning to implement change. In the context of change management, staff/members' lack of acceptance and negative reactions hinder change programmes (Hwang et al. 2016). In addition, Cunningham et al. (2002) suggest that organisational readiness for change refers significantly to employees' mind-sets. Analysing the construct of Perceived Organisational Readiness to Change (PORC) within organisations, they posit that organisations require a certain degree of readiness to successfully implement change. The PORC concept has contributed to the study of change management since the 1970s (Cinite et al. 2009). Its central argument is that the change process may be better understood through the way members of an organisation itself see the change, rather than through the assessment of other organisational infrastructures. Another construct associated with change management and reform is the organisational ability to do change; this can be understood as the requisite capacity to deliver change (Altmann & Lee 2015; Schweiger et al. 2015; Sune & Gibb 2015). Linking both concepts, we designed a set of interviews for the multi-level managers in Yogyakarta City, seeking to identify what dimensions shaped the bureaucracy's perceived readiness for the smart city programme.

Yogyakarta is a relatively small city, covering only 32.50 km² (1.02 per cent of the Yogyakarta Special Region Province). Administratively, it consists of 14 sub-districts and 45 villages, 616 community groups and 2,532 neighbourhood groups. A population projection by Statistics Indonesia gives the city's population as 422,732, giving a population density of 13,007 people per km²

(BPS Kota Yogyakarta 2018). Despite its relatively small area and population, Yogyakarta enjoys considerable advantages in the fields of education and tourism. Well known as a city of education, Yogyakarta hosts hundreds of institutions of higher education and attracts thousands of students from throughout Indonesia. The city is also recognised for its many tourist destinations, many of which involve cultural tourism.

Yogyakarta's smart city policy, thus, is constructed around education and tourism. It gives strong emphasis to smart people, with educating citizens through the education and tourism sectors one of its main objectives. One of its key projects has been the development of the Jogja Smart Tourism (JST) mobile application, which provides a list of tourism services (such as restaurants, hotels, and resorts), recommends sites and packages, and estimates travel costs. Currently, this application has been integrated into the Jogja Smart Service (JSS) application. This application has broader functions, covering five categories: (1) information and complaints covering all regions of Yogyakarta; (2) data and information services covering upcoming events, tourism sites, problems and job vacancies; (3) partnerships with other sectors; (4) general information about Yogyakarta, from the district to the village level; and (5) emergency services, consisting of medical and firefighting services.

Yogyakarta City was chosen as the locus for this study because it is one of the most successful smart city adopters in Indonesia, as indicated by its receipt of several awards. In 2017, Yogyakarta City was recognised by Rating Smart City Indonesia (RKCI) as the best smart city adopter (medium city level). The following year, it received the Best Smart Governance Award from City Asia Inc., as well as the Indonesia Smart Nation Award (ISNA) from the Ministry of Communication and Informatics, Republic of Indonesia. Yogyakarta City has also received awards for its smart city scheme and ICT infrastructure.

The Yogyakarta government has envisioned its civil servants as smartly serving the residents of the city. The bureaucracy, as a machine of the local government, has been directed towards promoting smart environment, smart mobility, smart government, smart economy, people-based orientation, smart living and smart disaster management. Prominent is smart culture, supported by the tourism and educational sector; this lies at the foundation of the Yogyakarta government's smart city design. This was mentioned by the Head of the Office of Communication and Informatics:

> smart culture is our spirit of public service and bureaucracy. Smart culture is our vision, one that should be supported by other smarts, such as environment, mobility, government, economy, etc. Smart tourism and education are the keys to bridging these smarts to our vision. This will support our initial steps towards the grand smart city design in Yogyakarta. It is still ongoing, due to our need to synchronise with regulations ... (TH, 27 May 2018)[1]

This statement clearly shows that the bureaucracy has a significant role in smart city implementation. It should not be regarded as a mere static machine, but as dynamically immersed in the operationalisation of smart city policies and programmes.

DIMENSIONS OF PERCEIVED BUREAUCRATIC READINESS

We conducted qualitative research by interviewing key civil servants at the Office of Communication and Informatics, where the smart city concept was pioneered. These interviewees represented the managerial-level staff of the Yogyakarta City government. Through these exploratory in-depth interviews on how civil servants perceived smart city readiness as transformative processes inside the bureaucracy, we discovered several aspects that constituted the concept of bureaucratic readiness. As perceived by the civil servants interviewed, perceived bureaucratic readiness could be promoted through political and administrative commitment, legal support, information technology infrastructure and expertise, and governance structures and strategies.

Political and Administrative Commitment

One of the most important aspects of smart city implementation is the political will of the mayor, as the top executive leader of the city government. Also necessary is the administrative and managerial commitment of upper-echelon administrators. Respondents believed that the commitment of their superiors was central for ensuring bureaucratic readiness in smart city implementation. One informant from the middle echelon stated that:

> the mayor has a commitment to improve the quality of public services, as indicated by integrated, simple, transparent and accountable service delivery. How can that be improved? By utilising information and technology, with a special focus on the smart city programme. It means that smart city should not just be jargon, but entrenched in the culture of Yogyakarta City. (JM, 6 May 2018)

Such political commitment is to the principle of integrated, simple and transparent service delivery, and must be given to ensure public service quality. 'Smart city' should not be mere jargon, but culture, something that unites civil servants in smart city implementation. This was evident in a statement by the Head of the Office of Communication and Informatics of Yogyakarta, who said that culture is a pivotal framework for formulating the grand design of the smart city. This also shows that the government is currently formulating a grand design using the spirit of smart culture.

Other informants stressed not only the need for political commitment, but also managerial support, meaning that their superiors (i.e., those who lead their departments or smart city projects) should be committed to smart city implementation. One informant felt that his manager had taken an important role in the smart city programme. He stated:

> while my manager is rather technical in action, he is a spearhead working with the mayor for the smart city and city branding programme. Further, Echelons 3 and 4 are reminders of the smart city roadmap, especially in Yogyakarta, which seeks to achieve smart culture through its education and tourism objectives … (ITH 8 May 2018)

This narrative emphasises the role of the upper echelons as the translators of the mayor's vision to the bureaucracy. In other words, these upper echelons are the ones who transform policy into action through strategic plans and operational actions. The two quotes above have indicated that commitment from political and administrative officials is highly beneficial to the concept of bureaucratic readiness. The commitment of the mayor is political in nature, while public managers provide support through technical involvement. Administrative commitment can be observed, for example, in the public servants' statements that their managers direct the detailed and operational aspects the of smart city initiative. This provides a strong indication that bureaucratic readiness is a main factor in the successful implementation of smart city initiatives. Mayors offer political commitment, while managers give managerial commitment.

Legal Support

The alignment of regulations was the second theme evidenced during interviews, where informants voiced a view that every government action must be backed up with regulations. An interview with a civil servant at the Office of Communication and Informatics supported this premise. The informant said:

> the regulatory framework for smart city should be comprehensive, from the national level to the sub-national level. It gives us a standing point to execute the policy using local government authorities … (AR, 10 May 2018)

Such legal support is necessary to ensure that the bureaucracy can do its duties. In Indonesia, a legal framework provides an umbrella for government actions and policies. However, there are currently no specific regulations dealing with the implementation of smart city programmes. The vague concepts of electronic government and smart cities may result in implementation being unclear. One informant stated that the Yogyakarta smart city programme's main legal foundation is Regulation of Yogyakarta Mayor No. 15 of 2015, which regulates e-government implementation in the city. He said:

> we have a regulation on e-government as an information management system that supports efficient, effective, transparent, accountable and participative public service provision. Moreover, we also have a five-year roadmap for developing an e-government implementation for our city, particularly city of Yogyakarta ... (SG, 12 May 2018)

The above quote indicates that civil servants in the Yogyakarta City government still do not really understand the distinction between smart city and e-government. One civil servant stated that the regulatory framework should be comprehensive, reaching from the national level to the sub-national level, while another civil servant said that the government does have regulations, but only for e-government. The Yogyakarta City government is encouraged to clearly distinguish between the smart city and e-government concepts. The current lack of distinction may be attributed to the practicality of the smart city programme, which relies on information technology as its core process. A smart city is more than just an electronic city, as seen in city government's smart city framework. This indicates that comprehensive regulations on smart cities are needed to guide the bureaucracy as an important agent in smart city implementation. Paskaleva (2009) found that legal support is needed to provide a comprehensive framework for policymaking and regulations. By this logic, it can be stated that the legal support of the Yogyakarta City government is necessary to ensure the bureaucracy's readiness for the smart city initiative.

IT Infrastructure and Expertise

IT infrastructure is very beneficial for implementing smart city programmes. As technology develops, the IT infrastructure supporting smart cities must also be updated. One representative from the technical department at the Office of Communication and Informatics stated:

> we are committed to continuously building IT infrastructure because it is a main value of our city's smart city programme, as is civil servants' skills. Technological developments must receive an expert response. Technology is evolving; human skills must also evolve ... (SRW, 15 May 2018)

This narrative offers a constructive concept of infrastructure, involving two elements: physical IT infrastructure and civil servants' technical expertise. It is not only hardware that can improve smart city implementation; software and technical proficiency are also needed. Another opinion highlighted the drawbacks of insufficient IT maintenance and expertise. One informant stated:

> the most important point to be considered in the smart city project is the availability of IT experts. Why? Because the government, as a contractor, requires project sustainability. If experts come from the outside, we will lose this sustainability as experts leave the project. I analogise it like this: the civil servant should hold the 'key' to computer programmes instead of non-civil servant experts ... (SS, 13 May 2018)

The Yogyakarta City government agrees that infrastructure not only designates hardware and software (computer/machine learning), but also the availability of IT experts capable of maintaining them. Both will play pivotal roles in the sustainability of Yogyakarta's smart city initiative, as the smart city concept not only involves the digitalisation of governmental operations but also their use by city residents. In other words, the infrastructure provided by the city government is meaningless if citizens do not want to use it. Indeed, research by Florida et al. (2008) shows how 'technology, talent and tolerance' combine to affect regional development and highlights how certain occupations such as computer science, engineering and management services play a key role in the use of technology.

Governance Strategies and Structure

The last element mentioned in interviews was the strategies and structures through which the city government runs its smart city project. A strategy is the way the city government sets its targets and acts to achieve its smart city objectives. Structure, meanwhile, refers to the units that operate the smart city programme in the city government. In this regard, one high-ranked individual stated:

> strategy deals with vision and mission statements, as well as target designs and action plans. The way the government sets its strategy is very crucial, as it is the soul of the policy and programme. In Yogyakarta, strategy is linked to city branding as a counterpart of the smart city initiative. This gives a strategic direction and value to smart city implementation ... (EM, 21 May 2018)

In terms of structure, Yogyakarta's smart city programme is led by the Office of Communication and Informatics, which is supported by the secretariat of the city government in strategic and regulatory aspects. At the operational level, the Yogyakarta smart city programme is supported by all offices with smart city features. One informant argued that clear authority is one of the most vital aspects of policy implementation. She stated:

> we need a strong and a clear leader who can lead us, guide us, and evaluate us in smart city implementation. Indeed, the mode of administration recently changed, where less structure is preferred. However, clear structure and authority still matter … (SY, 21 May 2018)

POLICY AND ADMINISTRATIVE IMPLICATIONS

The perceived readiness and perceived capability of civil servants in harnessing bureaucratic resources are very important in smart city implementation. Andrews et al. (2016) mention that readiness in a public-sector context is a pivotal yet challenging matter due to its environmental differences from the private sector. Political and legal aspects must also be considered in smart city programmes. Smart city implementation needs to consider the following four aspects of perceived bureaucratic readiness. First, the political commitment of politicians and senior civil servants as the main supporters of the agenda-setting processes is crucial in setting policy agendas. Evidence from this study shows that political commitment necessarily shapes how state apparatuses perceive smart cities and how they respond to smart city policies. Second, legal support is regarded as the foundation of the administrative actions that can be taken by the local bureaucracy. The bureaucracy needs a legal standing to engage with the smart city initiative. Third, the local government bureaucracy requires both sufficient IT infrastructure and expertise. Infrastructure (as hardware) and expertise (as software) are regarded as strong modalities in bureaucracies' implementation of smart city programmes. Last, civil servants perceive well-defined strategies and structures as the cruxes of administrative operational mechanisms. In other words, clear visions and attainable outcomes must be integrated into comprehensive action planning, and this must be supported by an appropriate organisation and operating procedures.

Drawing from these findings, there are two implications that need to be taken into account in future endeavours. First, the Indonesian government is encouraged to provide a smart city concept to guide the country's cities and regencies towards well-designed smart city implementation. As adopters of the continental law tradition, Indonesia requires clear laws and regulations on smart city initiatives, which will provide local governments with the legal standing to execute smart city policies. Another issue is the synchronisation of policies at multiple levels of government. The decentralised regime in which Indonesia's administrative system operates requires a well-managed policy coordination system, one with synchronised policies and synergy among the involved stakeholders. In the case of Yogyakarta's smart city initiative, policy coordination is still ongoing, as the grand design is currently being formulated by the Yogyakarta government in conjunction with local legislatures.

Second, the institutional, networking, and human resources of stakeholders must be arranged to ensure the institutional and professional management of the smart city's organisational structure and design, regulations, norms, ethics and cultures. In this sense, Yogyakarta Smart City needs a sort of Smart City Council that can act as a partner of the Yogyakarta government in its smart city development. The importance of stakeholder interactions in smart city development has been mentioned by Chourabi et al. (2012). They note that stakeholder relations cover four main issues: stakeholders' ability to cooperate, support leadership, structure alliances and work under different jurisdictions. These interactions should be embedded within the brain-ware, software and hardware aspects of the smart city. Brain-ware management might consider the expertise, competency and skills of operational staff and organisational bodies that can promote successful smart city implementation, while software and hardware management is likely to uphold the principles of interoperability, scalability and reliability in urban services.

CONCLUSION

We conclude that smart city programmes in Indonesia are mainly initiated by local governments; other stakeholders, including the central government, civil society entities, private corporations and academia do take part in smart city development, but in relatively minor positions. Indonesia lacks specific regulations on smart city programmes, and as a consequence various concepts have been interpreted differently by local governments. Unsurprisingly, every city has its own understanding of the 'smart city' concept, which affects their agenda setting and smart city policymaking.

Given the nature of Indonesia's smart city policies and actors, we argue that bureaucratic readiness lies at the crux of smart city implementation. Building on this mini study in Yogyakarta City, the bureaucracy's perceived readiness for smart city initiatives must be considered in smart city projects. This perceived readiness can be demonstrated in four dimensions: political and administrative commitment; legal support; information technology infrastructure and expertise; and governance strategies and structures. These proposed dimensions are tentative in nature, and as such need to be tested in terms of validity and reliability. Further studies, thus, could attempt to develop and test this proposed scale.

NOTE

1. Each interview quote is followed by the initials of the interviewee and the date of the interview in parentheses.

REFERENCES

Altmann, P & Lee, C 2015, 'Cognition, capabilities, and resources: Developing a model of organizational change', *Journal of Management & Change*, vol. 34/35, no. 1/2, pp. 76–92.

Andrews, R, Beynon, MJ & McDermott, AM 2016, 'Organizational capability in the public sector: A configurational approach', *Journal of Public Administration Research and Theory*, vol. 26, no. 2, pp. 239–258.

Angelidou, M 2017, 'Smart city planning and development shortcomings', *TeMA. Journal of Land Use, Mobility and Environment*, vol. 10, no. 1, pp. 77–94.

Anthopoulos, L 2017, 'Smart utopia vs smart reality: Learning by experience from 10 smart city cases', *Cities*, vol. 63, pp. 128–148.

Armenakis, AA, Harris, SG & Mossholder, KW 1993, 'Creating readiness for organizational change', *Human Relations*, vol. 46, no. 6, pp. 681–703.

BPS Kota Yogyakarta 2018, *Kota Yogyakarta dalam Angka: Yogyakarta municipality in figures 2018*. Yogyakarta: Kantor BPS Kota Yogyakarta.

Chourabi, H, Nam, T, Walker, S, Gil-Garcia, JR, Mellouli, S, Nahon, K, Pardo, TA & Scholl, HJ 2012, 'Understanding smart cities: An integrative framework', in *Proceedings of the Annual Hawaii International Conference on System Sciences*, pp. 2289–2297. Maui, HI: IEEE Computer Society.

Cinite, I, Duxbury, LE & Higgins, C 2009, 'Measurement of perceived organizational readiness for change in the public sector', *British Journal of Management*, vol. 20, no. 2 pp. 265–277.

Cunningham, CE, Woodward, CA, Shannon, HS, Macintosh, J, Lendrum, B, Rosenbloom, D & Brown, J 2002, 'Readiness for organizational change: A longitudinal study of workplace, psychological, and behavioural correlates', *Journal of Occupational and Organizational Psychology*, vol. 75, pp. 377–392.

Florida, R, Mellander, C & Stolarick, K 2008, 'Inside the black box of regional development: Human capital, the creative class and tolerance', *Journal of Economic Geography*, vol. 8, no. 5, pp. 615–649.

Hollands, RG 2015, 'Critical interventions into the corporate smart city', *Cambridge Journal of Regions, Economy and Society*, vol. 8, no. 1, pp. 61–77.

Hwang, Y, Al-Arabiat, M, Rouibah, K & Chung, JY 2016, 'Toward an integrative view for the leader–member exchange of system implementation', *International Journal of Information Management*, vol. 36, no. 6, pp. 976–986.

Ibrahim, M, El-Zaart, A & Adams, C 2018, 'Smart sustainable cities roadmap: Readiness for transformation towards urban sustainability', *Sustainable Cities and Society*, vol. 37, pp. 530–540.

Kemkominfo 2018, *Langkah menuju '100 smart city', Sorotan media*, viewed 13 September 2018, https://kominfo.go.id/content/detail/11656/langkah-menuju-100 -smart-city/0/sorotan_media.

Mayangsari, L & Novani, S 2015, 'Multi-stakeholder co-creation analysis in smart city management: An experience from Bandung, Indonesia', *Procedia Manufacturing*, vol. 4, pp. 315–321.

Napier, G, Amborski, D & Pesek, V 2017, 'Preparing for transformational change: A framework for assessing organisational change readiness', *International Journal of Human Resources Development and Management*, vol. 17, no. 1, pp. 129–142.

Oreg, S, Vakola, M & Armenakis, A 2011, 'Change recipients' reactions to organizational change: A 60-year review of quantitative studies', *Journal of Applied Behavioral Science*, vol. 47, no. 4, pp. 461–524.

Paskaleva, KA 2009, 'Enabling the smart city: The progress of city e-governance in Europe', *International Journal of Innovation and Regional Development*, vol. 1, no. 4, pp. 405–422.

Pratama, AB 2017, 'Bureaucracy reform deficit in Indonesia: A cultural theory perspective', *Journal of Public Administration and Governance*, vol. 7, no. 3, pp. 88–99.

Pratama, AB 2018, 'Smart city narrative in Indonesia: Comparing policy documents in four cities', *Public Administration Issues*, vol. 2, no. 6, pp. 65–83.

Sasono, MEN, Purwitaningsih, S, Yusuf, L & Navastara, AM 2016, 'Surabaya smart subway development as an alternative mode in Ahmad Yani corridor Surabaya by TOD concept application', *Procedia – Social and Behavioral Sciences*, vol. 227, pp. 132–138.

Schweiger, C, Kump, B & Hoormann, L 2015, 'A concept for diagnosing and developing organizational change capabilities', *Journal of Management & Change*, vol. 34/35, no. 1/2, pp. 12–28.

Skou, M & Echsner-Rasmussen, N 2015, 'Smart cities around the world', *Geoforum Perspektiv*, vol. 25, pp. 61–67.

Sune, A & Gibb, J 2015, 'Dynamic capabilities as patterns of organizational change: An empirical study on transforming a firm's resource base', *Journal of Organizational Change Management*, vol. 28, no. 2, pp. 213–231.

Thuzar, M 2011, 'Urbanization in Southeast Asia: Developing smart cities for the future?', in Montesano, MJ & Lee, PO (eds.), *Regional Economic Outlook*, pp. 96–100. Singapore: ISEAS–Yusof Ishak Institute Singapore.

Todnem, R 2007, 'Ready or not …', *Journal of Change Management*, vol. 7, no. 1, pp. 3–11.

Vakola, M 2013, 'Multilevel readiness to organizational change: A conceptual approach', *Journal of Change Management*, vol. 13, no. 1, pp. 37–41.

Wihantoro, Y, Lowe, A, Cooper, S & Manochin, M 2015, 'Bureaucratic reform in post-Asian Crisis Indonesia: The directorate general of tax', *Critical Perspectives on Accounting*, vol. 31, pp. 44–63.

Wiseli, D, Tanusetiawan, R & Purnomo, F 2017, 'Simulation game as a reference to smart city management', *Procedia Computer Science*, vol. 116, pp. 468–475.

Yunianto, A 2016, 'Telkom Indonesia untuk akselerasi ekonomi digital Indonesia (melalui smart city Nusantara)', paper presented at the 2016 Annual Meeting of Researchers. 28 July, Bogor.

10. The smart city as a complex adaptive system: the ebbs and flows of humans and materials

Ora-orn Poocharoen, Poon Thiengburanathum and Kian Cheng Lee

INTRODUCTION

The concept of the smart city has spread across the world in the past decade. The notion of a smart city implies a city's extensive use of technology to enhance governance, improve livelihoods and ensure sustainability. However, a city is a complex system and cannot simply be changed through technology adoption alone (Deakin 2013).

This chapter aims to make two contributions to the literature and practice related to smart city projects around the world. First, there is a deep-seated contrast between the linear thinking of smart city plans and the reality of cities as complex systems comprising unexpected events, decision-making points, behaviour of actors, and sub-economic, sub-social and sub-cultural systems. This contrast, in practice, contributes to the challenge faced by bureaucracies and public-sector organisations in planning and implementing smart city projects.

Second, we suggest an approach to define and design smart city projects that can possibly reconcile such a contrast. The approach takes into account the reality that cities are not static but rather made up of dynamic parts that interact continuously in a complex system. With innovative ways to utilise technology – in particular, the Internet of Things (IoT) and big data – this chapter proposes capturing and analysing the ebbs and flows of a city as the fundamental goal of smart city projects. The ebbs and flows refer to humans, material things, stories and histories of the city. With the help of technology, these ebbs and flows can be stored, interpreted, analysed and utilised to assist the city's evolution in an effective manner. This approach, we argue, can help take the smart city movement to a higher ground that is more than a mere adoption of technology for the sake of immediate efficiency.

The structure of the chapter is as follows. In the first part, we discuss the concept of complex adaptive systems (CASs) in relation to urban systems. Next, we introduce our concept of multiplicity of urban systems (MUS). Based on CAS and MUS, we suggest implications for smart city development stages. We propose an approach for smart city projects as the endeavour to document flows that make up the city's complex system. In the second part, we describe the case of Chiang Mai as a smart city to illustrate the characteristics of CASs. Drawing on the suggested approach, we analyse where Chiang Mai is today and recommend ways forward for the city's smart city development strategy. The chapter concludes with key analytical points that contribute to theory and practice of the smart city movement.

CITIES AS COMPLEX ADAPTIVE SYSTEMS (CASS): MULTIPLICITY OF URBAN SYSTEMS (MUS) AND THE IMPLICATIONS FOR SMART CITY PROJECTS

This section introduces the authors' framework of analysis, which focuses on cities as CASs. First, we introduce the general view of CASs that has been applied in the field of public management and the field of urban planning. We then explain a new approach called MUS, which has been adapted from CAS. In the last part of the section, we apply this new approach to suggest a pathway to measure and plan for smart city projects. The pathway is illustrated as smart city development stages.

In the field of public policy and public management, CASs have six core elements: (1) system; (2) environmental factors; (3) environmental rules; (4) agents; (5) processes; and (6) outcomes. Important characteristics of CASs are frequent fluctuations, interdependency of parts in the subsystems, the absence of universal rules about behaviour and action and initial conditions that can set path dependencies (Klijn 2008; Cairney 2012). The system's behaviour, according to the theory, is based on a set of key concepts: adaptation, emergence, non-linearity, self-organisation, path dependency and bifurcation (Rhodes 2008). Small interventions or incidents can create the 'butterfly effect', whereby the context might radically change (Bovaird 2008). Unknown developments might occur due to external forces or through internal interactions between entities, because elements, such as people, communities and organisations, have self-organising capacities (Teisman & Klijn 2008). A key management challenge is adaptability, because linear causal effects, top-down planning and command and control do not work in the world of complexity (Teisman & Klijn 2008). Thus, this requires new ways to reconcile the need to accept emergence and adaptation, while also having effective control over development directions through the planning process (Roo et al. 2012).

There are three different approaches in systems thinking: hard system, soft system and evolutionary system. The hard system is justifiably quantified and is a problem-solving approach in systems science. The soft system refers to problems as ill-defined or not easily quantified, or a methodology for systems that cannot easily be quantified, especially systems that involve people holding multiple and contradictory frames of reference. The evolutionary system is a form of system reproduced with mutation, whereby the fittest elements survive and the less fit do not survive.

In the field of urban planning, the concept of an urban system refers to a set of interacting or interdependent component parts forming a complex or intricate whole (Kim 1989). Various techniques can be used to study, analyse and provide better understanding of the levels of interaction between different components of the urban system. Once the urban area is viewed as a system, there are several ways that the area can be modelled. The system concept can be further classified into three levels:

1. **Static system:** At this level, it is more structural than behavioural, while a dynamic model is a representation of the behaviour of the static components of the system. It cannot be changed in real time.
2. **Dynamic system:** This level consists of sequences of operations, state changes, activities, interactions and memory. It can be changed in real time.
3. **CAS:** This level refers to dynamic and collective behaviour mutation and self-organisation that correspond to a change-initiating micro-event or collection of events.

Cities and urban areas correspond to the CAS Level. Cities cannot be created from the top down by the imposition of simple rules. CASs are hard to understand. Disorderly and complicated, they cannot be broken down into smaller components. It would be simplistic to ignore smaller parts, or just leave them as unsolvable problems. We maintain the view that the static and dynamic levels can also be useful for expressing subsystems of the city, particularly at the conceptual level or in the early state of modelling. Adaptation in complex systems happens at the level of individuals or of types. The system does not adapt; the parts do. When enough parts have altered their behaviours, system-level adaptation then occurs.

Based on the above, urban planning that does not respect the spectrum of diverse behaviours and that aims for simple concepts such as fresh air, more public space and large private space will hinder the ability of an urban system to adapt accordingly to suit the residents. This ability to adapt is the cornerstone of complex systems. Hindering the adaptive property can lead to the collapse of the system.

We argue that new analytical frameworks and practical tools are needed to model, understand and manage urban transformations. Despite the increasing availability of urban (big) data and methods of analysis with the potential to allow an evidence-based understanding of socio-spatial change in different geographical contexts, these approaches do not always guarantee the analysis of cities as CASs. Although smart cities are seen as offering solutions to pressing global challenges, there is room for more in-depth understanding of correlations and causalities between different urban systems and to address the links between 'soft' (economic, ecological and social) and 'hard' (engineered, physical) systems. The ability to link and model different kinds of urban data and systems is indispensable for a holistic understanding of cities as CASs.

MULTIPLICITY OF URBAN SYSTEMS (MUS): MULTI-LAYER, MULTI-AGENT, MULTI-FLOW, MULTI-ACTIVITY AND MULTI-DIMENSION

Based on the above discussion, we have coined the concept of the MUS. It is used to comprehend the complexity of smart city projects. It acts as a thinking tool for classifying urban/city subsystem elements. The interactions between people, spaces, land-use, economic activities, employment and cultural and traditional practices are all a part of, but not the whole of, an urban system.

The MUS model differentiates five categories of urban subsystems (Figure 10.1):

1. **Multi-layer:** land-use and natural resources in each geographic location – e.g., building, water body, transport, air quality, noise, roads, etc.
2. **Multi-agent:** key stakeholders – e.g., government, social groups, civil society, companies, etc.
3. **Multi-flow:** human and material flows – e.g., energy, people, finance, information, traffic, history, etc.
4. **Multi-activity:** rest, transport, work, entertainment, recreation, eating, studying, praying, etc.
5. **Multi-dimension**: economy, environment, social life, political life, spirituality, physical health, mental health, community life, etc.

The subsystems can be modelled by capturing various forms of data, particularly the data that reflects change or the dynamics of the system. In other words, it would be the study of the adaptation of the complex system. This framework can be applied to both macro and micro levels of a city. For example, it can be used to model regional areas, provincial areas, urban zones

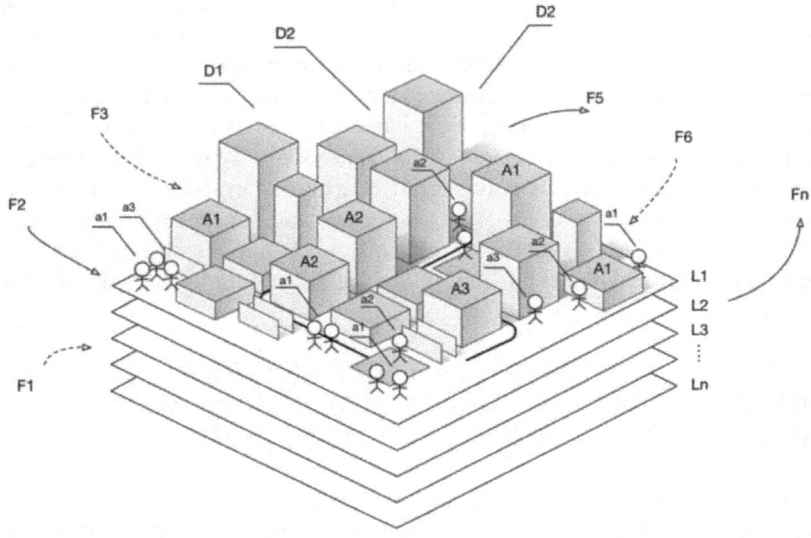

Source: Authors.

Figure 10.1   Framework of multiplicity of urban systems (MUS)

and small neighbourhoods. To use big data and other types of data to analyse such change of urban spaces is what we define as capturing the ebbs and flows of information of urban subsystems. Smart city projects have great potential to achieve this endeavour.

We propose that, in the case of a smart city, this framework can be applied to identify critical agents in different activity layers in the smart city development context, as shown in Figures 10.1 and 10.2. From a broader perspective, Figure 10.2 illustrates the four main layers of an urban system: (1) natural resources and geography; (2) infrastructure (hard and soft); (3) urban activities (investment and construction, operation and maintenance, and service and utilisation); and (4) people actors (e.g., public and private actors). These layers give us a holistic view of human settlement.

Figure 10.3 gives us more details about the activity layers. In smart city processes, agents can be simplified into four groups: (1) people/citizens; (2) institution (public and private); (3) infrastructure; and (4) service actors (e.g., public, private and other investors). Also, there are relationships or flows among the agents. The relationships include investing, supporting, regulating, servicing, contributing and utilising. Smart city development

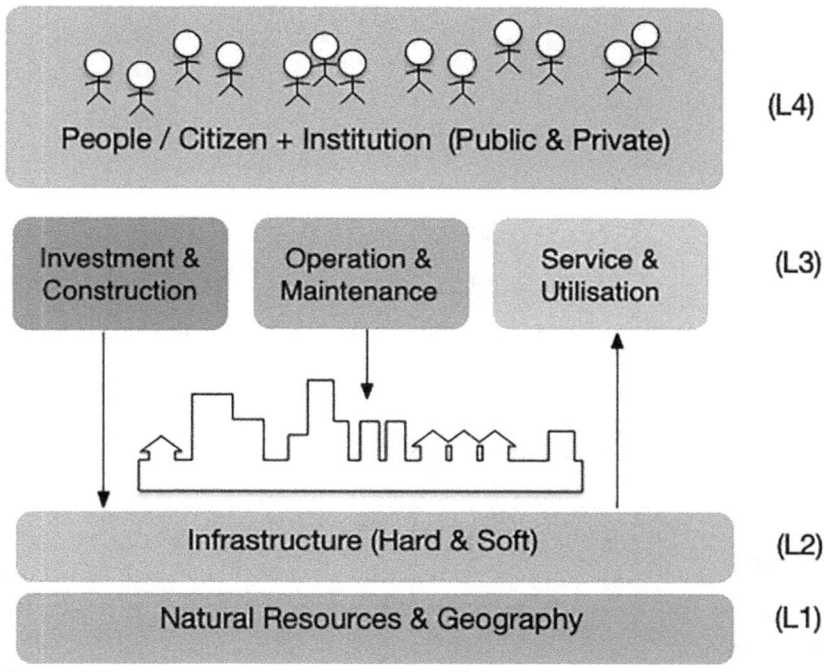

(L4)

(L3)

(L2)

(L1)

Source: Authors.

Figure 10.2   Four levels of human settlement dimensions

projects cannot focus exclusively on the investment perspective but must also consider all possible types of relationships. To achieve a high level of smart city development, other relationships or flows are required as part of the policy analysis.

In brief, CAS can be applied to the design of smart city goals and projects. Smart city development is not only about installing information and communication technology (ICT) and better Internet connections, as it is actually a complex set of urban policy designs and deployments that involve many stakeholders to achieve a better quality of life (Musa 2016). The MUS model allows us to digest the complexity as system elements. Some elements are critical success factors for the system. Success, here, is defined as a better society built on new information and communication infrastructures.

Based on the frameworks of CASs and MUS, we would like to suggest a series of smart city development stages that can be followed by practitioners.

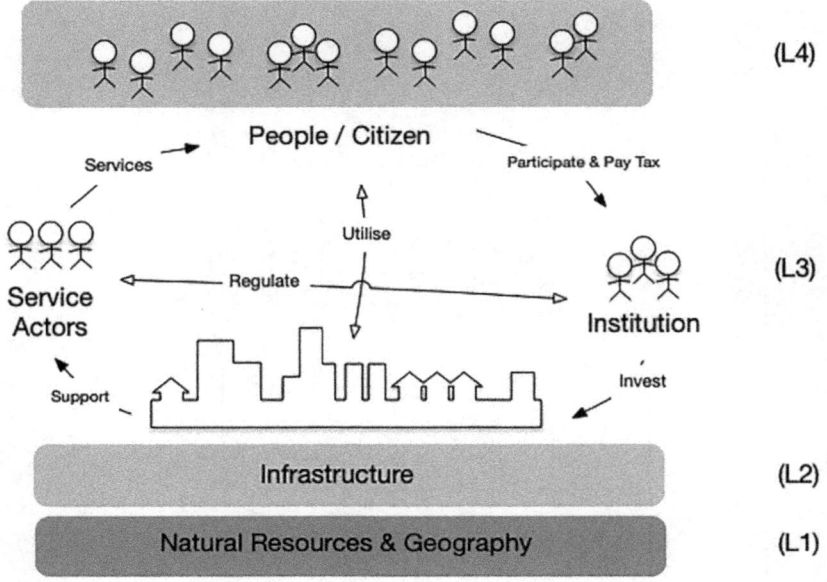

Source: Authors.

Figure 10.3   Four elements of an urban development system

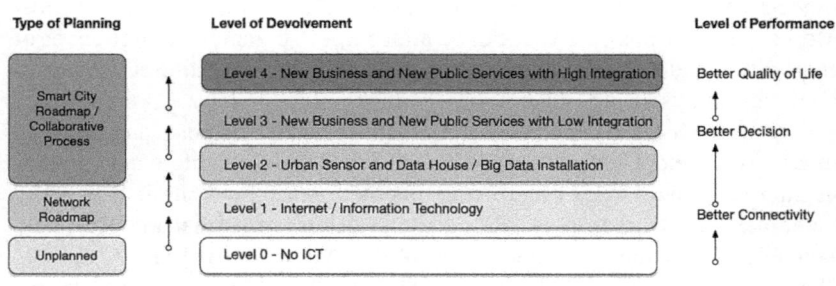

Source: Authors.

Figure 10.4   Levels of smart city development and their performance

Figure 10.4 reflects a 'smart city' as a process of city transformation towards a place that offers a better quality of life.

We suggest that there are five levels of smart city development:

1. **Level 0 – No ICT:** No modern connectivity and no intention of smart city development.
2. **Level 1 – Internet/Information connectivity:** This level of development aims at creating better connectivity. The city should have at least some form of city communication network planning. The Internet system is a required technology at this level.
3. **Level 2 – Urban sensor and city data house:** This level of development aims for big data installation (e.g., city open data platform). This data-harvesting phase includes the start of basic data analytics. However, there might not be significant value-added services for residents. Some stakeholders might be able to make better decisions based on the provided data. The IoT is usually a required technology at this level.
4. **Level 3 – New business and new public services with low integration:** This level of development indicates value-added activity to the public via new services and technological advancements. However, there is no significant integration among entities or segments. Business innovation and reasonable cost of ICT investment and operation are key factors in initiating new businesses and services in the city. There will be higher productivity and better quality of life in the city. At this level, there are enhanced civil activities, but they are not fully integrated with the smart city platforms.
5. **Level 4 – New business and new public services with high integration:** This level of development is similar to Level 3 but with enhanced governance sophistication. New values and services are generated based on comprehensive data analytics and mature civic-level activities. Strong collaboration among all stakeholders is required to enable new public values at this level.

CHIANG MAI SMART CITY: AN ILLUSTRATION OF A COMPLEX ADAPTIVE SYSTEM (CAS)

In 2016, the Thai government announced the Thailand 4.0 policy as the new direction for the country in line with the nation's 20-year Strategic Plan. Thailand 4.0 is a new economic model designed to develop Thailand into a value-based economy with an equilibrium between the environment and society while pulling itself out of the middle-income trap (Naprathansuk 2017). Within this national roadmap, the government's smart city development serves to build a high-capacity digital infrastructure with digital technology utilised to boost the economy, create an equitable society, transform government practices, develop the workforce and build trust and confidence (Naprathansuk 2017). From a macro perspective, the smart city project is embedded in Thailand's Digital Economic Policy, through which it seeks to develop a better smart digital service and industry while supporting the bigger plan to become a digital services hub of ASEAN (Association of Southeast Asian Nations) in connection with global digital business (Digital Economy Promotion Agency (DEPA) 2018; Khianmeesuk 2017, p. 82).

In February 2017, DEPA (under Thailand's Digital Economy and Society Ministry) announced investments for developing Chiang Mai into an innovation-driven smart city. Chiang Mai was among the first cities in Thailand, after Phuket and at the same time as Khon Kaen, to be developed using the smart city model. The model aims to capture and populate multiple levels of information (including building, social, environmental, governmental and economic data) from sources like sensors, real-time traffic information and social forums for access by managers, governments and citizens using mobile apps, tablets and dashboards. The smart city outlook (integrating ICT with IoT) is viewed as critical for secondary cities with a burgeoning urban population, like Chiang Mai, as well as for Thailand's move to become the digital hub of ASEAN. Secondary cities are driving forces for decentralisation efforts. Development of secondary cities should help redistribute the unbalanced wealth generation in Thailand, which has been concentrated in Bangkok.

Chiang Mai was founded in 1292. It was the capital of the Lanna kingdom before merging with Thailand and it is now the second most important city in the country. It is the centre for trade in the north of Thailand as well as with the border cities of Myanmar and, more recently, southern China. The city sits along the Ping River, a major tributary of the Chao Phraya River. It currently has a population of about 1 million people. Chiang Mai is an example of a complex city that has rich and poor sections co-existing together, with modern and traditional values mixed in all aspects of the city, such as culture, tradition, architecture, festivals, entertainment and the retail economy. As a result of Chiang Mai being a popular tourist destination and city of choice among long-term visitors and residents, it has undergone rapid change, becoming cosmopolitan and international. The city has experienced rapid urbanisation in recent years (McGrath et al. 2017).

Based on the CAS framework introduced in the previous section, awareness of the density and economy of the city is insufficient for designing smart city plans and projects, as it is also important to capture its complex nature. After one year of direct collaboration with the Chiang Mai authorities as consultants, we observed the following complex characteristics of Chiang Mai's smart city initiatives:

1. **Multiple projects:** During 2017–2018, numerous projects labelled as 'smart' emerged. They include smart farming, smart traffic, smart mobility, smart security, smart university campus, smart tourism, smart health, smart office and smart energy. Most of these projects aim to create applications and datasets to improve decision making and public services. Many projects aim at improving one-way communication (information provision) with citizens via websites and applications.

2. **Multiple actors:** It is nearly impossible to draw an exact line to identify exactly who is involved in the Chiang Mai Smart City projects. The public agency involved depends on the project. Agencies include the Ministry of Agriculture, the Ministry of Science and Technology, the police and universities. There are also representatives of central government agencies and local governments. Some foreign visitors are also involved, such as leaders of smart cities from Japan and France, international companies with smart city experience and multilateral organisations such as the World Bank. Among citizens, we found a number of groups related to smart city projects, such as the Chiang Mai Citizen Jury, the Green-Beauty-Scented group, Chiang Mai Cycling, the Crypto Currency group, Trash Hero, Creative Chiang Mai, TEDx Chiang Mai and the Chiang Mai Conscious Community.

3. **Multiple dimensions/policy domains:** Based on our scenario-planning exercises, we were able to identify key directions for Chiang Mai that can be considered contradictory.[1] The directions include being a city of mobility, innovation, arts and temples. Some actors and projects are aiming for preservation of historical heritage, while others aim to build physical infrastructure such as expansion of the international airport and a light-rail system. Policy domains that stand out are tourism, health and wellness, food, digital startups and the creative industry.

4. **Multiple layers:** Initiatives were found to represent many layers of data and information. The environment and physical layers include information on traffic, air quality, water quality, housing, parking spaces, and so on. The digital economy layer includes projects such as applications for tourists. The layer that represents deliberation was not prominent in Chiang Mai. It is this layer of data and information that will move the city to Level 3 or 4 (Figure 10.4) of new businesses and public services with active citizens.

5. **Emergence:** All of the projects were planned, budgeted and executed independently of each other. Each project emerged out of the occasion, circumstance and window of opportunity that was present. The citizen-led

projects were mostly spontaneous and focused on the environment. Some projects were initiated by the private sector, some were top down from the central government, some by local governments, some by international development agencies and some by universities.

6. **Mixing of the endogenous and exogenous environments:** Elements and agents operating in the city are influenced by both the endogenous and exogenous environments of the city simultaneously. For instance, the DEPA and private agents of the city worked together to create a network of startups and small businesses that can accelerate the digital economy. The DEPA represents the central government and thus is exogenous, but the people working in the DEPA have links to Chiang Mai and can also be considered part of the endogenous environment. In addition, the private agents that joined the network are based in Chiang Mai, as well as having businesses in Bangkok and other cities. Thus, we see the mixing of the endogenous and exogenous environments and actors as inevitable.

7. **Multiple jurisdictions:** One of the main obstacles for developing a comprehensive smart city plan is the fact that the scope of the 'city' is not fully defined. There is no governing unit for the Chiang Mai metropolitan area (in contrast to Bangkok and Pattaya). Municipalities tend to submit requests for very small projects (of less than US$33,000) that focus on improving basic infrastructure and household income via small businesses and simple skill sets. In contrast, the provincial and regional levels focus on larger projects – for instance, the Northern Food Valley – that are less focused on bringing immediate positive benefits to targeted groups in the urban community and instead have a provincial overview. The Smart Farming project created by the Ministry of Agriculture and Mae Jo University is an example of a successful project utilising digital technology to enhance water management for farmland. This project has indirect links to city management.

Now that the complex nature of smart city initiatives has been described, in the next section, we assess whether Chiang Mai is ready for smart city development.

A useful tool for assessing the smart city development process is the International Data Corporation's (IDC) Smart City Maturity Index of (Clarke 2013). This model outlines five stages of maturity and ways to progress to more comprehensive, organised systems:

1. **Ad-hoc:** This stage is the traditional government modus operandi with ad-hoc projects, department-based planning and discrete smart city projects.

2. **Opportunistic:** At this stage, opportunistic project deployments result in proactive collaboration within and between departments. Key stakeholders start to align around strategy development, a common language is developed and barriers to adoption are identified.
3. **Repeatable:** At this stage, recurring projects, events and processes are identified for integration. Formal committees are formed and documents are written up to define strategy, processes and technology investment needs, with stakeholder buy-in. Sustainable funding models and governance issues become a focus.
4. **Managed:** At this stage, formal systems for data flows and leveraging technology assets are in place and standards emerge. Performance management is based on outcomes of a shifted culture, new budgets, information technology (IT) investments and governance structure of a broader city context.
5. **Optimised:** A sustainable, citywide platform is in place. Agile strategy, IT and governance allow for autonomy within an integrated system of systems and continuous improvements. Superior outcomes deliver differentiation.

We combine the levels of smart city development (Figure 10.4) and the IDC Smart City Maturity Index to analyse Chiang Mai's position. The model allows a fairly quick assessment of the city's strengths and weaknesses, also enabling benchmarking of Chiang Mai against similar cities and identifying other cities with which it can partner to tackle similar challenges. Chiang Mai is at Levels 1 and 2 – Ad-hoc and Opportunistic – demonstrating that it needs tactical service delivery to ensure buy-in from stakeholders.

1. **Assess the current 'as is' situation across all smart city dimensions:** In smart city transformation, the assessment scale is specific and concrete to allow easy and definitive assessment of the current status of a city.
2. **Define and learn more about the challenges and success of other areas of city life, including city characters – e.g., constraints and opportunities:** If the model is completed in the suggested manner – i.e., by a group of city stakeholders working together – then all will gain a much better understanding of how transformation is being tackled throughout the city.
3. **Articulate the future 'to be' situation:** Once completed, the maturity model makes it easier for the city to set clearly defined smart city goals, outlining where it needs to be on a two- to five-year horizon.

Figure 10.5 shows Chiang Mai as being located between Levels 1 and 2, starting some ad-hoc projects and forming collaborations, as well as starting to acquire urban data. There is a citywide closed-circuit television (CCTV)

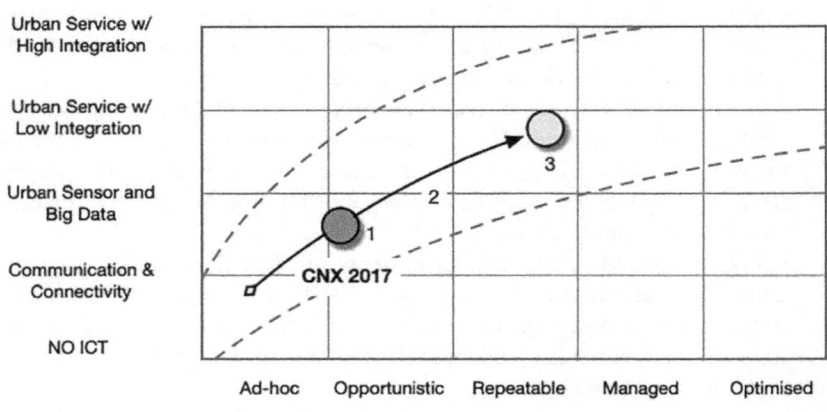

Source: Authors.

Figure 10.5   Applying a maturity model for Chiang Mai

system to ensure safety and a GPS tracking system to monitor urban transit movements. Air-quality sensors are installed across the city and these are especially crucial during the 'haze' season of February–May. There are also water-level gauges to monitor the Ping River and provide flood warnings. However, these systems operate separately, as there is no integration for a city-wide central warning system.

We observed that international and national private sectors are the first movers to see the smart city projects as investment opportunities. These include Mobike for bicycle sharing, Grab and Uber applications for on-demand taxi services, the BeNeat application for cleaning services and the CM Transit by RTC bus application.

The public sector has been taking part in smart city development, with stake-holder consultations conducted to formulate Chiang Mai Smart City planning. The DEPA has appointed a Smart City Committee to further collaborate with the national and local agencies for Smart City Planning and Development. The DEPA has set up its branch office in Chiang Mai and declared the city as one of its model smart cities alongside Phuket and Khon Kaen. The DEPA is utilising its financial instruments to match private-sector funding for smart city development. However, there is much more need to integrate all the projects to pursue a common goal for the city. In order to plan and execute a programme of continuous improvement in the pursuit of a smart city strategy, a method of assessing various dimensions of attributes at the current and target maturity levels would be a useful approach for stakeholders to continue integrating their ideas and projects.

POLICY RECOMMENDATIONS: THE BETTER THE FLOW, THE SMARTER THE CITY

In this section, drawing on ideas from CAS and the tools discussed, we offer policy recommendations and a roadmap for Chiang Mai Smart City. From the concept of MUS, a flow view can be used to analyse the performance of a city (Figure 10.6). From the logistics domain, there are three types of flow: (1) information; (2) finance; and (3) materials. The logistics domain can also include products and services. It differs from urban metabolic definitions of flows in a city.[2] For this context, a city is a fabric of different supply chains. It is also connected to other cities around the world. These views give us an insight into the complexity of a city and how to leverage them via better flows and better decision making along the flows.

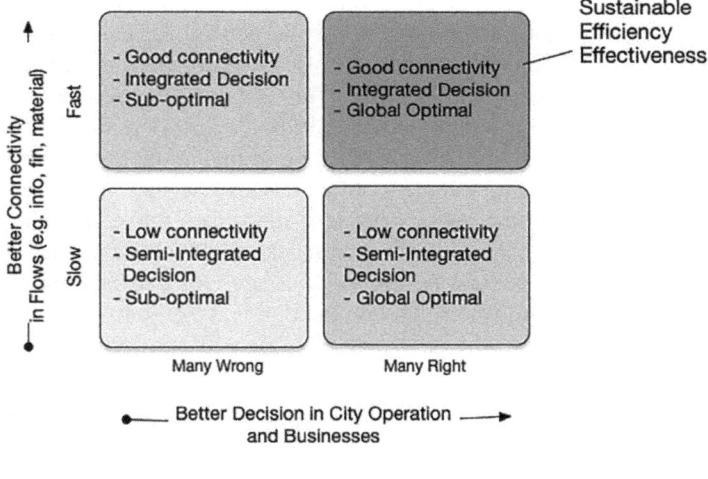

Source: Authors.

Figure 10.6   *Performance of flows (e.g., information, finance and materials) in terms of speed and quality of decision making in smart city development*

ICT infrastructure enables better speed and quality of connectivity. Furthermore, it allows us to make better decisions in terms of four main factors:

1. Data and information accessibility.
2. Integration and communication.
3. Analytic and synthesis systems.
4. Governance – e.g., transparency and accountability.

As with any living entity, the smart city development scheme allows a city to respond to opportunities. Because the scheme improves connections and communication among entities, it allows for better and integrated decision making. Due to complexity and limited resources, it is impossible to do everything, but the flow view gives us insight into the consequences of each flow, as well as tipping points that disable the entire flow.

In sum, Chiang Mai should become completely familiar with its urban flows, including human, capital, history, mobility, energy, waste, water and air. Data (big or not) should be captured in such a way that it can provide useful information for making decisions for the city. These flows can be depicted as visuals, thus requiring expertise from the geo-spatial community (mapping, imagery), data scientist community (big data analytics) and official statistics community (structured indicators). This requires empathy, or the effort to care about and understand people and things. Let us take a simple example of a plastic cup. The flow of a plastic cup begins in its production factory and then moves through its shipment, its use and its disposal. The same process can be captured for all manmade materials. If we care (have empathy) about the flow of the plastic cup, we will want to make sure that it is not left in unwanted places like rivers or oceans. Flows of nature can also be captured, such as air, temperature and water. The production and usage of energy can also be captured as a flow. This idea matches well with the technology of the IoT, whereby devices and sensors are easily installed and connected to one another.

Furthermore, it is recommended that smart city projects be divided into three levels:

1. **Physical platform:** Hard tangible spaces, infrastructures, transportation and environment.
2. **Digital platform:** Soft aspects, creative economy, e-governance and digital services.
3. **Deliberative platform:** The integration or blending of digital and non-digital worlds where there is active citizenry and meaningful collaboration in a dynamic world.

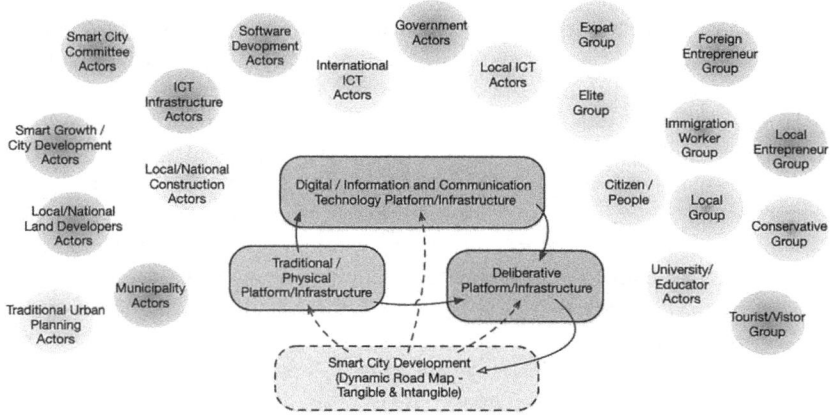

Source: Authors.

Figure 10.7   *Interaction among three platforms of smart city*

The three platforms help practitioners to focus on the technological aspects of smart city projects while also considering how all platforms can be integrated to make sense of the whole (Figure 10.7). This provides a rich dynamic roadmap to support different stakeholders in the smart city ecology. It also allows for parts or subsystems of the complex setting to adapt, which enables the flow of information to be efficiently captured, analysed and used for decision making.

In addition, in order to have a clear goal for becoming a smart city, it is recommended that aims align with the United Nation's Sustainable Development Goals (SDGs). Given that Chiang Mai is a highly diverse city with a mix of the local population, people from Bangkok, long-term foreign residents, tourists, investors and a growing Chinese population, together with a variety of socio-economic statuses, we recommend that Chiang Mai Smart City should focus on being inclusive and just. The third platform on deliberative processes will help Chiang Mai move in the right direction.

CONCLUSION

This chapter has outlined how a city is a CAS. By using Chiang Mai Smart City as an illustration, it described smart city initiatives with the intention of illustrating the non-linear nature of the projects, the unplanned emergence of events and the adaptive nature of actors. This chapter articulated the notion of a smart city being made up of a MUS – i.e., multi-layer, multi-agent, multi-flow, multi-activity and multi-dimension. These are fabricated into a complex system. This perspective provides us with a richer insight into the process of a city's evolution with respect to the intervention of the notion of a smart city. We also introduced the framework of levels of smart city development to guide practitioners in designing smart city projects.

The smart city development of Chiang Mai is still in its early stages. The model analysis concludes that it has a high potential for rapid growth because the key elements for development are readily available in the city. This includes geography and natural resources, basic infrastructure and key actors in smart city society, especially investors and citizens who have a high adoption rate for smartphone usage. However, there are basic limitations in Chiang Mai that may delay the process of smart city development: urban sprawl and the majority of citizens being caught in the low-middle-income bracket, limiting their ability to pay for smart services. Therefore, investment for ICT in the city has a slow return.

Chiang Mai is struggling between Levels 1 and 2 on the smart city development index. Smart city planning and roadmaps are critical tools for the deliberative process among Chiang Mai's citizens. The planning process needs to be a triple helix, involving private, public and academic sectors, ensuring balanced development between the top-down and bottom-up processes. The DEPA holds the key to reducing risk in smart infrastructure investment through its financial instruments.

The MUS model is a new framework that can be used to define and design smart city projects, taking into account multi-layer, multi-agent, multi-flow, multi-activity and multi-dimension. The new framework acknowledges the reality that cities are not static but are instead made up of dynamic parts that interact continuously in a complex system. There are innovative ways to utilise technology – in particular, the IoT, big data, blockchain, etc. The flows of the city can be captured within this model, helping the smart city movement to rise to a higher ground where it does not simply adopt technology for the sake of efficiency. It also improves decision making within the flows of the city.

Furthermore, a linear smart city roadmap cannot cope with this type of complexity. A combination of the three platforms of a smart city – digital, physical and deliberative – is a cornerstone of an adaptive roadmap, which can support different stakeholders of a smart city ecology, particularly in the era of transition.

NOTES

1. We conducted a series of scenario-planning exercises for the city in 2018. The conventional techniques of scenario planning were used. This includes brainstorming of uncertain driving forces, labelling of plausible scenarios and identifying key variables for each plausible scenario.
2. Urban metabolism is a model used to facilitate the description and analysis of the flows of materials and energy of cities, such as is undertaken in a material analysis

flow of a city. It provides researchers with a metaphorical framework to study the interactions of natural and human systems in specific regions.

REFERENCES

Bovaird, T 2008, 'Emergent strategic management and planning mechanisms in complex adaptive Systems', *Public Management Review*, vol. 10, no. 3, pp. 319–340.

Cairney, P 2012, 'Complexity theory in political science and public policy', *Political Studies Review*, vol. 10, pp. 346–358.

Clarke, YR 2013, 'Business strategy: IDC government insights: Smart City Maturity Model — assessment and action on the path to maturity', *IDC Government Insights: Smart Cities Strategies* #GI240620, viewed 15 June 2018, http://www.icd-gi.com.

Deakin, M 2013, 'From intelligent to smart cities', in Deakin, M (ed.), *Smart cities: Governing, modelling and analysing the transition*. Oxon and New York: Routledge, pp. 15–32.

DEPA 2018, *Thailand 4.0 pathway forward* (published in Thai), viewed 15 December 2018, http://www.depa.or.th/en/home.

Khianmeesuk V 2017, 'A conceptual framework to develop smart cities in Thailand', *International Conference on Literature, History, Humanities, and Interdisciplinary Studies* (LHHISS-17) Bangkok, Thailand, July 1112, viewed 15 December 2018, https://eares.org/siteadmin/upload/ED0717041.pdf.

Kim, TJ 1989, *Integrated urban systems modelling: Theory and applications*. Dordrecht, the Netherlands: Kluwer Academic Publishers.

Klijn, EH 2008, 'Complexity theory and public administration: What's new?', *Public Management Review*, vol. 10, no. 3, pp. 299–317.

McGrath, B, Sangawongse, S, Thaikatoo, D & Corte, MB 2017, 'The architecture of the metacity: Land use change, patch dynamics and urban form in Chiang Mai, Thailand', *Urban Planning*, vol. 2, no. 1, pp. 53–71.

Musa, S 2016, 'Smart cities – A roadmap for development', *Journal of Telecommunication Systems & Management*, vol. 5, no. 3, article 1000144.

Naprathansuk, N 2017, 'A national pilot project on smart city policy in Thailand: A case study on Phuket, Khon Kaen, Chiangmai Province', *European Journal of Multidisciplinary Studies*, vol. 6, no. 1, pp. 337–346.

Rhodes, ML 2008, 'Complexity and emergence in public management', *Public Management Review*, vol. 10, no. 3, pp. 361–379.

Roo, DR, Hillier J & Wezemael JV 2012, *Complexity and planning: Systems, assemblages and simulations*. Farnham: Ashgate Publishing.

Teisman, GR & Klijn, EH 2008, 'Complexity theory and public management', *Public Management Review*, vol. 10, no. 3, pp. 287–297.

11. 'Green' and 'smart' in South Korea: conceptions from the state to the citizen

Michael Manning, Jill L. Tao and Jae-in Noh

INTRODUCTION

Context matters. Although the city is a ubiquitous construct, that construct is still embedded within a web of social, political and economic environs. Thus, while we may consider the species *urbanus magnus* as one that flourishes and thrives no matter where it is planted, in truth, cities thrive because of their uniquely adaptive qualities. Just when we think we have the city defined, its qualities categorised and its values quantified, it morphs into something unexpected. The best laid plans often go astray, and this is probably a blessing, as those who live in the perfect implementation of a planned city have consistently testified (Jacobs 1961).

Yet plan we do, especially when trying to improve the city's prospects for the future. The smart city, the ubiquitous city, the green city: all are visions of an urban future that holds the promise of a better life in metropolitan areas. Much has been written on what such creatures are, the qualities of the smart city (Angelidou 2015), the strengths of the ubiquitous city (Shwayri 2013) and the inevitable weaknesses of green cities (Ko et al. 2011). But the broader national context and its influence on these entities have gone missing. How have we missed the forest for the trees?

The creation of New Songdo International City (hereafter referred to as Songdo) and its identity as a 'smart city' is tied to its beginnings as a 'U-city', and examining its creation and development as part of its association with 'green growth' helps illustrate how local governments in South Korea behave as national policy shifts. In this chapter, we examine how the local context shifts as national priorities shift. We examine the way in which the creation of Songdo leaned heavily on private sector implementation in an attempt to decouple the political undertones from the project. This supplements in an important way recent examinations of Songdo that offer purely fiscal incentives for the shift to private sector partnerships (Kim & Choi 2018). By creating a city that runs on only market values, the question of who gets to live in a smart, ubiquitous, green city, becomes a relevant question. Are 'smart' and 'green' cities only open to those who can pay? We survey a number of Songdo residents and non-residents to see what differences there may be between those who consider Songdo desirable, and to assess what they consider to be the characteristics of a smart, green city. We find significant differences between respondents' conceptions of 'green' and the national government's policy components, indicating a gap between national policy priorities and citizen expectations. We also compare the citizen and national expectations of the Songdo model to international models in order to highlight some of the uniquely South Korean conceptions of the 'smart, green' city and discuss the implications for the national government's plans to export the model to other nations.

SOUTH KOREA'S SMART CITY MODEL AND POLITICAL PRIORITIES

Defining the 'Smart City'

Smart cities, as evidenced by a vast and growing literature, are far more diverse in terms of what they offer in a practical sense than they are as a concept (Komninos 2015; Hard & Misa 2010). The idea of a 'smart city' in South Korea is somewhat different from the models often found in Europe or in the United States because of its incorporation into new cities, or cities that have new parts (Lee et al. 2016a, 2016b, 2016c). Thus, rather than retrofitting older sections of urban areas, South Korea is much more likely to take a new suburb and use it to display the benefits of smart city technology. This can provide local government officials with a powerful justification for redeveloping older sections of cities by literally building support around such areas (Lee et al. 2016a).

Good examples of the typical advantages touted by South Korean smart cities are the following: use of 'smart grid' technologies to integrate public transportation and private transport, so that traffic flow can be streamed live to commuters who may be deciding what route to take to work (for example, if there is excessive traffic on the roads, more commuters may decide to use the metro, thus preventing more congestion); the installation and centralised use of CCTV (closed-circuit television) to reduce the incidence of crime; and the use of information screens to market local amenities, and to provide timely and useful information to residents and warnings in the event of natural disasters, such as regional flooding, earthquakes, or fires. But as these examples illustrate, these uses of technology focus more on general provision of information to citizens, rather than incorporating data provided by citizens into other, more comprehensive, long-term policy goals. Thus, South Korean smart cities are not yet truly interactive, nor are they meant to be. If there were a word that sums up what a smart city in South Korea is meant to provide to its residents, that word would be convenience.

The other striking characteristic of smart cities in South Korea is cost. Most smart city development is done at surprisingly cheap rates. For example, in Anyang, a city to the southwest of Seoul, the local government installed a comprehensive CCTV system and a bus/subway notification system to integrate with traffic flow information for the same cost as building 'a 1–2-kilometer road' (approximately US$33 million) (Lee et al. 2016a, p. 1). In Namyangju, a city to the northeast of Seoul, the local government generates roughly US$74,000 annually through advertising revenue generated through its smart screen system (Lee et al. 2016b). And in Songdo, located on the west coast of Incheon, the creation of a public–private partnership for investments in the construction of smart city infrastructure allows for a cost-sharing model that provides companies that invest with a way to share in the returns generated by their investment (Lee et al. 2016c). By focusing its smart city efforts on the use of existing technologies to augment what South Koreans are already doing on a daily basis, local governments are able to sell the smart city idea as a highly attractive concept, and therefore something that citizens will begin to demand. But who demands it and how these new manifestations of urban life are perceived is still a matter of some contention.

Although the national government has pursued digital urban growth policies since 1984 (Shin & Kim 2012), the interest in smart cities in South Korea began in earnest in the early 2000s. In 2004, the Korea Ministry of Information and Communication established a new initiative called the IT389 Strategy. This strategy was meant to capitalise on the widespread installation and usage of broadband technologies, and the links to urban economic growth and development were clearly laid out (Shin & Kim 2012). This policy was supported by key private sector actors, such as Samsung, POSCO and LG, who sought out government support for these initiatives (O'Connell 2005). The central government was supportive of testing the ideas in different cities where the local government officials were willing partners and as long as the economy was doing well. But in 2007, due to the global recession, initiatives were tabled and the funding for several of the national government's technology-driven policies disappeared (Shin & Kim 2012). However, the U-city components of the IT389 Strategy escaped the funding axe and continued on when the Lee Myeong-Bak administration entered office in 2008, incorporated into a new policy that focused on creating smart infrastructure, largely through the use of wireless technology and RFID (radio frequency identification) tags (Shin & Kim 2012).

The South Korean model for the U-city/smart city was, therefore, a centrally driven one, following in the footsteps of the country's industrial development model of previous generations, with heavy participation by the technology and construction sectors. As some observers noted, 'A centrally coordinated technology project has generated a supply-push strategy, which has led to the risk [that] domestic demand or export market potential was not carefully considered' (Shin & Kim 2012: 5). Since the central government was the driving force behind such initiatives, transitions between administrations often resulted in shifts in focus. This was evident with the 'green growth' models that came to prominence in the late 2000s, which have been described as uniquely Korean (Shwayri 2013). In recent years, environmental variables have been added to growth policy. Some observers have said this change was due to a shift towards sustainability, notably by the United Nations, which was then led by South Korean Ban Ki-moon as Secretary General from 2007 to 2016. Domestically, the industrial policy underwent an evolution at the national level with the equation of a demand for a cleaner environment with an increase in the wealth of a nation (Lee 2010).

This was by no means a grassroots movement, pushing for a cleaner environment or a switch due to a rising public consciousness with respect to sustainability issues. Research on the level of concern for environmental issues and economic growth at the time show that South Korean citizens placed emphasis on the latter; therefore, 'green growth' models attempt to incorporate both 'green' qualities and economic growth (Kim & Kim 2010). Much like the smart city strategies promoted by the national government, the green city strategies promoted by green growth policies focus squarely on increasing efficiency to improve quality of life variables that are considered important to citizens. But what makes the Korean model unique?

Green growth as a general urbanisation strategy counters population growth and increased energy demands by increasing population density in order to decrease transportation impacts and the related increase in greenhouse gases (GHGs). This is done while simultaneously investing in renewable energies and enforcing energy efficient building standards (Lee 2010). Green growth requires the preservation of natural resources, use of renewable energy sources and, ideally, low-carbon output (Lee & Kim 2015). These desired outcomes, as outlined by the United Nations, help mitigate the effects of climate change, and under ideal circumstances, square nicely with the efficiency and convenience goals of smart cities (Lee et al. 2016b). However, such models are not as aggressive as and should not be confused with sustainable green city models. Arguments favouring green growth models are often associated with developmental states, based on the pressing need to improve living standards through growth while at least paying lip service to sustainability (Lee 2010). The problem with sustainability prioritisation is that it is ambiguous, with no consensus on which elements need to be 'sustained' (i.e., energy expenditures, quality of life, or environmental conditions) (Mori & Christodoulou 2012). This means it is open to co-optation as a model for urban development.

Political Opportunism and the Convergence of 'Smart' and 'Green'

While the U-city initiative continued through national political transitions, South Korea's 'green growth' models were adopted at the national level by the Lee Myeong-Bak administration. The original green growth policy was crafted by the Presidential Committee on Green Growth, which brought together key government administrators from multiple agencies who pushed the National Assembly to adopt the 'Framework Act on Low Carbon, Green Growth' in 2010 (Park et al. 2011). The implementation of the act was to be carried out in each province of the country, where each major city in each province would be charged with creating its own low-carbon green growth council and plan for execution of the green growth goals. As noted by evaluators of the early attempts at implementation, this approach suffered from a lack of understanding of the goals of the new law at the local level. This was illustrated by the remarkable consistency across cities in their proposed 'green growth' plans (Park et al. 2011, pp. 16–18).

Part of the reason for the timid response by local governments was the generally abysmal state of the economy. South Korea was suffering from the tightening of world markets and the only way to generate any kind of growth, green or otherwise, was to borrow (Lee & Tao 2012). The Lee Administration decided that the best way to allocate the potential benefits from green growth would be to let each region of the country, as represented by a major city, pursue different strategies that exemplified the policy focus. Thus, each city came up with its own strategy for 'green growth'. These were better tailored to cities' strengths, which meant that cities might continue with whatever plans they had originally devised, but with accommodation for the new emphasis on low-carbon emissions. Incheon Metropolitan City, for example, began a Green Growth through Sports campaign (Park et al. 2011). Since it had just been chosen as the site for the 2014 Asian Games, Incheon was already starting its plans to build several stadiums to host the competition. Each stadium's construction and operations were re-examined with a new eye to green compliance (see Table 11.1).

This perhaps exemplifies the inherent contradictions in the 'green growth' models. Scrapping the plans for construction would have reduced carbon emissions, but it would also have caused real economic harm to the local economy at a time when it could least afford the loss. Nonetheless, compliance with new national priorities had to be incorporated into the projects. Thus on-site practices, such as recycling of materials used in construction of temporary

*Table 11.1   Incheon Metropolitan City plans for 'green growth' Asian
              Game stadiums*

Stadium	Construction costs (million KRW)	Plans for eco-friendly measures
Gyeyang Stadium	14,275	Facilities linked to the Central Square, construction of ecological space
Namdong Stadium	21,159	None
Sipjung Stadium	14,845	Efficient reduction of human traffic and construction material
Ganghwa Stadium	12,854	None
Sunhak Stadium	10,067	Reducing manmade design by increasing the context of nature in placement and construction plans
Songlim Stadium	16,950	Connecting the flow between people and green belt via 3-D image
Munhak Stadium	16,665	Remodelling into a green stadium
Total	**106,815**	

Note: KRW: South Korean won (currency).
Source: Metropolitan City of Incheon 2009.

structures (scaffolding, administrative buildings and safety zones), were incorporated to cut down on waste and cost, with an added carbon footprint reduction compared to standard practices (Park et al. 2011). Politically, the provision of steady jobs during an economic slump was also appealing. The then Mayor of Incheon, Ahn Sang-Soo, saw this as an opportunity to grow while costs were low.

These green elements were often laid on top of existing smart elements. Construction started in the early 2000s under the Roh Administration, which hailed from the opposition Uri (Democratic) Party. This was another way for the new administration to lay claim to any benefits realised that were begun by previous regimes.

THE ROOTS OF 'GREEN' IN SMART CITIES

Defining 'Green'

What exactly do we mean when we say 'green' elements? The definition of 'green' is not universal (Ko et al. 2011). Defining 'green' through a political lens presents numerous problems due to both its wide-ranging policy scope and its relative isolation to the industrialised democracies in the 1980s and 1990s. Green political parties first surfaced in the 1970s in the UK and New Zealand, but were not called 'green' until much later (Frankland et al. 2008). The early parties were the outcomes of grassroots movements, incorporating many different concerns that members saw as interrelated. Environmental groups have addressed green issues by pressuring businesses to decrease carbon emissions, focusing on conservation efforts related to excess lumber extraction and maintaining natural habitat, and monitoring the contamination of water commons (Kleiner 1991). Given the breadth of policy goals, the shift from political to administrative structures for implementation reduced the apparent necessity for political movements and the focus on green political parties de-intensified over the early part of the twenty-first century (Frankland et al. 2008).

Other examples of government approaches to defining 'green' can be found in the early embracers of green political movements. One notable example that survives today is the Finnish organisation Sitra (see Table 11.2). First created by the Bank of Finland as a fund in 1967 to promote 'stable and balanced development, economic growth and international competitiveness and co-operation', its mission has changed over time. As Finland's economy has grown and developed, its focus has shifted to building a model that incorporates social, political and private sector changes to enhance renewable energy usage and reduce carbon output among all countries. The initiatives being promoted by this organisation provide additional context for defining green because they identify successful environmental policies utilised in different countries and encourage adoption of these policies internationally. The carbon saving solutions encouraged by Sitra offer a benchmark for comparison with those policies incorporated into South Korea's green growth agenda. The policies cover a broad range of public sector approaches to reducing carbon emissions and improving sustainable use of resources (Sitra 2017).

These approaches provide a benchmarking system for green policies that can be compared to South Korea's 'green growth' approach. The strategy for South Korea's 'green growth' was three-pronged: (1) focus on climate change adaptation and energy independence; (2) create new engines for economic growth with an emphasis on low-carbon or reduced carbon approaches; and (3) improve the quality of life through green measures to enhance international standing (Lee & Kim 2015, p. 21; Jones & Yoo 2011). Of these three, the first strategy received the lion's share of attention and funding from the central government. South Korea's legacy of energy infrastructure and its heavy reliance on fossil fuels explains this emphasis. In 2014, 85.7 per cent of South Korea's energy supply was reliant on fossil fuels (coal, petroleum and liquefied natural gas (LNG)) (Lee & Kim 2015). Since South Korea has no domestic source for such fuels, this has proven to be an expensive energy legacy. Such infrastructure has translated into a heavy reliance on coal and nuclear energy for power generation, with a constant eye towards improving renewable energy options. However, since South Korea's economic miracle is premised on cheap energy provision, even with 'green growth' as a national strategy, there has been little

Table 11.2   *Comparison between Sitra green and Korean green growth*
                policies by policy goal type

Policy goal (Korea green growth)	Policy (Country)	Benefit
Mitigation of climate change and promoting energy independence	Afforestation and reforestation (Costa Rica)	Encourages forest growth to cut carbon emission through government funding.
	Reduction of deforestation (Brazil)	Extends the life of forests to conserve biodiversity and help fight carbon emissions.
	Reduction of food waste (Denmark)	Lowers food demand, eases burden on low income families and lowers carbon emissions.
	Low-carbon agriculture (Brazil)	Preserves ecosystem and lower carbon emissions.
	Reduction of methane from oil and gas production (United States)	Maximises resources, saves money, lowers carbon emissions.
	Industrial electric motors (United States)	Reduces energy costs, need for imported fuel, decreases fossil fuel emissions.
	Industry energy efficiency through increased government standards (China)	Job creation, reduced carbon emissions and energy consumption.
	Appliance efficiency through increased government standards (Japan)	Cuts utility expenses for households and businesses, reduces energy consumption and fossil fuel imports.
	Improved cook stoves (China)	Reduces fuel consumption, increased safety, reduces need for firewood, charcoal and other biomass.
Creating new engines for economic growth	Environmental tax on diesel, LPG Butane and heavy oil (for industrial use) (**South Korea**)	Imposed a 3 per cent tax on heavy fuel for industrial purposes and other heavy GHG producing fuels to improve air quality.
	Clean Development Initiative and Carbon Market (**South Korea**)	Creates a domestic voluntary carbon market to provide incentives for GHG reduction by industry.
	Shift to low-carbon industry and production (**South Korea**)	Create new jobs and training opportunities in the energy efficiency and sustainability sectors.
Improving quality of life and international standing	Expand the high-speed train system (**South Korea**)	Improves transportation efficiency and access to more areas of the country.
	Solar water heating (China)	Improves air quality in densely populated areas.
	Four Rivers Project (**South Korea**)	Improves water management, provides alternative transportation networks, improves flood control.

Policy goal (Korea green growth)	Policy (Country)	Benefit
Combination	Building efficiency (Mexico and Germany)	Reduces fuel imports, lessens utility costs, improves air quality.
	Bus rapid transit (Colombia)	Reduces travel times, traffic congestion and carbon emissions, increases property value on bus lines and improves air quality.
	Vehicle fuel efficiency (European Union)	Improves air quality, enhances energy security, reduces demand for fossil fuels.
	Bioenergy for heating (Finland)	Creates employment opportunities, reduces import of fossil fuels, enhances energy security.
	Wind power (Denmark and Brazil)	Reduces reliance on fossil fuels, creates employment opportunities, provides a source of income for landowners.
	Off-grid solar power (Bangladesh)	Provides a modern electrical energy source for citizens living in rural, developing areas.
	Grid solar power (Germany)	Reduces reliance of fuel imports, potentially cuts energy costs for consumers, creates and secures employment and improves air quality.

Source: Sitra 2017 for Sitra data; Republic of Korea 2009 and Jones & Yoo 2011 for Green Growth data.

movement until very recently towards shifting to more sustainable, but also more expensive, options (the new Moon administration has plans to phase out coal and nuclear plants) (Proctor 2017).

When comparing these national strategies to those identified elsewhere (Table 11.2), the focus elsewhere is somewhat different. Although cost is certainly a concern for many of the examples compiled by Sitra, a more consistent focus is improved air quality and lower carbon emissions. This is a rather notable difference with 'green growth' strategies, especially the lack of concern with air quality. Thus, South Korea's 'green growth' model as funded and implemented does not quite comport with 'green' approaches used elsewhere. At the city level, this presents challenges. For Songdo, constructed on the border of Incheon, the green growth concept was initially added to its other characteristics to help boost its international visibility (Kim & Choi 2018) and because of a new policy direction at the national level.

Characteristics of Songdo: Green and Smart Overlap

Songdo exhibits smart and green city characteristics that overlap with a green city due to an inclusion of sustainability and energy efficiency in the design of the city (Angelidou 2015; Clark et al. 2002). Energy efficiency standards were enforced early in Songdo's construction at increased cost and were installed to ensure energy cost stability in case of rapidly increasing energy prices, consistent with national concerns and green growth policy (Copiello & Bonifaci 2015). Energy efficiency was a key component of Songdo's presentation as the exemplary green growth model that South Korea wished to export (Shwayri 2013). Smart green features of the city include an advanced recycling system (approx. 75 per cent of residential waste); storm water collection centres that feed into the 100-acre Central Park canal and the manmade river that separates the New Songdo island from the mainland; extensive open spaces which are enhanced by underground parking structures constructed literally beneath the green spaces; an advanced transportation systems that includes multiple bus lines, five subway stations and extensive bike paths; LED traffic lights are primary green features; and, Leadership in Energy and Environmental Design (LEED) buildings have been constructed (Kamal-Chaoui et al. 2011; Anderson 2015). Additionally, all residence areas operate in somewhat autonomous fashion, with public schools, grocery stores and small businesses incorporated into the structures of the high-rise residential buildings. This cuts down on the need for automobile use for short distance errands and the availability of delivery by low emission vehicles also helps reduce the overall carbon footprint of the city.

Songdo also has a smart waste system, which acts a renewable energy source, and this has been cited as the main 'smart' sustainability amenity provided by the city (Anthopoulos 2016). The city provides high-speed Internet accessibility, 'integrated building and facility management, security and hazard management, e-learning, remote healthcare facilities and automated traffic control', which are additional smart city features (Kim 2010). However, Anthopolous' study notes that 'smart' features of Songdo are minimal when compared to other 'smart' cities, so its defining features do not lie solely in its being smart; they also lie in its being 'green'.

RESEARCH DESIGN AND TEST OF CONCEPT

Assessing Local Green Sensibilities

As outlined in Table 11.2, the South Korean Government's view of what constitutes a green growth project or policy is fairly broad. But how might we identify the prevailing view of 'green' for ordinary South Korean citizens? Lee and Kim (2015) identified community participation as a key explanatory variable in the success of green initiatives nationwide. Lack of citizen involvement in green initiatives can be due to the absence/ineffectiveness of outreach programmes or failures of the central government to enact initiatives that are consistent with community views. The inconsistencies of local governments in fully investing in green growth initiatives can be linked to a lack of grassroots support (Lee & Kim 2015). There is little literature detailing the conceptions ordinary South Koreans have of green growth, but extant research shows that gender, age, level of education and income level all play a role in the level of environmental concern (Kim & Kim 2010) and may help explain perceptions of what constitutes 'green'. Thus, it may be useful to see what features South Korean citizens might consider critical elements of green city models.

We crafted a questionnaire using variables identified by Kim and Kim (2010), along with variables from cities often used to benchmark 'green' approaches and amenities (Sitra 2017) and included 'red herring' items that differentiate amenities South Koreans consider upscale, e.g., things associated with 'new' and/or 'clean' imagery (Anderson 2015) from things that may be considered 'smart' or 'green'. The questionnaire was designed to determine whether South Korean citizenry identifies more with 'mainstream' green city features or green features highlighted by the South Korean Government.

The survey was administered to participants within Songdo during June of 2016 to ensure that respondents had some knowledge of the city. Responses were collected at two shopping centres (Hyundai Premium Outlet and Canal Walk Shopping Centre) and at Incheon National University. The respondents were targeted randomly at each location using a 'man in the street' poll with two surveyors asking people to participate. Each location was chosen based on the high level of pedestrian traffic in those areas to increase the likelihood of a representative sample when compared to the surrounding city of Incheon.

Respondents' Characteristics

A total of 134 South Koreans participated in our survey, with slightly more women responding than men (76 to 58). Of those who participated, about a quarter were already residents of Songdo (33), with another 28 per cent (38) interested in moving to Songdo. The remainder resided outside of the city and had no interest in relocating (63). Most of the survey respondents (93) had a university degree. This is not unusual, however, in South Korea, where two-thirds of the population between the ages of 25 and 34 has a college degree and of those between the ages of 35 and 44, 60 per cent have a college degree (OECD 2017). Income distribution was evaluated based upon monthly earnings, where respondents were asked to report their family income. According to the Korean Statistical Information Service (KOSIS) in 2015, the average monthly income for South Koreans was 3,980,000 KRW. However, those who resided or wished to relocate to Songdo did have higher average incomes than those who did not. The average income for residents was 4,300,000 KRW,

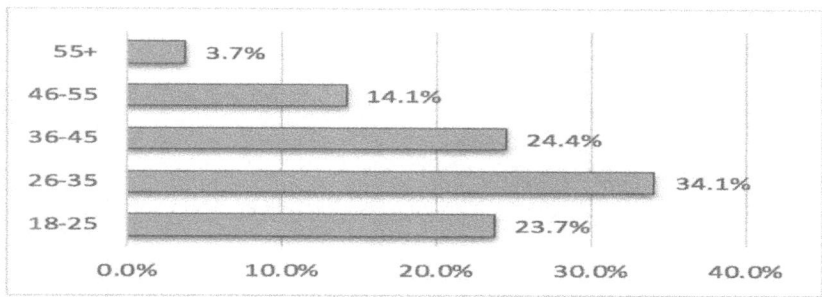

Figure 11.1   Age range of survey respondents by percentage

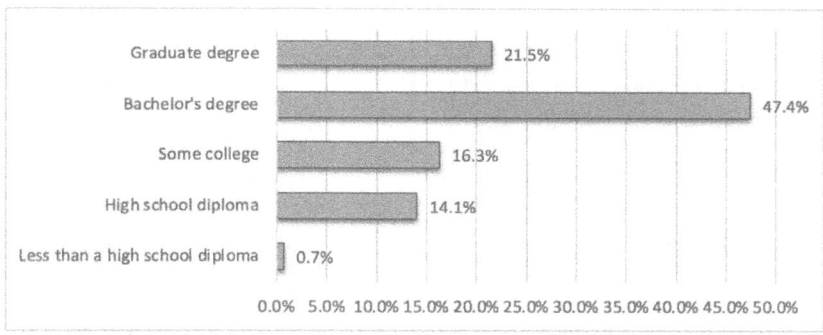

Note: The sample from our survey represents the educational achievement trends within Korea's population with a majority of participants earning of minimum of a bachelor's degree.

Figure 11.2   Educational achievement of participants

while the average for those outside was 3,325,400 KRW. The breakdown across variables can be seen in Figures 11.1–11.3.

General Perceptions of Green

Participants were provided with a list of 15 characteristics derived from the literature on green cities, including 'green' features actually present in Songdo, with other features not present but benchmarked in other 'green' cities outside South Korea. As noted previously, additional 'red herrings' (asterisked in Table 11.3) were included to control for perceptions of desirable but 'non-green' amenities. This provides clarification as to whether residents

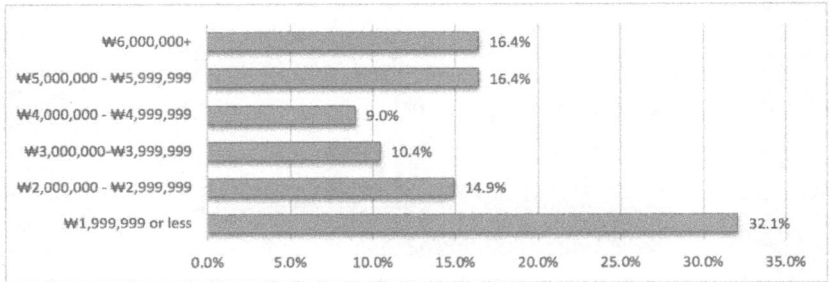

Note: One survey participant did not provide an answer to this question.

Figure 11.3   *Breakdown of monthly household income among survey respondents (KRW)*

conceptualise 'green' in a manner consistent with both the Korean national government and groups in the West such as Sitra (Table 11.3).

Respondents were also asked to identify characteristics that brought to mind the image of an ideal green city (Table 11.4). They were told to choose only one characteristic that best captured a 'green' city image. Five options were provided: parks and natural spaces; clear air and clean environment; solar panels and renewable energy sources; vast public transit; and farms and gardens. Parks and natural spaces received 68 responses (50.4 per cent). Interestingly, farms and gardens were not chosen by any of the participants, despite the prevalence of public vegetable gardens throughout Songdo. This illustrates the urban nature of the green city in South Korea, where farming is still thought of as a rural activity, despite ample evidence to the contrary.

Solar panels and renewable energy sources (which are staples of the Green Growth Initiatives) only received 7.4 per cent of the responses. This finding is interesting for two reasons. The first is that it is a relatively unpopular choice, indicating that perhaps citizens do not view renewable energy as a green activity. This may be due to the mixed messages local governments and citizens received: the national government's initiatives focused primarily on energy security and independence, with a side of environmental concern. The second point of interest is that people expected to see the use of renewable energy in Songdo at all. As a new city, Songdo has had multiple opportunities to incorporate new technologies into building designs. China has certainly done this with gusto in new urban areas, fitting both new and older developments with solar technology whenever feasible. But Songdo's 'green' emphasis has been elsewhere, literally buried underground in waste-to-energy features from some of the high-rise apartment complexes. This feature, perhaps more than any other, illustrates the disconnect between Songdo as a showcase city for

Table 11.3   *Characteristics of a 'green city'*

Green city characteristic	Location of feature	Frequency	Percentage
Smart grid development	Songdo	25	18
Parks and open spaces	Songdo Benchmark	110	81.5
Express trash collection	Red herring*	24	17.8
Areas for urban farming and agriculture	Benchmark	0	0
Conservation areas for indigenous animals	Benchmark	20	14.8
LED lighting for traffic lights	Songdo	21	15.6
Recycling mandates	Songdo Benchmark	23	17.0
Well-developed public transit system	Songdo Benchmark	28	20.7
Prioritisation of renewable energy resources	Benchmark	46	34.1
Building regulations requiring strict energy efficiency standards be met	Benchmark	40	29.6
Solar panels on public buildings with incentives for citizens	Benchmark	28	20.7
Broadband Internet technology	Red herring*	10	7.4
Promotion of local food consumption	Benchmark	10	7.4
Animal species protection training	Benchmark	7	5.2
Collection and reuse of storm water and waste water	Songdo Benchmark	24	17.8

Table 11.4   *Image of a 'green city'*

Image description	Frequency	Percentage
Parks and natural spaces	68	50.4
Clear air and a clean environment	47	34.8
Solar panels and renewable energy sources	10	7.4
Vast public transit	10	7.4
Farms and gardens	0	0
Total	**135**	**100**

national green growth initiatives and the features it has incorporated to attract residents. The design by Gale International has clearly been adapted to better meet the local renditions of what constitutes a 'green' city.

SOUTH KOREAN VISIONS OF 'GREEN': GRASSROOTS VERSUS ENERGY SECURITY

The original architects of Songdo were the international real estate development group Gale International, POSCO Engineering and Construction and government officials at the local and national levels. The primary focus in the initial plan for Songdo focused on ease of access to amenities (Whitman et al. 2008). Aside from conservation areas demanded by environmental groups when construction displaced indigenous animals (Ko et al. 2011), the primary green initiatives focused on construction of a city that reduced dependency upon automobiles for transportation. Infrastructure systems linking buildings, bike paths connecting residential areas, the adherence to compact city principles, at least in some parts, and linking amenities and residential areas all contribute to increased energy efficiency. These features are viewed favourably by the South Korean population, but because they are prized for their increased convenience rather than for their intrinsic environmental value.

In Songdo, energy efficiency was prioritised over renewable energy resource models. Initial designs showed that LEED standards would be used as a benchmark to ensure that high-energy efficient standards were met; however, the adherence to these standards is limited to few complexes within the city. Additionally, solar panelling, which was also highlighted in early promotions (Whitman et al. 2008), is largely absent, despite the emphasis on alternative energy investment by the Lee Myung-bak administration. However, as the results show, citizens are not concerned with alternative energy sources.

Criticisms of Songdo emphasise its shortcomings when compared to alternative green city models that focus more on renewable energy sources. Models outlined by Sitra embody aggressive strategies nations are using to address climate change concerns. When comparing South Korea's Songdo model to such strategies, it is clear that the exportable green growth model desired is not in itself a green model recognisable to the outside world. Nevertheless, for a nation whose citizens are still primarily concerned with economic stability and growth (Kim & Kim 2010), this model mixes sustainability elements and economic development deemed suitable to South Korean sensibilities. These results indicate that Songdo's plan is largely representative of the South Korean viewpoint of what constitutes a 'green city' at the grassroots level.

The evolution of the planning of Songdo has been in constant flux since the initial stages. A shift from smart city features to an emphasis on green growth has allowed the incorporation of green growth characteristics envisioned by President Lee's administration, at least on a limited basis. However, the disconnect between the national focus on energy security and the more pressing concerns of average South Koreans highlights the difficulties in implementing a national strategy when there are mixed understandings of what that strategy means. Our study has concluded that the vision for a green city represented by the South Korean national government does not closely represent the views of the population. Inclusion of green spaces (parks and recreation areas) and cleanliness (express trash removal) are the most commonly cited attributes of green cities by ordinary South Koreans surveyed, regardless of whether they resided inside Songdo or not.

CONCLUSION

Songdo's inclusion of ideal green features and their integration with the smart city features for its residents is evident; however, alternative amenities not associated with green cities are offered (e.g., cleanliness, open space and parks), which may motivate people to relocate to Songdo, but primarily those who have means. As a real estate venture, this is good news, but as a model for export, especially as a 'smart and green' city, Songdo delivers a different interpretation of green that may not fit what the national government intends to market. Additionally, the national rendition of 'green' cannot be used as an accurate representation of local perceptions. A more honest rendition might be 'modern and convenient'.

This mismatch between the product that the central government is marketing abroad and what actually exists in an experiment such as Songdo brings up the issues raised by the analysis of the smart city: a top-down, centrally driven model that is tone-deaf to what citizens need runs the risk of co-optation by the industries that may profit from the government's investment. This has been a constant critique of the developmental state and recalls the mantra that innovation does not necessarily thrive in environments that do not foster a true competition of ideas. However, such a critique oversimplifies what has happened in Songdo and overlooks some critical points of contention.

The first, and perhaps most important, is that Songdo, as a city, was established to be a number of simultaneous experiments, with some that have clearly initially failed (energy efficiency incorporated into construction, for example) while others have demonstrated success (the smart transportation infrastructure has spawned multiple apps and there is widespread use of the RFIDs by residents, including highly useful updates when businesses relocate). This has been possible because of the central government support, not in spite of it. With a buffer for failure, these kinds of experiments do not have to suffer the 'automatic' punishing recoil of the market (Lindblom 1982, p. 324). As one observer stated, 'Much of this technology was developed in U.S. research labs, but there are fewer social and regulatory obstacles to implementing them in Korea ... There is an historical expectation of less privacy. Korea is willing to put off the hard questions to take the early lead and set standards' (Anthony Townsend, quoted in O'Connell 2005). Second, because the city has been allowed to serve as 'one big Petri dish' (B.J. Fogg, quoted in O'Connell 2005), now the government has a much better idea of what has worked and what has not, so its final marketing push, which just started in 2018, may be more accurate than would have been the case otherwise.

In the years to come, as this model is borrowed by places as diverse as the Maharashtra state in India, Duqm in Oman, Tempere in Finland and Barcelona in Spain, the ability of the model to adapt to changing circumstances has already been proven by the Songdo experiment (European Union 2018; Global Construction Review 2018; Property Report 2018; Smart City Hub 2017). One of the great weaknesses of centralised planning and implementation is regime change. When a new regime comes in, it often cleans house and this can lead to disruptions of policy experiments, muddying links between cause and effect. However, in South Korea, there are vested interests on all sides, so even as policies morph into new versions of older ideas, the original architects document what has changed and what is new. So the links, although much more complex, remain evident.

If South Korea could make one modification of its model based on our analysis, we would recommend a stronger focus on who will benefit from the ubiquitous technologies that are evident in Songdo. As inequity in urban areas continues to be a concern in the developing world, the lack of acknowledgement that such issues have not been considered in the testing of these technologies should be addressed. Since the target market for this model is often the developing world, as evidenced by the examples outlined above, we conclude that although residents of Songdo have confirmed that the city does, for the most part, represent their expectations, additional citizen participation in the planning process is needed to achieve a more inclusive vision for South Korea's urban future.

REFERENCES

Anderson, C 2015, 'The scene of New Songdo', *Asian Journal of Cultural Policy*, vol. 2, no. 2. pp. 1–17.

Angelidou, M 2015, 'Smart cities: A conjuncture of four forces', *Cities*, vol. 47. pp. 95–106.

Anthopoulos, L 2016, 'Smart utopia VS smart reality: Learning by experience from 10 smart city cases', *Cities*, vol. 63. pp. 128–148.

Clark, TN, Lloyd, R, Wong, KK & Jain, P 2002, 'Amenities drive urban growth', *Journal of Urban Affairs*, vol. 24, no. 5, pp. 493–515.

Copiello, S & Bonifaci, P 2015, 'Green housing: Toward a new energy efficiency paradox?', *Cities*, vol. 49, pp. 76–87.

European Union 2018, 'World cities project encourages smart-city initiatives in Busan, Seoul, Suwon, and Gwangju', Press Release, 18 March, viewed 18 July 2018, https://eeas.europa.eu/delegations/south-korea/43604/press-release-world-cities -project-encourages-smart-city-initiatives-busan-seoul-suwon-and_en.

Frankland, EG, Lucardie, P & Rihoux, B 2008, *Green parties in transition: The end of grass-roots democracy?* Farnham: Ashgate Publishing.

Global Construction Review 2018, 'Korea signs up for plan to build smart city in Oman', Chartered Institute of Building, viewed 3 August 2018, http://www .globalconstructionreview.com/news/korea-signs-plan-build-smart-city-oman/.

Hard, M & Misa, TJ 2010, *Urban machinery: Inside modern European cities.* Cambridge, MA: MIT Press.

Jacobs, J 1961, *The death and life of great American cities.* New York: Random House.

Jones, RS & Yoo, BS 2011, 'Korea's green growth strategy: Mitigating climate change and developing new growth engines', *OECD Economics Department Working Papers No. 798*, viewed 8 May 2017, https://dx.doi.org/10.1787/5kmbhk4gh1ns-en.

Kamal-Chaoui, L, Grazi, F, Joo, J & Plouin, M 2011, *The implementation of the Korean green growth strategy in urban areas.* Paris: OECD Publishing.

Kim, C 2010, 'Place promotion and symbolic characterization of New Songdo City, South Korea', *Cities*, vol. 27, no. 1, pp. 13–19.

Kim, S & Kim, S 2010, 'Comparative studies of environmental attitude and its determinants in three east Asia countries: Korea, Japan and China', *International Review of Public Administration*, vol. 15, no. 1, pp. 17–33.

Kim, YJ & Choi, MJ 2018, 'Contracting-out public–private partnerships in mega-developments: The case of New Songdo City in South Korea', *Cities*, vol. 72. pp. 43–50.

Kleiner, A 1991, 'What does it mean to be green?', viewed 7 May 2017, http://hbr.org/ 1991/07/what-does-it-mean-to-be-green.

Ko, Y, Schubert, DK. & Hester, RT 2011, 'A conflict of greens: Green development versus habitat preservation: The case of Incheon, South Korea', *Environment: Science and Policy for Sustainable Development*, vol. 53, no. 3, pp. 3–17.

Komninos, N 2015, *The age of intelligent cities: Smart environments and innovations-for-all strategies.* Regional Studies Association: Regions and Cities Series. Oxford: Routledge.

Lee, J 2010, *Green growth: Korean initiatives for green civilization.* Seoul: Random House Korea.

Lee, JS & Kim, JW 2015, 'South Korea's urban green energy strategies: Policy framework and local responses under the green growth', *Cities*, vol. 54, pp. 20–27.

Lee, JY & Tao, JL 2012, 'Fiscal stress and its impact on local expenditure autonomy', *Korean Journal for Local Government Studies*, vol. 16, no. 3, pp. 235–247.

Lee, SK, Kwon, HR, Cho, HA, Kim, JB, & Lee, DJ 2016a, *International case studies of smart cities: Anyang, Republic of Korea, Discussion Paper No. IDB-DP-458.* New York: Inter-American Development Bank, Institutions for Development Sector; Fiscal and Municipal Division.

Lee, SK, Kwon, HR, Cho, HA, Kim, JB, & Lee, DJ 2016b, *International case studies of smart cities: Namyangju, Republic of Korea, Discussion Paper No. IDB-DP-459.* New York: Inter-American Development Bank, Institutions for Development Sector; Fiscal and Municipal Division.

Lee, SK, Kwon, HR, Cho, HA, Kim, JB, & Lee, DJ 2016c, *International case studies of smart cities: Songdo, Republic of Korea, Discussion Paper No. IDB-DP-463*. New York: Inter-American Development Bank, Institutions for Development Sector; Fiscal and Municipal Division.

Lindblom, CE 1982, 'Markets as prison', *Journal of Politics*, vol. 44, no. 2, pp. 324–336.

Metropolitan City of Incheon 2009, *Plan for Renewable Energy Implementation in Incheon*. Incheon, Republic of Korea: Metropolitan City of Incheon.

Mori, K & Christodoulou, A 2012, 'Review of sustainability indices and indicators: Towards a new City Sustainability Index (CSI)', *Environmental Impact Assessment Review*, vol. 32, pp. 94–106.

O'Connell, PL 2005, 'Korea's high-tech utopia, where everything is observed', *New York Times*, 5 October, viewed 5 October 2005, https://www.nytimes.com/2005/10/05/technology/techspecial/koreas-hightech-utopia-where-everything-is-observed.html.

OECD 2017, 'Population with tertiary education', *OECD Data*, viewed 30 September 2017, https://data.oecd.org on.

Park, CS, Lee, CH, Choi, SJ & Lee, JW 2011, *Regional green growth in Korea*. Seoul: National Research Council for Economics, Humanities, and Social Sciences (NRCS).

Proctor, D 2017, 'South Korean president details phase-out of coal, nuclear power', *Power*, 1 August, viewed 1 August 2017, http://www.powermag.com/south-korean-president-details-phase-out-of-coal-nuclear-power/.

Property Report 2018, 'Inside the futuristic blueprints for South Korea's first smart cities', viewed 18 July 2018, http://www.property-report.com/detail/-/blogs/inside-the-futuristic-blueprints-for-south-korea-s-first-smart-citi-4.

Republic of Korea 2009, *National strategy for green growth and five-year plan*, Seoul: Presidential Committee on Green Growth.

Shin, DH & Kim, T 2012, 'Enabling the smart city: The progress of U-city in Korea', in *Proceedings of the 6th International Conference on Ubiquitous Information Management and Communication*, pp. 1–7. DOI: 10.1145/2184751.2184872.

Shwayri, ST 2013, 'A model Korean ubiquitous eco-city? The politics of making Songdo', *Journal of Urban Technology*, vol. 20, no.1, pp. 39–55.

Sitra 2017, 'GreentoScale', viewed 8 May 2017, http://www.greentoscale.net/en.

Smart City Hub 2017, 'Smart city as a service: Made in South Korea', Urban Planning and Building Section, 13 April, viewed 8 May 2017, http://smartcityhub.com/urban-planning-and-building/smart-city-as-a-service-korea/.

Whitman, CT, Reid, C, von Klemperer, J & Roy, A 2008, 'New Songdo City: The making of a new green city', paper presented at the *8th World Congress Council on Tall Buildings and Urban Habitat (CTBUH)*, Dubai, UAE.

Index

Charles Darwin's Looking Glass

TRANSATLANTIC STUDIES IN BRITISH AND NORTH AMERICAN CULTURE

Edited by Marek Wilczyński

VOLUME 11

PETER LANG EDITION

Dominika Oramus

Charles Darwin's Looking Glass

The Theory of Evolution and the Life of its Author in
Contemporary British Fiction and Non-Fiction

PETER LANG
EDITION

Bibliographic Information published by the Deutsche Nationalbibliothek
The Deutsche Nationalbibliothek lists this publication in the Deutsche Nationalbibliografie; detailed bibliographic data is available in the internet at http://dnb.d-nb.de.

Library of Congress Cataloging-in-Publication Data
Oramus, Dominika, 1972-
  Charles Darwin's looking glass : the theory of evolution and the life of its author in contemporary British fiction and non-fiction / Dominika Oramus.
    pages cm -- (Transatlantic studies in British and North American culture ; volume 11)
  Includes bibliographical references.
  ISBN 978-3-631-65870-3
1.   English literature--21st century--History and criticism. 2. Evolution (Biology) in literature. 3. Natural selection in literature. 4. Darwin, Charles, 1809-1882--In literature. 5. Literature and science--Great Britain. I. Title.
  PR488.E96O73 2015
  820.9'36--dc23

                          2015015472

The book was written as part of a project financed by the National Science Center, on basis of funds awarded by decision DEC-2011/03/B/HS2/03570.

ISSN 2364-2882
ISBN 978-3-631-65870-3 (Print)
E-ISBN 978-3-653-05206-0 (E-Book)
DOI 10.3726/978-3-653-05206-0

© Peter Lang GmbH
Internationaler Verlag der Wissenschaften
Frankfurt am Main 2015
All rights reserved.
Peter Lang Edition is an Imprint of Peter Lang GmbH.

Peter Lang – Frankfurt am Main · Bern · Bruxelles · New York · Oxford · Warszawa · Wien

This publication has been peer reviewed.

www.peterlang.com

Acknowledgements

Earlier versions of three of the essays included here were presented at the conferences *From Queen Anne to Queen Victoria. Readings in 18*th- *and 19*th*-century British literature and culture*, organized by Professors Emma Harris and Grażyna Bystydzieńska in 2011 and 2013, and at the 2013 conference of the Polish Association for the Study of English organized by Professors Małgorzata Grzegorzewska, Dorota Babilas, and Paweł Rutkowski.

Some of the other essays were published in the periodicals *Anglica*, *Acta Philologica*, and *The New Review* as well as in *We the Neo-Victorians*, a volume of essays edited by Dorota Babilas and Lucyna Krawczyk-Żywko. Some were written originally for this book. The references are included in the bibliography.

I am very much indebted to Philip Earl Steele of Podkowa Lesna who read the manuscript at various stages of its development and saved me from many foolish errors.

My research for all the essays was supported by a grant from the National Science Centre.

I am grateful for all the above support.

Table of Contents

Introduction: Charles Darwin's Looking Glass

This book undertakes to introduce a new and important context of Darwinism-inspired popular science, a context which has been rather neglected by literary studies to date. I believe that my tackling of this issue allows literary scholars to gain new perspective in describing contemporary civilization, which turns out to be the product of post-Darwinian ideology, as in popular understanding Darwinism is now the single most important theory explaining the workings of the universe and humanity's place in it. It is 'the Theory', with a capital T, the epitome of science. Thus Darwin is now the mass-culture icon of the ingenious scientist and the founder of modernity in science, an honor which until quite recently had belonged to Albert Einstein. Consequently, Darwin's life has become a mythic story repeated in his biographies (in the form of both books and films), although the biographical novels and fictive novels on him use historical and biographical detail with varying degrees of fidelity. And indeed, just as with other myths, Darwin's life has features of a canonical story whose every variant must contain certain well-known anecdote-like moments (among them the Alfred Wallace controversy; the journey of the HMS Beagle; the Galapagos discoveries; and doubts on whether to publish a heretical theory).

Darwin's life is everybody's property: writers and filmmakers freely translate it into stories which form a part of contemporary mythology in the meaning defined by Roland Barthes in his seminal *Mythologies*. One of the essays in my book, "The Voyages of Charles Darwin in Recent Fiction and Non-Fiction", attempts to describe the process of 'mythologizing' Darwin as seen in three books written in the last forty years and devoted to the young Darwin's voyage around the world. From Alan Moorehead's *Darwin and the Beagle* (1969) to Irving Stone's *The Origin* (1980) and Roger McDonald's *Mr. Darwin's Shooter* (1999), these works describing the voyage of the Beagle differ as far as their genre goes, but each of the writers adds more and more fictive details to the established facts, thus blending fiction and non-fiction. Analysis of these three books allows me to demonstrate the myth-making mechanism writers employ when they fantasize about Darwin's life. The naturalist's biography is reduced to a number of 'nodes', well-known moments, events or facts, such as his poor health, his quarrels with Robert FitzRoy, and his interest in finches. Such 'nodes' define Darwin as we know him, a figure of the 20th century's collective imagination. Each writer chooses from these nodes and narrates his own semi-imagined story, thereby producing diverse myth-like accounts of ostensibly one and the same 'Darwin', precisely in the way heroes and demi-gods in ancient mythologies feature differently in manifold myths. The name Darwin today denotes both a historical personage and a fictive character, and his biographies and biographical novels are 'faction' – combining fact and fiction.

This observation is further developed in two more essays: the first of which is concerned with Darwin's stay in the colonies – and the second with the contemporary biographies of his wife, Emma Darwin. The essay "History and Simulation in Thorvald Steen's *Don Carlos and Giovanni* and Roger McDonald's *Mr. Darwin's Shooter*" is concerned with presenting colonial history in these two novels. Referring to Jean Baudrillard's notion of history as simulation, as described in his famous *Simulacra and Simulations*, the essay discusses the books by Steen and McDonald in the context of postmodern poetics. These 'Darwinian fantasies' are told by unreliable narrators who refer to numerous classics as well as to other literary *Darwiniana*. The narrators mostly talk about books they read, the ones Darwin reads (and writes), and they presume that we readers have read them all. In reading these narratives we are closed up within a vicious circle of texts corresponding with one another, but having no relation to any extra-textual reality, past or present. History itself is a myth, a laboriously yet vainly re-produced 'faction' about our past. "Depictions of Emma Darwin in Recent British Non-Fiction" offers an analysis of the literary lives of Emma Darwin as myths. Referencing Roland Barthes, Mircea Eliade, and Edward Caudill, this essay looks at Mrs. Darwin's recent biographies from the angle of media studies, popular culture studies, and anthropology. Keeping in mind that 'myth' in the popular understanding denotes a tale which lacks literal truth and yet is a vehicle for a greater truth transcending the factual details, the numerous avatars of Emma Darwin we see in non-fiction written at the turn of the millennium serve to argue diverse ideological points. For some she is an embodiment of nineteenth-century wifely virtues who teaches us what true femininity is; for others she is a disappointed reader of Jane Austen's books whose life fails to resemble fiction; for yet others she is a fundamentalist Unitarian focused on her religion and blind to other people's ideas. Moreover, although the books analyzed in this essay are non-fiction, they make free use of the novelistic stock figures one encounters in Victorian literature: the happy wife, the dutiful mother, the skillful housewife, the shrewdly intelligent girl from the landed gentry who mocks her suitors mercilessly, and the devoted Christian widow. All in all, fact and fiction blur in the biographies of Emma Darwin, just as it is in the case of those devoted to her husband.

One cannot overestimate the impact of Darwin's theories on British literature of the 19th and the 20th centuries, particularly as regards science fiction. *On the Origin of Species* and the polemics the publication of the book provoked, made notions such as evolution, devolution, and anthropogenesis enter the popular imagination and find their way to 'penny-dreadful' novels. The very idea of evolution seemed uncanny at that time: if humanity has evolved from lower animals the line dividing what is human from what is not must be very tenuous indeed. The beast is hidden in each one of us, and can be easily awoken. The half-human hybrids we read about in ancient mythologies are therefore not just fantasies, but may become horridly real. The Gothic novels by Robert Louis Stevenson, Oscar Wilde, Bram Stoker, and most of all H.G. Wells, the author of the many-times filmed *The Island of Dr. Moreau*, fed on such fear and simultaneously prompted the emergence of a new literary genre: science fiction. Its authors speculated on the possibilities of devolution. They stipulated that if we have evolved over the eons, we may also devolve, as well – which, as falling is to climbing, will in fact be all the easier. Thus Dr. Jekyll may one day be horrified to find Mr. Hyde actually hidden within his own self. Dorian Gray may live through a similar trauma seeing his own bestiality exteriorized in his wicked image. By the same token, Wells' narrator – who, marooned on Moreau's island, encounters human-bestial hybrids – first thinks that Moreau's horrid experiment involves the reversal of evolution, and that the scientist, by subjecting people to some cruel vivisection, exposes the pre-human beast we all carry within. However, Moreau in fact is attempting the opposite and trying to humanize the animals. The doubt concerning what the terms 'human' and 'bestial' mean adds to the uncanny appeal of this work.

When in the 1920s Hugo Gernsbeck created the first American pulps devoted to science fiction, he started by re-printing Wells and Verne. The pessimistic late-Victorian fantasies about mad scientists and the bestial nature of people served as models for even more pessimistic tales from the times of the Cold War and nuclear tests involving A and H bombs. Today, over one hundred years after Wells, Darwin's theory continues to have an enormous impact on culture. Since the creation in the 1940s and 1950s of the 'Modern Synthesis', the blend of evolutionism and modern genetics, Neo-Darwinism has been considered the latchkey to all natural history. British intellectuals from Richard Dawkins to David Attenborough stridently claim we are very near to understanding how nature works and contemporary writers feel obliged to comment on this supposition.

The most vivid contemporary attempt at describing devolution is to be found in J.G. Ballard's *The Drowned World*. Similarly to his other early novels

(*The Crystal World* and *The Drought*) *The Drowned World* is a catastrophic novel in which Ballard depicts a dying civilization and a passive, defeated human race. The end of the world as we know it is a good moment to study sundry human reactions to trauma and to describe a noble but resigned protagonist whose aim is to die in the way he is destined to die. Ballard enters intertextual dialogues with Freud, Darwin, and the surrealists, and his reader is expected to decipher and interpret allusions and be brave enough to draw the most pessimistic conclusions. In this book the catastrophe is due to the hyperactivity of the Sun, which has resulted in mutants resembling primordial organisms from archaic epochs. Gradually, as Earth's climate and geography go back to their state from millions of years ago, biological evolution is also reversed. Ferns and reptiles dominate the Earth, mammals cease to multiply, and the remnants of the human race (forced to move to the poles) are witnesses of the end of civilization. The waters of the melted ice-caps flood most of the Earth and the heat is unbearable. The new coast-lines resemble those from the very distant past; the remains of human cities are deluged and looted by all kinds of pirates and savages. According to Ballard, despite our human nature and mammalian anatomy all of us retain on the cellular level memories of previous stages of evolution. We 'remember' our ancestors who evolved into humans. One of the characters postulates our innate propensity for backward movement; he believes that deep in our souls are traces of the passage from the most primitive protozoa to Homo sapiens. Memories from the turn of the Paleozoic and the Triassic era are encoded somewhere in the hind-brain. These long-latent genetic recollections of our ancestors, the first air-breathing amphibians, are now awoken by external stimuli resembling those from millions of years ago.

Thus the theory of evolution still inspires diverse genres of fantasy, ones which this book also attempts to explore. In the essay entitled "Recent Fiction about Charles Darwin: Peter Nichols, Henry Thompson, and John Darnton", I closely read three recent novels: Peter Nichols' *Evolution's Captain*, Henry Thompson's *This Thing of Darkness*, and John Darnton's *The Darwin Conspiracy* in order to describe the relationship between the theory of evolution, fantasy, history, and science. The apocryphal biographies of Darwin's associates and the Neo-Victorian fantasies about the truth behind the official version of Darwin's story prove how prolific his biography still is, breeding, as it does, new stories and prompting subsequent generations of writers to generate their own Darwinian fantasies.

Similarly, my essay "References to the Theory of Evolution in the Novels of John Fowles, A.S. Byatt, and Hilary Mantel" is concerned with Darwinism and literature. Challenging the cliché that Darwinism is atheism, the essay juxtaposes three aspects of Fowles' *The French Lieutenant's Woman*, Byatt's *Morpho Eugenia*, and Mantel's *A Change of Climate* – namely: the way nineteenth-century naturalists are depicted in the contemporary novel; what do these writers understand by 'science'?; and is there really an inescapable conflict between the theory of evolution and Christianity? Are we dealing with a simple replacement of God with natural history? Why are the mid-nineteenth-century Darwinian naturalists still stock figures in Neo-Victorian novels? Why are the intellectual debates provoked by their discoveries still a very important subject for British novelists today?

My next article "Echoes of the Mid-19th-century spiritual crisis in selected contemporary texts referencing Charles Darwin", which is based on the critical writings of Michael Ruse, Asa Gray, and Edward Caudill, discusses three contemporary books of fiction and semi-fiction – namely, by Randal Keynes, Jenny Diski, and Graham Swift. They all are about the search for Victorian forebears by twentieth-century narrators who have just experienced some trauma. Their acute feelings of nostalgia and suicidal depression are linked to the theme of the loss of Christian faith their Victorian great-grandfathers suffered after having read (or, in one case, having written) *On the Origin of Species*. Each of the three books recreates the past by reading the retrieved nineteenth-century documents that stimulate the narrators' imagination, making them spin their own gloomy yarns and indulge in self-destructive fantasies – thus demonstrating that the theme of 'Darwinism and the Victorians' truly does imply crisis in the spiritual history of Britain.

The idea of "The Theory of Evolution and the Life of its Author", from the title of my book, breeds stories of both fiction and faction. One further essay, "Darwin's Problem with Human Ancestry as Reflected in Recent Fiction", deals with how popular culture reworks the implications of Darwinism, which at first glance are politically incorrect. The issue in question is that humans, especially human babies and savages, display features and behaviors also to be seen in young primates. Describing non-Europeans as ape-like and infants as little monkeys is provocative in times of racial equality and the pop-cultural sentimentalization of babies. My reading of William Irvine's *Apes, Angels, and the Victorians*, Randal Keynes' *Annie's Box*, and its film adaptation by Jon Amiel, *Creation*, lets me demonstrate how Darwinian controversies are 'tamed' by mass culture. Popular texts reinforce the vision of both Darwin the loving father and Darwin the proto-conservationist, thereby rendering his orangutans-babies-savages chain of associations harmless. Darwin the genius is pictured as a reluctant rebel against the Biblical paradigm of Creation who struggles with the implication of his own discoveries.

As the full title of this book, *Charles Darwin's Looking Glass. The Theory of Evolution and the Life of its Author in Contemporary British Fiction and Non-Fiction* suggests, the essays in this volume deal with a number of subjects: Darwinian fictions; Darwinian non-fictions; the theory of evolution as reflected in both of them; and Darwin's life as reflected in both of them. I trust I have suitably addressed the above four issues in my essays, but one last aspect of my project still needs to be explained – namely, the titular 'looking glass'. Why and how does today's culture gaze upon the myth of Darwin, his theory, and his life in order to find its own reflection? What image does it find there? – what kind of narcissistic pleasure does it get? – are our times the era of Charles Darwin? – if so, then why?

The essays in this book were written in the strong belief that comparative analysis of diverse Darwinism-inspired discourses (post-modern novels, science fiction, nature films) can enrich literary studies. Such an analysis introduces new contexts to the standard ways of reading contemporary literature and, thanks to the interdisciplinary approach, texts written by scholars and journalists specializing in natural studies are discussed alongside works of fiction. The interdisciplinary approach allows me to demonstrate how deeply the diverse spheres of today's culture influence one another: Darwinian scholars use epic conventions to make their popular science papers interesting, and novelists who have read popular books by Dawkins or Wilson make their fictional characters behave in accordance with Darwinian theories. Moreover, the essays here show how the very notion of 'science' changed in meaning throughout the 20th century, and prove that for both novelists and filmmakers Darwin is now the mass-culture icon of the ingenious scientist and the founder of modernity in science. Thus, Darwinism in this respect has replaced the theories of relativity and quantum physics that were "fashionable" among literary scholars in the 20th century.

Darwinian paradigms (entities evolve in time via natural selection, survival of the fittest, and the spread of successful adaptations) are referenced in literature, popular science, show business, and education (e.g., the Discovery Channel's nature films) – which fact points to the homogeneous character of global culture. By comparing diverse uses of evolutionary discourse in current literature and films, my essays demonstrate how natural science influences the contemporary humanities and, conversely, how literary conventions are used in order to make scientific and popular science texts intelligible and attractive. *Charles Darwin's Looking Glass. The Theory of Evolution and the Life of its Author in Contemporary British Fiction and Non-Fiction* attempts to discover the common denominator of generically diverse Darwinism-inspired discourses and, additionally, to show how deeply fashionable scientific theses have infiltrated postmodern literature and popular culture. In the essay "Darwinism and the Humanities" I discuss the mid-1990s crisis of literary and social studies and demonstrate that by references to evolutionary biology scholars are re-introducing to their texts the human universals which in the heyday of postmodernism were deemed false and a product of a hegemonic ideology. With concepts such as a universal, culture-independent human nature re-entering the humanist debate by the Darwinian back-door, new and interesting approaches emerge. I prove this point by referencing Edward O. Wilson and his *Sociobiology*. Wilson's notion of 'consilience', the intellectual bridge between the sciences and the humanities, helps to understand recent critical attitudes in the fields of anthropology and literary and film studies, as pursued by Neo-Darwinism inclined scholars.

The essay "The Motif of Human Evolution in Selected Fiction and Non-Fiction" is an attempt to use the Darwinian vantage point in the analyses of novels and popular science films. I discuss on the same plane fictive accounts of early human societies by H.G. Wells and William Golding, and the educational TV films by Jacques Malaterre. My references include Charles Darwin, Geoffrey Miller, and Roger Lewin. The last essay in this volume, "Annie Dillard and Kurt Vonnegut on the Galapagos Archipelago as the Archetypal Darwinian Setting", compares the way Dillard and Vonnegut (who are both well-read in Darwinian criticism) use references to the naturalist's works and Darwinian paradigms in science to prove their very diverse points. Their Galapagos-set narratives deal with human nature, religion, the creation of the universe, the future of the human race, along with other grand issues. The ideological standpoints of both writers are very different and yet both find vehicles in the theory of evolution for their metaphors, which proves that ours truly is a Darwinian culture in which Darwinian paradigms serve all purposes. The life and oeuvre of Charles Darwin, as all my essays maintain, are a looking glass in which we enjoy gazing at the image we see: the image of ourselves as creatures who have evolved pleases us and gives comfort.

The Voyages of Charles Darwin in Recent Fiction and Non-Fiction

In the popular space comedy *A User's Guide to the Galaxy* the protagonist dressed up as Dr. Livingstone meets his future girlfriend at a fancy-dress party where everybody wears Victorian costumes. With a long white beard and a toy dog in her hands the girl is supposed to look like Charles Darwin, but much to her dismay people take her for Santa Claus. "I thought the beagle was a giveaway," she complains, pointing at her toy dog.

This movie, a standard product of contemporary mass culture, makes use of easily recognizable iconic associations, thus proving that in popular imagination the name 'Darwin' brings about the likeness of a stern looking bearded man and a beagle, the latter denoting of course the name of the ship on which Darwin traveled around the world and got his grand idea for the theory of evolution.

This chapter attempts to examine how the story of the voyage Darwin undertook in the 1830s entered popular culture about 150 years later, thereby becoming for late 20th-century readers and film-goers one of the icons of Victorian times. Numerous re-writings of this story have reinforced the following sets of associations: 'Darwin and the Beagle'; 'Darwin and the Galapagos'; 'Darwin and his finches' along with many others. In order to show how this happened I am going to analyze the motif of the Beagle voyage as described in three books published in the second half of the 20th century: Alan Moorehead's *Darwin and the Beagle*, which is non-fiction; Irving Stone's *The Origin. A Biographical Novel of Charles Darwin*, which tells the real story of Darwin's life, yet freely adds lots of fictive material to fill the gaps; and Roger McDonald's *Mr. Darwin's Shooter*, which is clearly a novel, though one that is based on a number of factual details. Thus, non-fiction has gradually been replaced with fiction, and the facts about the Beagle voyage have become generally known and sufficiently recognizable to serve novelists as their raw material. Therefore, the authentic story of the Beagle generates other stories which no longer belong to historical discourse, but rather to literature.

In 1969, Alan Moorehead, an Australian-born British journalist with "The Daily Express", and a former war correspondent during the Spanish Civil War and the Second World War in the Middle and Far East, published *Darwin and the Beagle*, a comparatively brief and beautifully illustrated account of Darwin's journeys. The Beagle voyage is presented here as a formative experience, and November 5th, 1831 – the day Darwin came to an interview with Captain FitzRoy and as a result was offered the post of the Beagle's naturalist – is held up as one of those pivotal events of which history is made.

Moorehead's style is ornamental, his sentences often longish, and the focus is on individuals and their lives, as in the following sentence introducing Captain FitzRoy to the reader:

> There was no room for compromise in his nature, no slack to be taken up and let go again; no real patience, and so he vacillated between moods of depression and elation, and by the time his interview with Darwin took place he was already giving way to those manic-depressive tendencies that were to end in his suicide thirty-four years later.[1]

Such a manner of writing renders history colorful: FitzRoy seems vivid and somewhat tragic and his maniacal sense of purpose (he insists that Darwin should look around the world for evidence of the Biblical Flood) is described as symptomatic of his mental problems. Darwin, according to Moorehead, is a young enthusiast dazzled by the vastness and diversity of nature: his visits to Brazil, Tierra del Fuego, the Pampas, Patagonia, along with his experiences of earthquakes, volcanic eruptions, and most of all tropical forests make him ecstatic. Once he realizes that life on Earth has existed for billions and not thousands of years his eyes open. Digging up fossilized bones of a prehistoric megaterium he swears to devote himself to natural science. In one of his letters he confesses that nothing compares to: "finding a fine group of fossil bones which tell their story of former times with a living tongue…." [2]

Yet it is only on the Galapagos Islands that the 'Eureka!' moment occurred, something comparable only to the inspiration Sir Isaac Newton is said to have experienced in the orchard when the apple fell. On these black, bizarre-looking volcanic islands Darwin famously noticed that species evolve if only transformation increases their chances of survival. In Moorehead's version it is there that Darwin examines finch beaks, realizing that these birds have evolved in directions that allow them to eat the various foods on the various islands. The 'Eureka!' moment happens on a sea journey: "just like a man might have a sudden inspiration while he is traveling in a car or a train"[3]. The word 'inspiration' is crucial here, as the entire story Moorehead tells leads to this moment and all that follows is a result of it. The chapter after the Galapagos section is entitled "Homeward bound" and recalls Darwin's return, the publication of his

---

1   Alan Moorehead, *Darwin and the Beagle* (London Hamish: Hamilton, 1969), 11.

2   Moorehead, *Darwin and the Beagle*, 86.

3   Moorehead, *Darwin and the Beagle*, 151.

bestselling journal, and the exploration of his ideas about evolution resulting in the publication of *On the Origin of Species* many years later.

Moorehead's book has a moral to it: progress results from the confrontation of clashing ideas. Two men in their early twenties, FitzRoy and Darwin go on a five-year journey around the world: the former is a devoted Christian and a firm believer in the Biblical Creation, the latter is a genius and a keen observer of nature. They continuously debate the beginnings of life on Earth aboard their ship, and these quarrels help the genius to formulate his grand theory of evolution. Thus the Beagle voyage brings about intellectual and scientific advancement for our rational civilization.

Irving Stone published his book on Darwin over a decade later. He was already then a well-known biographer, famous for admitting imaginative detail to his historical accounts. He also was careful in choosing the protagonists of his biographies: Vincent Van Gogh (in *Lust for Life*), Michelangelo (in *The Agony and the Ecstasy*), and Sigmund Freud (in *The Passion of the Mind*) are vivid individuals whose lives have provoked much debate. Stone's choice of Darwin suggests that by the late 1970s Darwin was already considered 'interesting' by the reading public.

The first part of Stone's book, entitled "Landsman", is entirely devoted to the Beagle voyage which made Darwin's name as a scientist and a man of letters, as if the whole previous twenty-two years of Darwin's life were insignificant. The rare retrospections usually concern the books Darwin had devoured as a small boy: memoirs of Alexander von Humboldt and Captain James Cook, Admiral Beechey's *Narrative of a Voyage to the Pacific and Bering's Strait*, William J. Burchell's *Travels in the Interior of Southern Africa*, and the children's book *Wonders of the World*, which was Darwin's favorite and which contained marvelous descriptions of exotic life in the colonies. He also read the diaries of Dr. Henry Holland, who was the Darwin family's friend and the author of *Travels in the Ionian Isles, Albania, Thessaly, Macedonia etc.- during the years 1812 and 1813*, and for his long journey on board the Beagle Darwin took with him Philip Parker King's *Narrative of a Survey of the Intertropical and Western Coast of Australia*.

The HMS Beagle was, according to Stone, a place where Darwin primarily read a lot and wrote even more: among his writings were numerous letters and sheets of notes later sent to his family and Cambridge teachers and hundreds of pages in his diaries. He copied passages of the journals and posted them to England where his sisters distributed the copies all over the country among friends, scholars, and intellectuals. Darwin's letters were also copied and posted to scientific societies and university people. While their author was still aboard the

Beagle the story of his journey and its scientific fruits was already famous. One of his sisters wrote back to Darwin on the attractiveness of his diary: "what a nice amusing book of travels it would make if printed"[4]. Irving Stone makes his protagonist comment on this authentic letter in a long passage of revelation written in free indirect discourse:

> He had never dreamed of publishing his diary (…) Over the years he had read many diaries of world travelers: it had never occurred to him that he might have even the faintest chance of adding to that lore. He found it a heady idea.[5]

Darwin grows into becoming a writer at the same time he grows ready to formulate his grand theory. In the 'Eureka!' moment on the Galapagos he realizes, in Stone's version, that animals of apparently the same species vary from island to island, and asks himself in an internal monologue the seminal question: "What causes these differences? Ay, there's the rub"[6], thus initiating the mental process that ultimately leads to generalizing his observations concerning nature's ways in his masterpiece published thirty years later.

At Galapagos, where the idea occurs, Darwin is very far from conceiving *On the Origin of Species*, but prompted by admiration of his friends and family he seriously thinks about editing his Beagle diaries for publication. Having read some of the text Captain FitzRoy exclaims: "Darwin, these two hundred pages are jolly well done!"[7]. He then asks Darwin to join with him in publishing their two accounts of the same voyage in one book of travels, and Darwin agrees. On entering the London harbour Darwin is already a writer with his first book commissioned. Therefore, in Stone's biography the immediate result of the five years on the Beagle is a book about that experience; in later years Darwin's part of the book (much better written than FitzRoy's account) was reprinted separately very many times under the title *The Voyage of the Beagle: Journal of Researches into the Natural History and Geology of the Countries Visited During the Voyage of H.M.S. Beagle Round the World*, becoming a huge bestseller and the origin of the myth of Darwin's inspiration for his theory.

In 1998 the Australian novelist Roger McDonald published *Mr. Darwin's Shooter*, a fictive story of the lives of people mentioned but in passing in Darwin's papers: sailors from the HMS Beagle and most importantly Syms Covington, the

---

4   Irving Stone, *The Origin. A Biographical Novel of Charles Darwin* (London: Corgi Books, 1982), 310.
5   Stone, *The Origin*, 310.
6   Stone, *The Origin*, 343.
7   Stone, *The Origin*, 359.

eponymous servant Darwin recruited from among the ship's staff. Darwin taught him to skin birds and prepare specimens for naturalists to examine in their laboratories back in England. The action of the novel takes place in Australia within a colonial society made up mostly of emigrants from England and the United States – and in the early 1860s, when Darwin finally publishes his *On the Origin of Species*. The colonials are all to some extant Darwinists and Australia as a whole is a Darwinian country, where all of nature boldly announces that the theory of evolution is true. On this continent: "there must have been a separate act of creation, it was maintained, and as Darwin had said on visiting there, to bring them into being. Swans were black. A mammal, the platypus, laid eggs."[8]

In the 1860s Darwin's first best selling book, *The Voyage of the Beagle: Journal of Researches into the Natural History and Geology of the Countries Visited During the Voyage of H.M.S. Beagle Round the World*, is read by everybody. Syms Covington, Darwin's ex-servant boasts his own copy and so does the novel's other protagonist, a young American doctor who had emigrated to Australia mostly because of his juvenile infatuation with Darwin. As a boy he had read *The Voyage of the Beagle...* along with novels by Herman Melville, and in his early twenties, being disappointed with America, he abandoned his Bostonian home and "shipped down the coast of South America as Darwin had, tasting Patagonian gales in his teeth and stopped at the Galapagos, and sailed the Pacific making his shell collection on the way"[9].

In a bookshop in Sidney everybody waits for *On the Origin of Species* to be shipped from London and wonders about its sure financial success: "If it was like Darwin's *Beagle's Voyage* there would be unaccountable numbers of copies, and in all the world's languages and libraries too"[10]. Darwinism is therefore presented as a cultural phenomenon connected to books and as a kind of fashion springing from reading about nature. Both the novel's protagonists are shaped by Darwin: the young doctor by reading Darwin's narratives, Syms Covington by accompanying him aboard the Beagle and copying passages of his writings to be sent to England. Covington is a historical figure, "the unacknowledged shadow behind [Darwin's] every triumph"[11], who most probably traveled with the young naturalist, killed rare animals for him, skinned them, preserved them and even edited Darwin's notes back in the 1830s. The servant must have been a born naturalist

8   Roger McDonald, *Mr. Darwin's Shooter* (Milsons Point and London: Anchor 1998), 74.
9   McDonald, *Mr. Darwin's Shooter*, 326.
10   McDonald, *Mr. Darwin's Shooter*, 389.
11   Janet Browne qtd in McDonald, *Mr. Darwin's Shooter*, 414.

himself, a dedicated and observant observer of wildlife. Yet the protagonist of *Mr. Darwin's Shooter* is very much McDonald's invention as factual data on Covington is scarce. Darwin mentions him several times in his letters to his sisters: "an odd sort of person (...) from this oddity very well adapted to my purposes."[12]. All the same, some of Darwin's diaries are in his handwriting.

Mr. Darwin's Shooter fills the gaps in the real Covington's biography with four hundred pages of fiction: Covington is brought up in England. A butcher boy and a devoted Christian he goes to the sea, travels under Captain FitzRoy on the HMS Adventure and HMS Beagle, waits on Darwin for over seven years, emigrates to Australia, gets rich and for years collects rare specimen of natural life, and sends them at his own expense to Darwin. On the news of Darwin's new book, the blasphemous *On the Origin of Species*, Covington suspects he had unwillingly helped to write a work which undermines the Holy Bible. He shares his frustration with his friend, the young doctor whom he tells his own story of the voyage of his youth. For him it was very much a divine expedition. With the orthodox Christian Captain and the initial agenda of finding evidence of the Biblical Flood they felt like God's crusaders:

> The Great Flood of the Bible was (...) a favorite story. The idea of a ship with animals aboard was a picture of a cozy creation. The ark in everyone's mind wasn't much bigger than their Beagle. It had (...) the same kind of Capt, precise in his rule and close to God.[13]

And yet even in the old Covington's perspective of thirty years the story of the Beagle voyage commences the anti-creation myth of evolution, a discourse which belies the Biblical version. The ship was in fact full of dead animals prepared for anatomizing. Just like in Noah's Ark they came in twos (Darwin asked Covington to always catch both the female and the male of each species), but they were definitely not going to breed and fill the Earth with life. The outcome of the journey is a new secular myth of rational society, or, as McDonald seems to be saying, two myths: the myth of evolution which explains nature, and the myth of Darwin on the Beagle experiencing the 'Eureka!' moment at the Galapagos.

Moorehead, Stone, and McDonald describe the voyage around the world with varying degree of historical exactness, yet in all three narratives it is not the journey, but the textual account of it that really matters. All three books show Darwin and his fellow naturalists as readers and writers. They are brought up on 17th- and 18th-century travel literature and then in turn they write down and publish

---

12 Charles Darwin qtd in McDonald, *Mr. Darwin's Shooter*, 413.
13 McDonald, *Mr. Darwin's Shooter*, 287.

accounts of their own adventures. In the case of Darwin the financial success of *The Voyage of the Beagle…* made him famous all over the world and prompted the sales of *On the Origin of Species*. Moreover, these two books correspond with each other and are interdependent in popular imagination: *The Voyage* describing the birth of the idea for *The Origin* and *The Origin* explaining and giving a theoretical frame to observations from *The Voyage*. As the theory of evolution serves as a secular equivalent of the Creation and the whole story of Darwin is mythologized in mass culture we may talk about the voyage of the Beagle as about the beginning of the beginning, the founding story of scientific progress, and the origin of *On the Origin of Species*. The juxtaposition of the three narratives depicting this seminal cruise shows no distinct border between fact and fiction, which makes them all a kind of narrative which Hayden White famously calls "historiographic metafiction".

History and Simulation in Thorvald Steen's Don Carlos and Giovanni and Roger McDonald's Mr. Darwin's Shooter

In his seminal *Simulacra and Simulation* Jean Baudrillard argues that late twentieth-century cinema is obsessed with history, which it tries to simulate[14] in response to the prevailing feeling of the lack of historical continuity, and of any reliable historical narrative bonding us to the past and explaining our world as resulting from past events.

> The great event of this period, the great trauma, is the decline of strong referentials, these death pangs of the real and of the rational that open onto an age of simulation [...] today one has the impression that history has retreated leaving behind it an indifferent nebula, traversed by currents, but emptied of references. It is into this void that the phantasms of past history recede, the panoply of events, ideologies, retrofashions – no longer so much because people believe in them or still place some hope in them, but simply to resurrect the period when at least there was history.[15]

"The phantasms of past history" Baudrillard describes are history-inspired scenarios which, without attempting to be accurate, play with the notions of the past, reenacting events and situations we know from history books and turning them into spectacle. History is not shown but simulated in such a way as to emphasize its fictional character. The spectators are continuously made aware that they are watching an impression of what might have happened in the past, with the past itself being inaccessible.

The wealth of contemporary novels which discuss history in a metafictional manner seemingly attempt to be historical novels just to show it is no longer possible. The way they playfully adopt a voice from the past allows the claim to be made that what Baudrillard wrote about the cinema thirty years ago now applies to literature, as well. Julian Barnes, Graham Swift, Peter Carey, A.S. Byatt, and John Fowles, to mention just the best known writers, obsessively come back to past epochs and past narrative voices, while their unreliable narrators deprive the reader of maintaining any illusion of authenticity. The nineteenth century is

---

14 By "simulation" Baudrillard famously means: "the generation by models of a real without origin or reality: a hyperreal." Jean Baudrillard, *Simulacra and Simulation*, trans. Sheila Faria Glaser (Ann Arbor: University of Michigan Press, 1994), 1.

15 Baudrillard, *Simulacra and Simulation*, 43–44.

recently the favorite period that novelists recall. Indeed, as Mark Llewellyn states in his article "What Is Neo-Victorian Studies?" such a literary fashion results in "the often perilously close to kitsch or clichéd engagements with the Victorian period"[16]. He adds, however:

> [I]t might also be argued that this is a fact of our contemporary culture; that in book-stores and TV guides all around us what we see is the 'nostalgic tug' that the (quasi-)Victorian exerts on the mainstream identification of our own time as a period in search of its past.[17]

Thus the propensity for the quasi-historical in general and the quasi-Victorian in particular seems to be deeply rooted in mass culture as well, much in line with Baudrillard's argument. Attempting to account for the historical fad in the English novel A.S. Byatt in the essay "The New Body of Writing: Darwin and Recent British fiction" makes an interesting point: very many neo-Victorian novels allude to the founding of natural history and the Darwinian fashion of the Victorian upper-middle class. Gentlemen-naturalists, fossil collectors, keen readers of mid-nineteenth-century essays on natural history by Lyell, Bates or Huxley are neo-Victorian stock figures. Certain neo-Victorian literary motifs or fictional situations are also recurrent.[18]

Byatt argues that the Darwinian references can be explained by the fact that nowadays, living in an agnostic culture, we look to natural history and sociobiology for our morality, adapting Edward Wilson's and Richard Dawkins' theses concerning humanity's in-born altruism and evolutionally-formed aggressiveness. This essay, which is an analysis of two novels featuring a young Charles Darwin – Thorvald Steen's *Don Carlos and Giovanni* and Roger Mc-Donald's *Mr. Darwin's Shooter* – aims at demonstrating a more Baudrillardian point. The novels in question simulate the history of Darwin's voyage round the world, reenact the well-known story, and at the same time continuously and systematically emphasize their own fictiveness, which is done by undermining the reliability of their narrators, intertextual referencing of the religious debate provoked by the theory of evolution, depicting overseas countries as Darwinian

---

16 Mark Llewellyn, 'What Is Neo-Victorian Studies?', *Neo-Victorian Studies* 1.1 <http://www.neovictorianstudies.com/past_issues/Autumn2008/NVS%201-1%20M-Llewellyn> accessed 12 November 2011, 168.

17 Llewellyn, 'What Is Neo-Victorian Studies?', 168.

18 Byatt's example is the discovery of a beautiful fossil on the Lyme Regis beach which is described in Fowles' *The French Lieutenant's Woman* and, 25 years later, in Hilary Mantel's *A Change of Climate*.

neverlands, and concentrating on an alleged pivotal point of history – its fable-like node: when Darwin conceived his theory.

"History is our lost referential, that is to say our myth. It is by virtue of this fact that it takes the place of myths"[19], claims Baudrillard. McDonald and Steen writing their novels in, respectively, the early 1990s and 2004 elaborate on one of the 'mythic' moments in all Darwin's biographies: his youthful voyage around the world and his adventures in the colonies. The voyage of Charles Darwin on the HMS Beagle in the early 1830s is considered a turning point in the development of the theory of evolution as it was during this journey that Darwin conceived his most important ideas: natural selection and the survival of the fittest. The fossils he and his men dug up in South America together with the specimens of contemporaneous fauna and flora he collected on Pacific islands prompted him in later years to write *On the Origin of Species*.

Thanks to bestselling popular science books by Wilson, Jared Diamond, Matt Ridley, Konrad Lorenz, and Richard Dawkins, in the second half of the 20th century Darwinism infiltrated contemporary culture via quasi-biographical books on him. Alan Moorehead's *Darwin and the Beagle* published in 1969 is a good example of such non-fiction. The book narrates Darwin's famous voyage in a light and entertaining manner, yet it follows historical data known from letters and diaries and is bibliographically very accurate. In the early 1980s Irving Stone's *The Origin. A Biographical Novel of Charles Darwin* was published, marking a new fashion in literary Darwiniana; it is both a fictional and non-fictional telling of the real story of Darwin's life that freely adds lots of fictive material to fill the gaps. It is significant that Stone's book begins not until Darwin is already mature and is preparing for the HMS Beagle voyage which made his name as a scientist and a man of letters. Not only does it ignore the whole previous twenty-two years of Darwin's life, but also gives much prominence to the formative experience of the naturalist's stay in the colonies: all that happens there is described as significant to the nascent theory.

Stone is the first to overtly mythologize Darwin's voyages, McDonald and Steen apparently follow his lead, but writing from within post-modern culture they are aware they do not depict history but simulate it by reenacting the story which their readers probably already know from numerous biographies, for instance: Jon Amiel's feature film *Creation, Darwin and the Tree of Life*, the BBC series by David Attenborough, or popular accounts of how modern science was born by writers such as Bill Bryson. Thus they narrate the story of the Beagle

---

19 Baudrillard, *Simulacra and Simulation*, 43.

voyage from carefully chosen points of view: their narrators left Europe more or less voluntarily in the early nineteenth century and made their permanent homes in countries overseas. Their memories of having met or traveled with Darwin prove to be the most important experience of their youth, as it shaped their worldview and propelled them towards a religious crisis. However, in their later years when they recall Darwin's story they both have a grudge against him and yet are still deeply fascinated by him, which seriously hinders their reliability.

In both novels the simulation of history is rather peculiar, as McDonald and Steen choose one of the best accounted episodes in the history of science. As McDonald points out in his essay about the making of his novel: "Darwin's archive is an immense resource: he remains the most thoroughly documented scientific genius of the nineteenth century"[20]. Thus, there is much material available: nearly every episode of the voyage is accounted for in the letters and diary entries of Darwin and his shipmates. Moreover, later on those materials were summarized by biographers and then used as a building block by the historians of science constructing their grand narrative, to borrow Lyotard's famous term, of how evolutional thought was born.

By using obscure and 'low-born' narrators and choosing the period from just before Darwin formulated his own narrative of progress (the grand story of how nature evolved from baser forms to more advanced species), McDonald and Steen counterbalance the canonical story with voices undermining its legitimacy. McDonald's half-real and half-fictional protagonist, Syms Covington is a butcher boy in a provincial English town brought up in an orthodox Puritan family, who at eleven becomes a protégé of a travelling preacher and at fifteen, together with a group of other God-fearing young acolytes, joins the British navy. His second ship is the HMS Beagle, a vessel famous for its Christian Captain, Robert Fitz Roy, who is devoted to studying the Bible and finding evidence of its exact accurateness all over the world. Darwin is one of his passengers and, once he realizes that Covington is very skillful at capturing all sorts of creatures and skinning them, he hires the young cadet.

The boy stays in his service during the HMS Beagle's long voyage and then for an additional two years spent on classifying and describing the specimens gathered during the journey. Only when Darwin gets married does Covington leave his position: he then emigrates to Australia where he becomes a successful

---

20   Roger McDonald, 'Evolution of a Novel: *Mr. Darwin's Shooter*', *Australian Humanities Review* 12 <http://www.australianhumanitiesreview.org/archive/Issue-December-1998/mcdonald.html> accessed 12 November 2011, 3.

businessman and the father of a very large family. For the remainder of his life he exchanges letters with Darwin every few years and at his own expanse he sends his former master specimens of marine life he catches in Australia. In the late 1850s, having heard that the texts Darwin publishes promote agnosticism and even prove the Bible wrong, Covington feels deeply wounded and betrayed. In such a difficult moment he tells the story of his travels as a young man to the young Boston-born doctor who is his neighbor on the Australian coast. McDonald comments on his sources:

> Covington's archive by comparison with Darwin's is tiny. It consists of a contested birth-date, a scrappy diary held in the Mitchell Library, Sydney, a few watercolours, a photo-graph, and scattered mentions in Darwin's letters and diaries [...] Midway through the voyage of the Beagle Darwin wrote to his sister back in England:

> "Tell my father how much obliged I am for the affectionate way he speaks about my having a servant. It has made a great difference in my comfort [...] My servant is an odd sort of person" Darwin continued, "I do not like him much; but he is, perhaps from his very oddity, very well adopted to all my purposes."[21]

The very existence of Covington, whom McDonald calls in the afterword to the novel an unacknowledged shadow behind Darwin's success, renders the canonical version of the HMS Beagle myth doubtful. The real-life Covington, as one can gather from less often published files in Darwin's archive, learned collecting, preserving, and shooting. He was much better than Darwin in slitting open bird's stomachs, digging up bones of prehistoric creatures, carding, and sorting. He was also Darwin's valet, secretary, copyist, and house-keeper. Moreover, flying in the teeth of the standard story of the young naturalist's epiphany on the Galapagos, it was Covington who noticed that finches of historically the same species had developed differently on different islands. It was not Darwin's practice to catalogue the collected specimens by island, "but the real Covington had labeled by islands the birds he had shot for his private and potentially saleable collection. When they were back in London Darwin called for these birds to be examined"[22].

In giving the fictive Covington voice McDonald does not try to "unearth" the real and buried story of how the theory of evolution was really formulated, but rather simulates one of the possible scenarios of these events. He suggests that the eventual scientific effect of the Beagle voyage, the text of *On the Origin of Species*, can be only hazily related to its causes: different things happen during

---

21 McDonald, 'Evolution', 4.
22 McDonald, 'Evolution', 5.

the HMS Beagle years, but the cause-and-effect chain is neither simple nor knowable[23]. Equally unknowable are the answers to the questions at the very crux of McDonald's novel:

> Had Darwin on their voyage found proof of natural selection as a theory able to explain life on earth as completely as creationism? More importantly, had Covington himself handed the proof over to Darwin – willingly and blindly? Had he thus committed, as he puts it to himself, a crime against God and his own good nature?[24]

Steen's *Don Carlos and Giovanni* is narrated by a fictitious personage, Giovanni Graciani, formerly of Genoa, Italy and now of Buenos Aires, Argentina. He is a casual laborer in the shipyard and an émigré forsaken by his family. In the early 1830s, during a time of political upheaval in Argentina, Giovanni accidentally meets a young rich Englishman, Darwin (known in Buenos Aires as Don Carlos) and takes some money from him. In return he helps in capturing and transporting to the harbor carcasses of a few South American animals.[25] The two young men talk a lot and Giovanni, who is very inquisitive and scientifically-minded, grasps Darwin's points very quickly and is able to comprehend the basics of the now-emerging theory of his interlocutor, which in turn makes him question the Christian dogmas he was brought up to believe.

They say goodbye to each other and the reader sees him again not until three years later, when Darwin is back in Britain. Giovanni is by then an atheist and a seriously-ill political outcast wanted by the Argentinean secret police. He receives a letter from Darwin who is in London, immersed in his scientific projects

---

23 In *The Postmodern Condition* Lyotard describes such hazy relations of causes and effects using the term paralogism.

24 McDonald, 'Evolution', 6.

25 Darwin's biographers often depict his stay in South America in terms of a romantic adventure full of impromptu expeditions, similar to these described by Giovanni, camping in the pampas and getting to know tough local men. Compare Irving Stone: "Charles stretched out on the ground, nestled his head on his saddle, covered himself with his saddle blanket. He was happy and comfortable [...] There was a death-like stillness on the plain, the gauchos' dogs were keeping watch. He thought, 'This scene will live in my mind a strongly marked picture which I will not soon forget." (Stone, *Origin*, 268–269). And Moorehead: "he never seems to be tired, never loses his curiosity or his sense of wonder. Finally after forty days in the wilderness we find him riding into Buenos Aires through orchards of quinces and peaches, and with his beard, his wide hat, his worn clothes and his sunburnt face he must have looked like some cowboy, or perhaps a gold prospector coming into town after a hard spell on the trail. He was leathery and as horny as gauchos themselves." (Moorehead, *Darwin and the Beagle* 101).

and working busily to formulate his grand theory. Reading the letter Giovanni thinks about his excursions with Don Carlos and ponders the philosophical implications of the theory of evolution. Finally, feeling the approach of death, he decides to describe in one more letter both the story of his acquaintance with Darwin and the ideas that sprang to his own mind from having talked to Darwin. This letter, addressed to his brother who had renounced him some years earlier, is the very text of the novel we read and at the same time the new extra-canonical version of the Darwinian 'myth' of the HMS Beagle and the formulation of the theory of evolution. It is a version surprisingly modern and full of anachronistic insights concerning the impact of evolutionism on the newly emergent world order with independent post-colonial countries such as Argentina and the European fatherland. And yet we are made to believe that the brother will never get the letter, and even if he does get it he will destroy it like all the previous ones from his outcast brother. The narrative situation is thus bizarre: the truth, (or rather a truth) about young Darwin will never make its way into the official histories, though we ourselves somehow read it. The novel paradoxically narrates its own destruction. The implied suggestion is that whatever really and objectively did happen is unknowable, the historical discourse as we know it is dependent on certain events narrated in discourses never classified as history and never preserved. Thus no outside party can know why what happened did happen, and we are left with gaps filled up by historians according to agendas of their own.

Both Covington and Giovanni are unreliable narrators and their stories suggest that our official knowledge about history is far from exact. Yet they do not provide us with any revelation, as their own accounts are erroneous and hazy. They are but simulations of history, and this is all we can hope for as, to quote Linda Hutcheon's essay on recent quasi-historical novels entitled "Historiographic Metafiction. Parody and the Intertextuality of History": "After all, we can only 'know' (as opposed to 'experience') the world through our narratives (past and present) of it, or so postmodernism argues. The present, as well as the past, is always already irremediably textualized for us"[26].

Moreover, it is significant that both narrators come from social strata deprived of the power to write down official histories: the canonical story of how modern science was born thanks to one of its paragons, Charles Darwin, is told by people from outside the milieu of early Victorian gentlemen-naturalists. They lack proper education (despite very quick minds and, in later periods of their life,

---

26 Linda Hutcheon, 1988, 'Historiographic Metafiction. Parody and the Intertextuality of History' <http://hdl.handle.net/1807/10252> accessed 12 November 2011, 8.

having read much natural history), their attitude to religion puts them outside the mainstream, and in the moment they spin their yarns they are at the world's ends in, respectively, the former penal colony of Australia and the politically unstable newly-born Argentinean regime of general Rosas.

Both Covington and Giovanni were born in provincial European towns and both died in the colonies as citizens of newly-emerging centers of civilization. Their attitude toward Europe in general and Darwin in particular is ambiguous. Covington considers his years of service to Darwin the most important period of his life: he felt privileged to be chosen by the naturalist and he was proud to have proved very skillful at his job. Nevertheless, towards the end of his life he feels betrayed as he realizes that his efforts to help Darwin indirectly prompted his master to work out a theory at variance with the Bible. He is angry at Darwin's affluence, his English gentry background, and his country squire ways. Once they had said farewell to each other Darwin immediately forgot him and was genuinely surprised by the former servant's letters and his small gifts of fossils.

Giovanni's anger is even more pronounced. An outcast, and in the last months of his life a crippled beggar, the young Italian suffers the contempt of the colonial upper classes and the persecution of the government's secret service agents. Darwin's optimism and eagerness to learn seem to him the upper class frivolities of a naïve youth able to afford being generous and open-minded. Giovanni's mother's and brother's plebeian Catholicism makes him suspicious towards every religion and the fact that Darwin does not see the contradiction between his Anglican faith and his nascent theory adds to Giovanni's anxiety.

Steen and McDonald emphasize the importance of the religious upbringing their narrators underwent and systematically juxtapose these dogmas with the religious debate surrounding the publication of Darwin's texts in the years after his return from the HMS Beagle's voyage. They do so primarily by referencing canonical religious works. The Covington/Darwin and Giovanni/Don Carlos theological conflicts are thus intertextual in nature and not reports on "real" people's quarrels. Commenting on his protagonist McDonald confesses that he created this character by conscious imitation of religious discourse:

> He was born obscurely in Bedford, the hometown of John Bunyan and religious non-conformity. Building from this lone early established fact, I created him imbued with trusting faith from childhood, coming from an older England, a stranger to the Anglicanism of the ruling order.[27]

---

27 McDonald, 'Evolution', 4.

He also adds that his quest for theological veracity was a textual endeavor: he started by looking for literary sources of Covington's religiousness: "Seeking a language for Covington to represent an older, more trusting religion, and to stand against Darwin's "modern" pattern of thought, I delved into *Pilgrim's Progress*" (McDonald 1998b, 3). The repetition of "older" in both above mentioned quotes is significant: Darwin's intellectual attitude toward dogmas is contrasted with the myth of the old and true religion of lower social orders – by the same token Steen describes the Catholicism of Giovanni's family as superstitious and fanatical. This is emphasized when Giovanni compares Darwin's intellectual courage to look for unorthodox interpretations of natural phenomena with what he and his brother were told in their childhood in Genoa: "Do you remember when we each found a fossil in the cliff face at home? Mother told us that fossils were relics of the Flood [...] Was the Creation anything like that?"[28].

Equally textual is the science/religion conflict which lies at the core of Darwin's myth in contemporary culture[29]. Since the days of Darwin modern science has been contradicting religious dogmas, and people who witnessed its birth, such as Covington and Giovanni, have to feel this. Covington is unable to accept the implication of his friend and master's discoveries. The Biblical story of the Flood is his favorite chapter of Scripture and he tends to interpret all the fossils they find in terms of this story. He does not accept the fact that species can be changing in time, as for him this would suggest that the original act of Creation was not perfect and his idea of God precludes such a possibility. Darwin's easy assumption is that it is not him who suggests that the Bible is wrong, but "the nature of beings and their station in life that would speak the blasphemies if there were any"[30].

For Giovanni, who had lost his Catholic faith long before meeting Don Carlos, Darwin's reluctance to take responsibility for the theological implications of his theory is hard to accept. After the short period of his work for Don Carlos,

---

28 Thorvald Steen, *Don Carlos and Giovanni*, trans. James Anderson (København and Los Angeles: Green Integer, 2004), 56.

29 Jon Amiel's film *Creation* is a good example of this: narrating a period in Darwin's life just after the death of his daughter and just before the publication of his theory it is concerned with Darwin's inner struggle whether to publish a 'heretical' book or not. Christianity is presented as the religion of both a stupid village vicar punishing children for making precocious remarks and Darwin's subtle and sophisticated wife. In one of the flashbacks we see Darwin talking to God and promising Him to destroy his notes and love Him blindly if his daughter lives.

30 McDonald, *Mr. Darwin's* 358.

Giovanni started reading Charles Lyell, George Buffon, and other geologists. He grew aware that the scientific discoveries of the early nineteenth century prove the Biblical Creation story wrong. To his dismay Darwin refuses to see this.

> Don Carlos is dependent upon some higher organizing power: God or reason. Don Carlos doesn't quite trust either of them. That is why he makes confessions to both God and reason. He is riding two ostriches that run in opposite directions […] Don Carlos is a liar. To attempt to reconcile belief and science is to remain uncommitted to either. Had he relinquished his faith, he would have had to trust in himself and his science. Is it fear of loneliness, cowardice, or sheer immaturity that enables him to accept this unholy alliance? I have only one explanation. His instinct always warns him to leave an escape route open to his naïve faith, like an ox with various boltholes.[31]

Giovanni calls Darwin "reasonably gifted"[32] and yet he cannot help feeling somewhat superior, or at least able to notice discrepancies in the young naturalist's theories and his behavior. For Giovanni even the choice of the three books Darwin took with him to Argentina is significant: a volume of Lyell's *Principles of Geology*[33] the Old Testament, and John Milton's Paradise Lost are contradictory. Milton's book, a Protestant classic, puts forward a literal understanding of the Bible, especially the Old Testament, while Lyell demonstrates that the Biblical story is inaccurate. Nevertheless, Darwin seems happy reading them all by the campfires. Thus Giovanni cannot understand Darwin's "pathetic anxiety about Noah's ark"[34], the vessel described in the Bible in detail which clearly could not have housed pairs of "thousands of species of mammal, six or seven thousand species of birds and fifteen hundred species of reptile"[35]. How can Darwin worry

---

31 Steen, *Don Carlos*, 178–182.

32 Steen, *Don Carlos*, 182.

33 This is again a critical cliché repeated by most of Darwin's biographers, who make him an intellectual disciple of Lyell. As early as in 1956 William Irvine, the author of a joint biography of Darwin and Huxley entitled *Apes Angels and Victorians*, writes: "The *Principles* are perhaps the most important link in the long, tenuous, precarious chain that leads up to *On the Origin of Species*. Lyell taught Darwin not only how to think about geology, but how to think. From him Charles learned observation in the higher sense of a thoughtful activity which suggests and tests hypotheses. Again, he learned how to construct hypotheses. In other words, he came to see nature as logical, regular, and self-explanatory." William, Irvine, *Apes, Angels, and Victorians* (London: Readers Union Weidenfeld and Nicolson, 1956), 37.

34 Steen, *Don Carlos*, 247.

35 Steen, *Don Carlos*, 247.

about the ark when his whole life is devoted to working out a theory contradict-ing the Bible?[36]

McDonald makes a similar point in an equally intertextual manner. In his discussion of the religious controversy surrounding Darwinism overseas he is indebted to Peter Carey's famous *Oscar and Lucinda*[37], a novel also set in Aus-tralia in the years of the publication of *On the Origin of Species* and concerned with how the newest scientific discoveries alter the traditional patterns of moral behavior inherited by the European emigrants from the more God-fearing past.[38]

Thus it is traditional texts: the Bible, *Pilgrim's Progress*, Milton's *Paradise Lost*, Lyell's Principles, Darwin's own later writings, books by his twentieth-century biographers[39], anecdotic Holy Scripture-inspired superstitions of Italian Catho-lics, and even famous contemporary post-colonial novels that are confronted in

36 Reading the book one cannot help sometimes having a feeling of Giovanni's intel-lectual superiority over Darwin. This results from the fact that the novel is but a simulation of history written with contemporary awareness, some of which is given to the narrative voice of Giovanni. Thus the narrator only apparently belongs in the early nineteenth century and the character of Darwin does. Giovanni and Darwin's conversations uncannily become the twenty-first century exposing and discussing its own nineteenth-century roots.

37 Contrary to Carey, McDonald refrains in the book from noticing aboriginal local culture which is based on story-telling abilities and gives oral narratives special status.

38 One of Carey's characters sums up the dilemmas central also to McDonald's book: "Our whole faith is a wager, [...] We bet — it is all in Pascal and very wise it is too [...] we bet that there is a God. We bet our life on it. We calculate the odds, the return, that we shall sit with the saints in paradise [...] It is true! We must gamble every instant of our allotted time span. We must stake everything on unprovable fact of His existence." Carey, Peter, *Oscar and Lucinda* (New York: Vintage International, 1988), 218.

39 Giovanni's and Covington's descriptions of the young naturalist echo those in Irving Stone's *The Origin* where we read about Darwin onboard the Beagle as a youngster whose head spins (Stone 1982, 343) with the multitudes of discoveries, ideas, and plans for the future. Just like Stone, McDonald and Steen depict a young energetic na-turalist much distinct from the sternly-looking elderly scientist whom we know from the portraits painted in Darwin's later years. Compare: "Don Carlos as presumably the only person on the planet who enjoyed the earthquake in Chile on 20th February, 1835... Don Carlos has just finished reading Charles Lyell's work on the history of the earth's crust. He managed to keep his feet as he pondered that the thing he was now experiencing might provide him with a theory about how the Andes came into existence" Steen, *Don Carlos and Giovanni*, 247. And: "Naked life was his subject in nature. Rocky islands and stones from beaches were living material to his eyes likewi-se skeletons of fishes and birds from which he conjured former existence." McDonald, 'Evolution of a Novel: *Mr. Darwin's Shooter*', 152.

a partly anachronistic simulation of the early nineteenth-century controversies. To quote Hutcheon again:

> [H]istoriographic metafiction... demands of the reader not only the recognition of textualized traces of the literary and historical past but also the awareness of what has been done – through irony – to those traces. The reader is forced to acknowledge not only the inevitable textuality of our knowledge of the past, but also both the value and the limitation of that inescapably discursive form of knowledge.[40] (Hutcheon 1988, 8)

Additionally, Darwin's discoveries change the way the colonial wilderness is perceived in Europe, which is the result of the popularity of his bestselling books, ones extremely influential in the cultural life of the mid-nineteenth century. Covington and Giovanni share the belief that this young naturalist "discovered" new continents for the European imagination. And again Steen's Giovanni anticipates this cultural phenomenon, though he dies much earlier, in 1836, when Darwin and his servant are still editing his diaries and notes. Yet Giovanni also intuitively reckons how very important Darwin's voyages are for the nascent identity of colonial states (Argentina has just become independent).[41]

Covington dies in the 1850s during the heyday of evolutionism. In the last months of his life he reluctantly witnesses the emergence of a true Darwinian cult: in the 1850s Australian colonists grow interested in natural history and consider Darwin's voyage around the world a cornerstone of the process of the European colonization of the antipodes.[42] His books are obligatory items in all

---

40   Hutcheon, 'Historiographic Metafiction', 8.

41   It is significant that Darwin considered all continents with their respective wildlife as equally interesting and all the species worth studying in the immediate context of their ecosystems and not just in relation to similar European forms; "In New South Wales in Australia Don Carlos came across an ant-lion that caught its prey just as predatory insects in other parts of the world do. For Don Carlos this was the great affirmation. One and the same hand had been active across the entire Universe. All things had not been created simultaneously, in the same place." Steen, *Don Carlos*, 235.

42   Both Steen and McDonald writing about Darwin from the perspective of European emigrants in the colonies emphasize how much people in the new world missed him after the Beagle voyage was over. Yet, when HMS Beagle entered the London harbor Darwin swore never to come back to the sea, he got married, bought a country house near London, thus becoming a country gentleman, a true member of the privileged upper middle class. From the vantage point of the London-based scientific societies and land gentry parlours, the colonies seem a far away country populated by drop-outs, runaway convicts and the like fit to be visited but by young adventurers hunting for new species to get a name.

Australian libraries, and bookshops have standing orders for everything new by him published in Britain. Australian colonists think about the fauna and flora of their new homeland in Darwinian terms, seeing creatures such as the platypus to be living proofs of the explanatory power of the theory of evolution. Australia and South America can boast superb wildlife and a separate history (natural history included), independent of the European perspective and yet comprehensible only from the point of view of modern Europe-born science and somehow related to Europe-born theological thought. Such a conclusion is markedly late twentieth- or early twenty first century-like, and the fact that McDonald and Steen have their historical characters voice it gives their texts the status of simulated history, a textual game with readers who know about history only from previously read and watched material.

Such narrative vantage points result in paradoxes, the major one of which is the fact that these post-modern narrators only simulate early nineteenth-century simpletons and are equipped with knowledge denied to their contemporaries, Darwin included. As a result Giovanni is stricken by Darwin's unawareness concerning many things, and by life in the colonies and their ambiguous attitude to the European motherlands.

> This baby of twenty-seven is in the process of writing a book about his journey round the world. Without the slightest compunction he has entitled it *Journal of the Researches into the Geology and Natural History of the Various Countries Visited by HMS Beagle under the Command of Captain Fitz Roy from 1832 to 1836.*[43]

In the quasi-historical novels of Steen and McDonald it is Giovanni and Covington who really got to know the overseas world and who sacrificed their lives to colonial enterprises and not the "baby" Darwin, a hard-working optimistic youth who, once having matured, decided to abandon voyages and become a respectable English squire. Yet paradoxically it is not them but he who wrote *Journey of the HMS Beagle* and later *On the Origin of Species* and it is after him that colonial towns, rivers, straits, and species got their names: it is he who has become a textual icon.

The juxtaposition of the two recent novels about young Charles Darwin shows a number of similarities. Both books simulate history in the Baudrillardian meaning of the term, they are products of post-Darwinian, post-modern culture very much concerned with myths of its own beginnings. In *The Postmodern Condition* Jean-Francois Lyotard argues that within contemporary culture our quest for knowledge cannot depend on "grand narratives" such as one canonical

---

43 Steen, *Don Carlos*, 197.

version of the history of the West, meta-discourse on humanity's gradual under-standing of the workings of nature, or in fact the Darwinian narrative of progress in Nature: the evolution of baser forms to more advanced ones. To quote Lyo-tard: "in contemporary society and culture — postindustrial society, postmod-ern culture — [...] the grand narrative has lost its credibility, regardless of what mode of unification it uses, regardless of whether it is a speculative narrative or a narrative of emancipation"[44]. What we have instead is constant cravings for the lost systems of reference we no longer believe in, but enjoy playing with.

The two quasi-historical novels discussed above provide us with stories set before Darwin wrote down his enormously influential grand narrative in one of those mythical nodes history is made of, at least according to popular science books and films. Both narrators are unreliable: partly equipped with their own time's awareness and partly talking in camouflaged quotes – the Bible, *Pilgrim's Progress*, Stone's *The Origin* and other Darwiniana are among their sources. They also talk about the books they read, those Darwin reads, those Darwin is going to write and McDonald and Steen's readers are supposed to have read, thus clos-ing us up within a vicious circle of texts corresponding with one another, but never in fact touching the extra-textual world of past or present. "History is a strong myth, perhaps [...] the last great myth. It is a myth that at once subtended the possibility of an 'objective' enchainment of discourse"[45] argues Baudrillard – and McDonald and Steen seem to agree.

---

44 Jean-Francois Lyotard, *The Postmodern Condition: A Report on Knowledge*, trans. Geoffrey Bennington and Brian Massumi (Manchester: Manchester University Press, 1984), 37.
45 Baudrillard, Jean, *Simulacra*, 47.

Depictions of Emma Darwin in Recent British Non-Fiction

The life and works of Charles Darwin have been the subject of numerous essays written in the last 150 years. Depicting virtually every moment of the Victorian naturalist's eventful life are his own journal and *Autobiography*, the diaries and letters of his children, friends, and colleagues, scholarly and popular biographies, and, in the last half a century, feature films and popular science TV series. A surge of new publications on the subject of Darwin flooded the book market in 2009, the Year of Darwin. Two new feature films based on the records of his family life added further detail to the picture of Darwin as represented in mass culture, while the darwin online project freely provides Internet users with a vast amount of useful material.

The explanation for this enormous popularity is far beyond the scope of this paper, though a fair guess would be that it is because Darwin is universally acknowledged as the creator of the modern scientific paradigm[46], and in the contemporary world science is the single most important method of generating knowledge about reality. Therefore, in the collective imagination Darwin has become one of the founders of modern civilization and so he is interesting. Moreover, although science in modern civilization is enormously successful, it does not transcend culture: quite the opposite – our culture bends science to its own purposes and makes free use of its paragons to create, sell, and disperse stories. Books and films 'translate' Darwin's life and the Darwinian oeuvre into, to borrow Roland Barthes' canonical term, "the collective representation"[47] of the scientist, whereby they form a part of modern mythology in the Barthesian meaning of the term.

The egalitarian character of contemporary culture makes authors prone to using simplifications and stereotypes, but at the same time the vastness of the global book and movie market promotes originality and the skill to show old truths in a new and attractive way. As a result, the recent non-fiction about Darwin both promotes cliché stories and subverts them, thus showing how erroneous stereotypes are. My aim in this chapter is to discuss the above mechanisms on

---

46 This honor was first given to Darwin by Thomas Kuhn in *The Structure of Scientific Revolutions* (1962). Numerous historians and theoreticians of science later shared this opinion, among them Karl Popper and Edward O. Wilson.

47 Roland Barthes, *Mythologies,* trans. Annette Lavers (London: Vintage, 1993), 9.

the basis of the recent non-fiction devoted to the presentation of Emma Darwin, the scientist's wife. Her life as a Victorian 'wife of fame', an accomplished upper-middle class lady, and a mother of ten who sacrifices her own happiness for the good of her husband and her ever-growing family, makes it very easy to ascribe her biography to the standard set of feminist assumptions concerning the Victorian era, and the woman's position then. Yet four decades after the heyday of the Women's Liberation Movement, the authors of historical books about Victorian women are now aware of the risk of becoming cliché. Following in the steps of fiction writers, today they apply the narrative strategies of historiographic meta-fiction and the postmodern novel in general to show the current awareness of the difficulties in describing the past.

Among quite a number of recent non-fiction works devoted to the Darwins that focus on Emma Darwin I have chosen for my analysis just four books, as each represents a distinct, broader subgenre of non-fiction texts. John Darnton's *The Darwin Conspiracy* adds so much fictive detail to the 19th-century history of the scientist's family that one should in fact call it a novel, albeit a thoroughly re-searched one. Edna Healey's *Emma Darwin. The Inspirational Wife of a Genius* is a standard biography: it is full of details, well-documented, and yet written passionately by someone who clearly adores Emma Darwin. Janet Browne's *Charles Darwin. Voyaging. A Biography*, often referred to as his definite biography, is a very scholarly account of the scientist's life that exceeds 1000 pages of annotated narrative referencing a vast array of sources. Finally, William E. Phipps' *Darwin's Religious Odyssey* is a serious philosophical treaty discussing the scientist's path towards agnosticism in reference to, among other documents, his well-known private life, with the emphasis falling on his relations with Emma Darwin.

From the popular to the scholarly, four depictions of Mrs. Darwin have been created in the last fifteen years. They all are products of turn-of-the-millennium, science-loving Western culture, which considers the discovery of evolution by natural selection one of the milestones in the history of modern thought. Be-cause Darwin's life as centered around this discovery is a contemporary myth, a story whose diverse variants circulate in culture, I am going to refer in my analy-sis of the depictions of Emma Darwin to selected twentieth-century criticism devoted to modern myths – namely to those of Roland Barthes, Mircea Eliade, and Edward Caudill. These three scholars approach myths from very different vantage points, and this allows us to look at Emma Darwin's representations from the angle of media studies, popular culture, and anthropology.

The very term 'myth' in the popular meaning denotes a tale which lacks truth or refers to some fantastic entities and thus is ambiguous: myths may or may not be literally true, and yet they always teach a lesson and are vehicles of a greater truth which transcends the immediate details. As Edward Caudill claims, "a person that has become mythic may have a foundation in fact, but the constraints of historical evidence have been broken by the significance of the symbol"[48]. Emma Darwin of course did exist, a fact best attested to by her numerous great-grand-children, and yet her story is prone to becoming a myth in Caudill's definition as it may symbolize, among other things, the Victorian values of silently-suffering femininity, a woman who in fact is a victim of her family, her class, and its decorum.

Some of the recent Darwinian non-fiction does reinforce the Victorian ideal wife stereotype by emphasizing Emma Darwin's alleged "sweetness" and her readiness to surrender her own good for the sake of her family. In *The Darwin Conspiracy* Lizzie, the narrator and the youngest daughter in the Darwin family, calls her "the sweet Mamma [who] organizes the household with quiet command"[49], always thinks foremost about the comfort of her husband, and is very much concerned with appearances, e.g., writing long letters to relatives, making the children behave in a proper manner, attending church services, and keeping in mind that the reward for all her toils awaits her in heaven. Pious and conventional in her tastes and opinions Emma Darwin as described by Lizzie is focused on her husband. She is more of a servant or a nurse than a wife: "Mamma has become a veritable Florence Nightingale sacrificing herself at all hours to bring him tea and rub his back and read aloud to sooth his nerves and distract him from his various ailments"[50]

William Phipps in his discussion of Charles Darwin's marriage underlines what a suitable choice Darwin made in proposing to Emma Wedgwood, an accomplished lady of his own age and of the affluent middle-class, a person whom he had known since childhood and who had the best education then available for women, having been tutored in French, Italian, and German. She was taught to dance during her trip to Paris at age ten, several years later she attended a school for girls in London, and at eighteen she studied in Geneva where for several months she received music lessons from Frederic Chopin and became "accomplished on the piano"[51]. She also developed "a many-sided interest in the world,

48 Edward Caudill, *Darwinian Myths* (Knoxville: The University of Tennessee Press, 1999), XII.
49 John Darnton, *The Darwin Conspiracy* (New York: Anchor Books, 2005), 54.
50 Darnton, *The Darwin Conspiracy*, 50.
51 William E. Phipps, *Darwin's Religious Odyssey* (Harrisburg: Trinity Press International, 2002), 40.

in books and in politics," and was always praised for her talent to "read aloud clearly and well"[52].

Phipps' description of the Darwin family is very favorable and the terms of praise he uses are conventional and in accordance with stereotypical Victorian values: it was the sound union of two well-matched and well-meaning people from very good families, with very good financial status and very good education. Emma Darwin's readiness to serve the scientific and social needs of her husband and make his life comfortable at the expense of her own needs and unrealized talents makes her conform to the ideal of the devoted nineteenth-century wife. To prove this point Phipps quotes a very conventional letter Darwin wrote about his wife in his old age:

> I marvel at the good fortune that she, so infinitely my superior in every single moral quality consented to be my wife. She has been my wise adviser and cheerful comforter throughout life, which without her would have been during a very long period a miserable one... the greatest blessing.[53]

Janet Browne also underlines the conventionality of their union explaining the patent happiness of the couple by Emma Darwin's ability to behave in the very way that was expected of her and by the fact that theirs was the union of two first cousins coming from families with a three-generation tradition of intermarriages – a kind of marriage everybody approved of and waited for. All the parties involved took it with relief: "both the Darwins and the Wedgwoods consolidated their financial interests by sensible marriages as much as any other means and were apprehensive about fortune hunters"[54]. The genius and his wife in Browne's opinion lived happily and the usual stages of connubial upper-middle class Victorian life – the appearance of numerous children, their infancy and education, purchasing a country house and managing the estate – came in due course. This life was in fact typical and only seemed different to its protagonists as it was modified by Darwin's special kind of perception: he watched the world around him, his family included, with the eyes of a naturalist while Emma Darwin accompanied him silently, making his research possible and taking care about all the mundane details of family life.[55]

---

52 Phipps, *Darwin's Religious Odyssey*, 40.
53 Phipps, *Darwin's Religious Odyssey*, 153.
54 Janet Browne, *Charles Darwin. Voyaging* (Princeton: Princeton University Press, 1996), 393.
55 Such is Browne's attitude towards the Darwins in her book, the best examples of which are the scenes when she describes Darwin playing with his infant children

Janet Browne praises Emma Darwin for her ability to manage the ever-growing family and their country house in a quiet and effective way and portrays this country life in the following manner:

> The household at Down was getting larger and noisier. Darwin and Emma had four children in 1846. By the end of 1850 there were three more, making the family running from William, the eldest at eleven years old, through the two girls, Anne and Henrietta, to George (aged five), Elizabeth (aged three), Francis (aged two) and the baby Leonard born in January 1850. Nursemaids, housemaids, ponies, dogs, gardeners, stablehands, and governesses increased to match.[56]

Childbirths, managing the servants, nursing the family in their numerous ailments, structuring her life around their needs – such a portrait of Emma Darwin complies with the depictions of patient heroines of the Victorian novel; the morally good women who tend to forget about their own selves and are much applauded for this.[57]

---

at the same time as he makes notes about human expressions and behavior having their origins in animal ancestors. He noted down the dates of the infants' first smiles, first attempts at communication, first tears – not as a doting parent, but as a cool observer: "ingeniously he discovered a part of Victorian daily life to investigate that other scientists ignored. The natural history of babies, he decided, was essential to his proposals about evolution" Browne, *Charles Darwin. Voyaging*, 422. Interestingly, in Jon Amiel's *Creation*, the filmed story of the Darwin children's young years, there is a scene showing Darwin bent over the cradle of his infant eldest daughter with a pencil and notebook in his hands, busily taking down some essential data concerning human communication skills. In the background a stern-looking beauty in a nightgown, Emma Darwin, plays the piano all the while watching her husband and her baby. *Creation* is a historically accurate picture of this family's life in the years just before the publication of *On the Origin of Species* based on the non-fictive account of Randal Keynes, the great-grandson of Charles and Emma Darwin, who was in the possession of the family papers and heirlooms.

56 Browne, *Charles Darwin. Voyaging*, 471.

57 Yet Browne notices, albeit in passing, the "inward-looking life" (Browne, *Charles Darwin. Voyaging*, 537) of the Darwins who with their children, devoted servants, and beloved relatives drew into a single close-knit unit hardly interested in other people and reluctantly going out. Emma Darwin's existence among all the havoc of the very big and vivid family was therefore, again in accordance with the Victorian ideal of mute female suffering, strangely solitary and her accomplishments (playing the piano with the utmost artistry and reading aloud beautifully) served for the entertainment of her husband and her children only.

Edna Healey in *Emma Darwin. The Inspirational Wife of a Genius* is very much interested in all the "forgotten women who shaped [Darwin's] life"[58], not only Emma Darwin – who is definitely not forgotten – but various women from the two previous generations of the interconnected Darwin-Wedgwood clans. In unearthing their stories Healey devotes a number of chapters of Emma Darwin's biography to the times before her subject was born. Her research is really impressive and done in line with a mildly feminist agenda of bringing into the limelight the forgotten females: acutely intelligent spinster aunts; pre-Victorian hard-working grandmothers; women who died young due to poor medical services; and female relatives whom mental solitude, intense post-natal pains, and overwork drove them to drink spirits and take overdoses of opium (which they called 'laudanum'). Some of them in fact might have been remembered[59] by their diary-writing offspring (for example Charles Darwin's own mother, whom he claimed not to remember at all though he was already eight when she died), and yet their histories are lost forever despite the best efforts of 21st-century feminist historians, because – Healey repeats this several times – they lived and suffered wordlessly, leaving no trace behind.[60]

When Healey finally describes Emma Wedgwood herself, she tends to use a whole set of stereotypes derived from Victorian novels in order to show the consecutive stages of her subject's life. As a small girl, the youngest of eight children, she was "the happiest being that was ever looked at"[61] and spent her days in the

---

58 Edna Healey, *Emma Darwin. The Inspirational Wife of a Genius* (London: Review, 2001), 4.

59 'Forgetting' and 'being forgotten' are the key notions here. Healey writes about the family papers dumped in the scrapyards and retrieved miraculously two generations later in "one of the most astonishing stories in the history of archives" Healey, *Emma Darwin*.4, which revealed the truth about the Darwin and the Wedgwood grandmothers and the whole gallery of other women living at the turn of the nineteenth century.

60 A good example is one of the Darwin women from the generation before Charles': "her story is a blank. Only once is her voice heard: her second daughter, Kitty, refused to marry on a Monday, because her mother told her it was bad luck; her own wedding, she said, had been on a Monday" Healey, *Emma Darwin*, 35, and yet, judging from the testimonies of her family voiced in letters and diaries, this silent or silenced woman all her life was "a credit to the belief in women's ability" Healey, *Emma Darwin*, 42. Thus, according to Healey, the affluent Victorian gentry in the late 18th century, when the generation of Darwin's parents was young, made their women "accomplished", but applauded them for renouncing their exceptional abilities.

61 Healey, *Emma Darwin*, 63.

fields and the garden, among flowers, doting aunts, and faithful servants. Riding ponies, fishing and swimming with her brothers, she was everybody's favorite and a very pretty little lady. Healey claims that "this childhood spent lapped in love gave Emma a stability and tranquility that marked her all her life"[62]. In the next stage she was an English girl on the continent – visiting galleries and concert halls, conversing in French, German, and Italian, attending balls, and making friends with the European intellectual elite of her day. Then, in her late twenties, she received numerous marriage proposals but "laughed at her friends' match-making efforts"[63], and the amorous men whom she refused. Indeed, her mother remembered once having found "a young curate walking round the lake, weeping with disappointment at his rejection"[64]. Finally, at thirty, when she accepted Charles Darwin, she momentarily became a paragon of wifely virtues decorating the newly-bought house, planting gardens, instructing the servants, and farmhands. It was she who created "the environment essential for Charles [where] he could direct the burning force which might have destroyed him into the work which changed the world"[65]. She fulfilled her duties, never complained, and even the continuous effort of child-bearing did not prevent her from being a cultured lady very much concerned with maintaining standards:

> Despite producing nine children in eleven years of marriage Emma did not allow her mind or her musical talent to atrophy. She played the piano every day, to please herself, to sooth Charles in the evenings, to amuse the children[66].

Not only was Emma Darwin mindful of her artistic and intellectual development, but she was also incredibly tolerant as far as her husband's work was concerned. Healey emphasizes that she never objected to the "experiments" Darwin and the children were always conducting, though some of them were very messy and very smelly as they involved cutting open animals' bodies, examining their vitals or the mud and excrement clinging to their feet and feathers, and surveying consecutive stages of their carcasses' decomposition. In one well-known biographical anecdote, Healey recounts how Darwin made his wife play the piano

62 Healey, *Emma Darwin*, 63.
63 Healey, *Emma Darwin*, 138.
64 Healey, *Emma Darwin*, 138.
65 Healey, *Emma Darwin*, 138.
66 Healey, *Emma Darwin*, 197.

with a jar of worms on the lid so that the creatures' reaction to music – if there was any at all – would show.[67]

It is precisely the devotion to her husband that motivated Emma Darwin: the thought of him and the consideration of his needs were behind most of the choices she made not only during his lifetime, but also after his death. In the last section of her book Healey very emotionally describes the separate graves of the spouses:

> The churchyard grass is rough. Emma's tomb is hard to find. There are no flowers, and the words on the gray headstone are fading... Charles's mortal remains are in Westminster Abbey. All her married life she had prayed that they would never be parted... It was her last sacrifice, and she made it willingly because she loved him.[68]

One must admit that the self-denying femininity represented by Emma Darwin as Edna Healey sees her does have some period charm. Despite her clearly visible compassion for the long-dead women of the Darwin and Wedgwood families, Healey does not suggest that the world and history were unfair or unjust to Emma Darwin. Much less does her book comply with some militant feminist agenda, though it freely makes use of clichés about Victorian females and finds the stereotypes attractive. The author seems proud to be old-fashioned and critical of the Women's Liberation principles. The unsaid assumption is that Emma Darwin lived happily and was (and still is) adorable just because she chose to be the "wife of fame" hidden behind her husband and yet indispensable to everybody.

The feminist angle of the story of Emma Darwin's marriage (or the decision to avoid it) depends on the ideological stance of her diverse biographers. Much

---

67 Healey also admits offhandedly that Emma Darwin was perhaps not so very perfect as far as the keeping of the ideal Victorian house was concerned: she never considered tidiness to be of much importance, did not understand why the servants insisted on polishing stairs and chair legs. Healey goes on to add that the reason was neither idleness nor neglect, but rather Emma Darwin's strong belief that the untidy house full of "experiments" in progress was ideal for children who could play and learn simultaneously and, moreover, the atmosphere free of any nagging helped Darwin conceive his theories concerning natural history, which of all the things made him the happiest. Yet Healey refrains from mentioning the well-known fact that later in life Emma Darwin was also atrociously indifferent to what she was wearing, at least in her daughter's opinion, who claims that her mother was the happiest in an old dress she wore working in the garden, and had to be earnestly persuaded to change when visitors arrived to talk to her "genius" husband.

68 Healey, *Emma Darwin*, 321 and 354.

more controversial is the question of Emma Darwin's religion, which makes her biographers feel rather confused. The most standard way of dealing with this is to contrast Emma Darwin's devout Unitarian religiousness with Charles Darwin's learned agnosticism and to underline that the spouses loved each other dearly[69] and were aware they could not reconcile their conflict about God so they preferred not to talk about it.

Some historians even explain that the delay in the publication of *On the Origin of Species*[70] resulted from Charles' reluctance to upset his wife. Other biographers seemingly describe the Darwins' conflicting opinions concerning the Bible, but in reality they refer rather to the contemporary American public debate between Darwinists and creationists, or to the conflict between the theory of evolution and Intelligent Design theory. In *Darwin's Religious Odyssey* William Phipps explains the aged Darwin's mistrust in Christianity in terms of the fact that he could not accept that God would make good people who do not believe in Him (like himself, his father, his brother, and most of his best friends) go to hell. Such an explanation was given by Darwin in his *Autobiography*, but edited out after his death by Emma Darwin, who wrote on the margin of the manuscript "it seems to me raw. Nothing can be said too severe upon the doctrine of everlasting punishment for disbelief, but very few would call that Christianity (tho' the words are there)"[71]. Thus Emma Darwin was both shielding her late husband from the accusation of being the Devil's Chaplain and looking for some common denominator of their beliefs: he and she did agree that what was damnable was not unbelievers but "the vindictive doctrine itself"[72]. Her person, and her words written on the margin of her husband's text, offer means to show the 21st-century reader the alleged reconciliation of Darwinism and religion, by proving that the very conflict is false and the result of verbal not ideological differences.

---

69 John Darnton, for example, does not deny Emma Darwin's great love for her husband and one of her children who, following her father's steps, turned atheist ("our household held two non-believers", Darnton, *The Darwin Conspiracy*, 191), but makes her despair about this and from time to time utter some cliché devout exclamations: "you have set foot on the unholy path" Darnton, *The Darwin Conspiracy*, 162; or "this is what comes from turning away from our Savior Jesus Christ" Darnton, *The Darwin Conspiracy*, 162.

70 The book devoted to the presentation of Darwin's theory was published twenty years after he had discovered that the species do "transmute", as he put it then.

71 Qtd in Phipps, *Darwin's Religious Odyssey*, 196.

72 Phipps, *Darwin's Religious Odyssey*, 196.

Janet Browne discusses Emma Darwin's religiousness a number of times in her voluminous book, but always in the context of Darwin's psychological dependence on his wife only. Browne suggests that the spouses had some sort of unspoken agreement: Darwin would never talk about his disbelief aloud, and that would allow Emma Darwin to always hope his faith was alive. To prove such a point Browne quotes from Emma Darwin's letter to her husband: "may not the habit in scientific pursuits of believing nothing till it is proved, influence your mind too much in other things which cannot be proved the same way, which truly are likely to be above our comprehension"[73].

Once his theory was intellectually ready – though physically it remained merely in the form of a sketchy draft – Darwin refrained from publishing it, and simply prepared the manuscript for publication. He added to it a letter[74] addressed to his wife to be opened after his death in which he commands her to have the paper published at her own expanse. Janet Browne comments: "he would prefer to be dead rather than deliberately hurt Emma's feelings… he would prefer someone else to put it before the public… he recognized the weightiness of his paper"[75].

The situation as Browne sees it is rather paradoxical: Darwin does not want to publish his book in order not to hurt his wife's feelings, and yet he appoints her – and not one of his learned naturalist friends who urged him for years to have his thoughts written down – to publish it posthumously and thus wants to make her his accomplice. The good wife who always chooses to obey her husband would undoubtedly do this, even if she felt that the book was vile and morally wrong, but Darwin being dead would not see her suffer, Browne is saying. Strange as these ideas about the Darwins' marriage may seem, Emma Darwin did not have to face such a loyalty test, thanks to the fact that Alfred Douglas Wallace discovered the mechanism of natural selection independently and urged Darwin to publish *On the Origin of Species* in his lifetime in order to ensure his primacy.[76]

---

73 Qtd in Browne, *Charles Darwin*, 411.
74 The Darwins were privileged members of a very literate class which left us with thousands of pages of diaries, letters, journals, and notebooks: writing was for them as spontaneous as talking and, moreover, being introverts, both spouses, when they felt they had something complex and difficult to communicate, wrote it down. Therefore, despite living for nearly forty years in the same house they did write letters to each other when they were bothered by important things, especially when the subject was emotionally charged.
75 Browne, *Charles Darwin*, 447.
76 Emma Darwin in the end did not have to choose between her dead husband's will and her duty as a Christian fearful of blasphemies and never was in the position to decide

The two women-biographers, Browne and Healey, adore Emma Darwin's complicity, her old-fashioned nuptial love, and un-feminist lifestyle. Describing such attitudes in all their touching details the two writers show how attractive the "good wife" myth still is and how much pleasure writing and reading about it can give. Healey tries to explain Emma Darwin's decision to help her husband in his endeavor to prove via natural history that the Bible is not true to the letter by claiming that she considered God too great and too good to mind the theory of evolution. She also allegedly thought that Darwin was deprived of the gift to believe:

> Conscious of their differing religious views she had to accept that her faith was beyond reasoning: she saw Charles's unbelief as that of a man who was color-blind. The God she believed in would not have rejected a person who was so clearly good.[77]

Yet by no means do all the biographers of the Darwins consider Emma to have always been a good and faithful wife who helps her husband to bring forward the great theory despite all the obstacles, her own worldview included. Edward Caudill in *Darwinian Myths* describes her as a kind of censor who bowdlerized her husband's notebooks, letters, and his *Autobiography* after he died. If she did not do this herself, she at least prompted the editing; "Darwin's son Francis editor of the *Autobiography* tempered his father's religious views in order to appease Emma Darwin, a Unitarian... the *Autobiography* minimizes the materialism in Darwin's early notebooks and eradicates [the offensive references]"[78] Caudill also suggests that Emma Darwin "might make a good suspect"[79] when he discusses possible sources of "the reconversion story". A few years after Darwin's

---

whether to destroy the theory of evolution or let it change the world. Nonetheless, the emotional tension of such a hypothetical moment of choice is psychologically attractive and it recurs in biography-inspired fiction devoted to the Darwins. In Jon Amiel's *Creation* Darwin, having read the letter from Wallace, sits down to finish his book, but once it is ready he does not send it to his London publishers but rather hands it to his wife saying that it depends on her whether the book will be destroyed or printed. Not only does she have to renounce her beliefs for the love of her husband and not prevent the publication, but also to actively help in it, sharing his sin. Next we see her reading the manuscript by candle light all through the night and then, after the cut, we see her in the morning light burning something in the garden. Darwin, his face horror-stricken comes near the fire to see last year's leaves burning and to learn that she mailed the manuscript to London herself: "Let God forgive us both", is all she says.

77 Healey, *Emma Darwin*, 342.
78 Caudill, *Darwinian Myths*, 55.
79 Caudill, *Darwinian Myths*, 48.

death an anecdote narrating his alleged conversion in his dying hour surfaced and was used to prove the theory of evolution evil. Presbyterian ministers in the US claimed in their sermons that Darwin in agony: "whined for a minister... renouncing evolution [he] sought safety in the blood of the Savior" [80], the patent nonsense of which Emma Darwin is to blame.

The issue of religion and how it is handled in the diverse non-fiction devoted to Emma Darwin's life and marriage shows in the most extreme way the diverse ideological standpoints of her biographers. Her religious persuasion and the consequent motivations behind what she did and said are a subject which is both ideologically and emotionally charged. Her biographers feel strongly about these issues and, as already stressed, try to interpret the very same set of data in diverse ways, making her the perfect wife who renounces her beliefs on the behalf of her husband's work, or the devout Unitarian fundamentalist whose shears censor her husband's texts once he is dead. For some contemporary writers her life can teach us a lesson on how forgotten wifely virtues make families happy; for others her religiousness proves how dangerous fundamentalism, especially zealous creationist fundamentalism, always is. Yet some essayists try to prove the unattainability of the past by showing that we may know what historical figures did, but we may never know why.

Each of the above attitudes makes use of the collective representation, the Barthesian myth of Emma Darwin which is a figment of contemporary culture and which transforms some particular person who lived in Victorian times into an embodiment of some universal truth.[81] Thus the story of Emma Darwin painted in a sentimental fashion by some of her recent biographers perpetuates the myth of the good old days when family values, romantic love, happy marriages, and the traditional ideals of femininity were respected and, consequently, people were happier.[82] The relevant context here is the thesis of Mircea Eliade,

---

80 Caudill, *Darwinian Myths*, 48.

81 Compare Roland Barthes: "by treating collective representations as sign-systems, one might hope to go further than the pious show of unmasking them and account in detail for the mystification which transforms petit-bourgeois culture into a universal nature." Barthes, *Mythologies*, 9.

82 Discussing the comfort-giving myths of femininity which women themselves enjoy – myths of women as universal mothers, women as men's best helpmates, women as personification of the bounty of nature and the fertile soil – Angela Carter famously calls myth "a consolatory nonsense" Carter, Angela, *The Sadeian Woman. An Exercise in Cultural History* (London: Virago, 1979), 9. She claimed that such myths are used to seduce women into submission in her essay written way back in 1979, one which

who discusses myths in popular culture as means for mental escape to some earlier, better times:

> Plays, books and movies are myth-like because they reveal the existence of another world alongside the everyday one – a world of extraordinary figures and events akin to these found in earlier, superhuman myths... Most of all moderns get so absorbed in plays, books and movies that they imagine themselves to be back in the world before their eyes.[83]

Yet other biographers aim their books at readers who are trained to demythologize history as historiographic metafiction which "demands of the reader not only the recognition of textualized traces of the literary and historical past but also the awareness of what has been done – through irony – to those traces"[84]. Thus, post-structurally oriented authors in recalling the life of Emma Darwin do two things simultaneously: they describe it in stereotypes we tend to use while thinking about the Victorians, and also expose the stereotypes by signaling they are aware of their inescapability. The most striking example of this approach may be found in *The Darwin Conspiracy*, where Darnton makes his narrator, Lizzie Darwin[85], aware of the conventionality of the type of representation she is making:

> The bustle provides the impression that the Darwins are a normal and contended family... I perceive a strangeness beneath the gaiety and manners... an astute observer sitting in our midst at the grand table might notice a forced quality to the laughter and, were he as perspicuous as some of the modern novelists from Maudie's Library like Mrs. Gaskell or Mr. Trollope, he might be able to detect the reason for it. We are not as we present ourselves to the unknowing outsiders, our attempts at hospitality and gaiety are more play-acting.[86]

---

seems today very much dated, as now writers such as Edna Healey clearly miss the pre-Women's Liberation Movement myths Carter was criticizing and clearly enjoy stories from within the heart of patriarchy.

83 Mircea Eliade, *Myth and Reality*, trans. Williard R. Trask (New York: Harper Torchbooks, 1968), 5–6.

84 Linda Hutcheon, 'Historiographic Metafiction. Parody and the Intertextuality of History' <http://hdl.handle.net/1807/10252> accessed 12 November 2011.

85 Lizzie is not like the John Fowles' postmodern narrator who describes the 19th century with the 20th century's consciousness in *The French Lieutenant's Woman*, nor is Darnton's book an attempt at a literary experiment, nor is it a philosophical treaty about the inaccessibility of the past. Passages like the one quoted above are rather signals that some sort of a literary standard has emerged at the turn of the 21st century – it is now conventional to recognize that we cannot escape the conventions we apply.

86 Darnton, *The Darwin Conspiracy*, 54.

Moreover, the authors of non-fiction who research the past are prone to notice in the texts written long ago traces of a similar anachronistic awareness, and as a result in Emma Darwin's biographies we come across passages showing that the Victorians comment on themselves being Victorian. Phipps quotes a letter about marriage Darwin wrote in jest to a friend before he started courting Emma Wedgwood: "as for a wife, that most interesting specimen in the whole series of vertebrate animals, Providence only knows whether I shall ever capture one or be able to feed her if caught"[87].

Darwin the naturalist joking about the absent-minded naturalists of the Mr. Pickwick kind brings to mind the social comedies and humorous prose of the Victorian period. The Darwins being landed gentry – a social group frequently described in Victorian novels of manners – they are often described as being aware they resemble characters in period novels. In fact, comparisons to the works of canonical writers of the period are commonplace in Emma Darwin's biographies. Browne's monumental life of Darwin starts in the following way:

> He was born into Jane Austen's England. Indeed the Darwins would have stepped straight out of the pages of *Emma*. The four girls sharply intelligent about the foibles of others, their father as perceptive as Mr. Knightley.[88]

Healey is also aware that in describing real-life 19th-century people she resorts to literary conventions; she even compares some of the Darwins' distant relatives (who are mentioned several times in various chapters devoted to different periods of Darwin's life) to "minor actors in a stage play [who] appear repeatedly in different disguises"[89]. Moreover, Healey – true to form – equipped her heroes and heroines with the self-awareness of the fact that they are like characters in a novel and makes them joke on the subject. She describes Emma Darwin reading Mungo Park's diaries and fantasizing about acting out the stories of adventure and discovery, and how Charles Darwin compares his sisters, Susan and Catherine Darwin, to "Jane Austen's Lydia and Kitty, always running after anything in trousers"[90].

According to Healey, Emma Darwin herself considered none other than Jane Austen's books to give a fair likeness of her own class and the people around her: "Emma and her friends often identified with Jane Austen's characters"[91].

87 Phipps, *Darwin's Religious Odyssey*, 42.
88 Browne, *Charles Darwin*, 3.
89 Healey, *Emma Darwin*, 61.
90 Healey, *Emma Darwin*, 79.
91 Healey, *Emma Darwin*, 75.

Therefore, it is customary to describe the Wedgwoods and the Darwins as, first of all, readers of fiction for whom real life merely imitates the art of the novel.[92] This is a very 20th-century-like metafictive realization and it remains unsettled as to what extent the biographers gave the people they describe their own sensibility and the prevailing feeling of exhaustion: all that happens in 19th-century biographies is a blurred image of what happens in the Victorian novel. Janet Browne at a certain point quotes a famous description of the Darwins' wedding by Emma's romantically inclined friend who claimed that "it is very much like a marriage of Jane Austen's"[93], and then goes on to show that it was not and that the protagonists of the event were aware they were poorly cast in the roles of "Miss Austen's" newlyweds: neither of the protagonists seemed too much in love at the moment; Emma was too much bewildered to feel any overwhelming sense of happiness… 'I believe' said Emma afterwards, 'we both looked very dismal'"[94]. Generally speaking, not only was the ceremony itself awkward, but "the congregation was subdued"[95] and "it probably seemed a little flat to the main participants"[96].

In the eyes of their biographers the Darwins are 'textual', they communicate via texts, think in reference to texts, and when they have to make some important decision they resort to writing texts, the best example of which is the alleged story of Charles Darwin deciding whether to marry or not. Even before he started to look for a suitable candidate to become Mrs. Charles Darwin – at least according to most biographers – Darwin had to choose between the two ways of life available to a member of his class and profession: that of a scientist bachelor living in a London flat with a devoted male butler and some servants, or that of a married country squire doing natural history in his spare time. In order to

---

92 It is significant that in most of the books describing the Darwins' life in Down House there is at least one passage devoted to the novels they read aloud: "the evening readings were an important part of the family life at Down… Scott, Thackeray, Dickens…, light novels (Charles liked a pretty heroine and a happy ending) were all read" Healey, *Emma Darwin*, 75, 198. This quote is just one of the very many references to Emma Darwin's ability to read aloud beautifully: in the house of her father and then of her husband. She is believed to be aware that the expectations raised by the novels concerning love, marriage, and adventure are difficult to be fulfilled, and the biographers equip her with a seemingly 20th-century-like disappointment with the literary conventions which are so misleading.

93 Browne, *Charles Darwin*, 392.

94 Browne, *Charles Darwin*, 391–392.

95 Browne, *Charles Darwin*, 400.

96 Browne, *Charles Darwin*, 400.

make up his mind he took a piece of paper[97] and wrote down two headings: "not to marry" and "to marry". Below them he listed the arguments for and against matrimony (if one is a Victorian gentleman). In the pro-marriage section he wrote: "the children (if it please God), constant companion (and a friend in old age) who will feel interested in me, object to be loved and played with – better than a dog anyhow"[98].

Discussing the literary 'lives' of Emma Darwin one cannot escape stumbling on numerous contradictions: John Darnton considers her to be "the sweet Mamma", an embodiment of nineteenth-century virtues; Healey depicts her as a thoroughly adorable woman who can teach us what true femininity is; for Browne she is primarily a reader of Jane Austen's books fashioning her life to resemble them and disappointed with the results; for Phipps Emma Darwin is most importantly responsible for purging Darwin's texts of atheistic remarks. Her biographers use cliché statements: from comparing her life to those of Victorian stock figures: a happy newlywed, a dutiful wife, a shrewdly intelligent girl from a landed gentry background – to casting her in an ideologically-charged role in some feminist or religious crusade. Her life, especially in recent years, marked as they are by the revival of interest in Darwin, has become quite well-known and in these times of mass-culture it has been turned into a 'myth', a simplified Barthesian "collective representation" of a Victorian wife of fame. Within the constraints of mass-culture, "it is extremely difficult to vanquish myth... for the very effort one makes in order to escape its stronghold becomes in its turn the prey of myth"[99]. This statement aptly describes the story of Emma Darwin's lives: it is now impossible to determine what Emma Darwin truly was like. We may only infer from her numerous recent biographies and other non-fiction devoted to her what she seems to have been like, or what her biographers would like her to be in order to suit their agendas.

---

97 Still kept in the Darwinian archives.
98 Healey, *Emma Darwin*, 146.
99 Barthes, *Mythologies*, 135.

Recent Fiction about Charles Darwin: Peter Nichols, Harry Thompson, and John Darnton

The story of Charles Darwin's life: his travels around the world, his family life, the long years of working on his ground-breaking book, *On the Origin of Species*, and his anxiety over whether to publish what most of his contemporaries would consider a heresy – this has been described in numerous books and is now among the best known biographies of the 19th century. A chapter on Charles Darwin is a must in popular science books that describe both evolutionary biology and Victorian science. Scenes from his voyages, primarily his exploration of the Galapagos archipelago, adorn many wildlife films, and his ideas concerning the order in nature have become cliché. The best known from the latter is the Tree of Life metaphor: all beings living and extinct are compared to an enormous tree with forking branches and twigs; the green sprouts at the ends of the twigs are the wildlife species we see in the world now, but lower in the branches there are also the withered remnants of contemporary organisms' long-dead ancestors and cousins, such as, say, dinosaurs.[100]

Thus, there is hardly a fact in the well-documented life of Darwin that is not reported in books and films. Along with Darwin's biographies and biographical novels such as Irving Stone's *The Origin*, in the last two decades Darwinian 'apocrypha' have been written recounting the story of his HMS Beagle adventures from the viewpoint of characters from the margins of Victorian society – including his servants and émigrés in the colonies he visited. Such a switch of the narrative perspective seems to reflect the changing attitudes of contemporary culture to colonization, the Antipodes, the British Empire, wildlife, and evolutionary biology among other issues. Moreover, a new tone is to be heard in today's fiction devoted to Darwin: an increasing number of books depict him as far from ideal

---

100 This metaphor is used, for example, by David Attenborough in his TV series *Darwin and the Tree of Life*, which starts in the Down House (the Darwins' residence) and depicts the conception of the theory of evolution as the beginning of modern science. Also in Jon Amiel's biopic *Creation* the passage from *On the Origin of Species* describing the Tree of Life is read aloud by the voice off in the final scenes when Darwin, after his years of agonizing internal struggle, finally does write his masterpiece, which suggests that this was the discovery he has been afraid to make public all this time.

and prove that the standard story of his life was manufactured by his enthusiasts, rather than being true.

The aim of this chapter is to closely read three of the above mentioned narratives: Peter Nichols' *Evolution's Captain*, Harry Thompson's *This Thing of Darkness*, and John Darnton's *The Darwin Conspiracy* and to show how their authors deconstruct the standard picture of the scientist. Each of the three are very far from reproducing a number of the 'Darwins' we encounter in popular culture: the young enthusiast on his adventurous trip to the New World, the eager student of wildlife, the middle class gentleman who has to suffer the changeable moods of his aristocratic captain, the loving husband of a fundamentalist Christian wife, and the benign father of a house-full of beetle-collecting children who grieves for decades after the death of the eldest and most talented daughter. My analysis is going to prove that the very same set of the narrative "nodes" – events from Darwin's life that are historically accurate – can be used to construct ideologically diverse stories which follow the now fashionable aesthetics of the novel of facts and the neo-Victorian novel. The major historical controversies of interest to Nichols, Thompson, and Darnton are: Christian missions in Tierra del Fuego; the moment the theory of evolution was conceived; Alfred Russell Wallace's discovery of the mechanism of natural selection at the same time Darwin discovered it; and the origin of the Tree of Life metaphor as opposed to the ladder or chain of being comparison derived from Linnaeus' texts.

Evolution's Captain by Peter Nichols is a non-fiction account of the life of Robert FitzRoy in the context of the discussion concerning evolution (or the transmutation of species, as it was then called) conducted by naturalists and philosophers in the 18th and 19th centuries. FitzRoy, who actually chose Darwin for the vacancy of the Beagle's naturalist and for five years sailed him all over the world, in most of Darwin's biographies is depicted as a tyrannical and hysterical fundamentalist keen on proving the Bible's inerrancy to the letter and on pursuing his fancies about civilizing the savages. Darwin's enthusiasts blame him for postponing the publication of the Beagle diaries, making Darwin's life miserable on many occasions, and writing ardent critiques of the theory of evolution in newspapers.

Nichols is far fairer in his attempts to picture FitzRoy not merely in the context of the Beagle journey, but on the broader canvas of 19th-century nautical reality. During his first voyages to Tierra del Fuego (when Darwin was still in Cambridge collecting bugs and reading Lyell) he had to face hostile natives and pursue the mission he received from the Admiralty without killing any of them. In the conflict he took native children hostage, but to his dismay their parents were reluctant to negotiate their release. So FitzRoy decided to take them to England, educate them, and then bring them back to their fatherland to establish a Christianizing mission in Tierra del Fuego. A few years later, when he was returning with them to South America, Darwin joined the crew of the HMS Beagle. In Tierra del Fuego FitzRoy's men built for the now-civilized natives three wigwams resembling English living-rooms and equipped with furniture, china, and chests of decent clothes and left them to their own devices in the company of but one white missionary, an Englishman named Matthews. The surrealist enterprise failed and when the Beagle sailed back to Tierra del Fuego on its way home they found the mission in ruins, and Matthews, whom they rescued, was barely alive. In the years to come the three English-speaking natives were purportedly sighted by many European sailors: the only girl, Fuegia, worked as a prostitute whenever the white sailors arrived and in the end became rich. The brightest boy, Jemmy, lived a life of changing fortune: a local princeling, a naked beggar, a participant in a revolt against the Europeans. Nichols attempts to answer the question why FitzRoy, an exceptionally intelligent person, failed to foresee the inevitable pitiable outcome of the mission, and he blames "the crazy myopia of FitzRoy's vision, fueled and abetted by Britain's expansionist aspiration and its own special relationship with God"[101].

While FitzRoy sought to prove that the Fuegians, if only given education, would not much differ from white Europeans, for Darwin they were "Adam and Eve"[102], though not in the meaning of half-angels but rather half-apes. In his conversation with FitzRoy he compared them to Jenny, an orangutan he had seen at the London Zoo and which was capable of expressing the emotions of greed, sulkiness, pleasure, and sympathy. The rules of the 'transmutation of species' (again, in these days Darwin was not yet using the word 'evolution') thus apply to animals and humans alike, and the people of Tierra del Fuego are a good sample to study: "nature by making their habitat omnipotent has fitted the Fuegian to the climate and production of his country"[103]. FitzRoy, who passionately opposed such views, was therefore furious after reading Darwin's part of the Beagle diaries (which were to be published as a joint venture). He could not edit Darwin's text but he counterbalanced it with his own part, which he made impeccably factographic, but utterly boring. FitzRoy was well aware that his decision to take Darwin onboard gave rise to the theory of evolution, which

---

101   Peter Nichols, *Evolution's Captain* (New York: HarperCollins, 2003), 189.
102   Nichols, *Evolution's Captain*, 235.
103   Nichols, *Evolution's Captain*, 216.

fact deeply embittered him and in the final years of his life made him obsessively criticize *On the Origin of Species*.

FitzRoy was a failure in nearly everything he did in his life. Yet Harry Thompson in *The Thing of Darkness* attempts to understand the personality of this tense, high-strung workaholic who was very talented, always tried very hard, and never knew why he did not succeed. The Admiralty took away the Beagle's command from him, he failed as a politician, as the governor of New Zealand, and as a chief of the first London metrological office; he lost his fortune, his books never sold, and history remembers him mainly as an obstacle on Darwin's path to fame. Thompson's book, a 600-page account of FitzRoy's life from the moment he is given the command of the Beagle until his suicide in his fifties, freely adds fictional detail to historical facts. It is a novel focused on FitzRoy, his ideas, hopes and fears; often we read interior monologues whereby we get to know FitzRoy's uttermost secret thoughts and thus we grow to like and pity him. The reader adopts FitzRoy's point of view as far as Darwin is concerned: he seems egoistic, shamelessly self-centered and cowardly, yet intelligent and favored by fortune.

FitzRoy (the nephew of both admiral FitzRoy and the Duke of Grafton, a direct though illegitimate descendant of Charles the Second) is depicted as a high-born perfectionist sent to a military school at six, and to sea at twelve, and as the best graduate ever produced by the Royal Naval College. In his early twenties he was already an experienced officer and a commander of his own ship. It is from the moment he met Darwin that his good fortune left him, though his fall, according to Thompson, was a result of a hereditary neurotic condition:

> [A] groundswell of dread in the pit of his stomach, that he knew would remain with him as long as he lived. Something primeval lurked inside him, something that frightened him because he did not know if he could ever exert authority over it. He had traveled to Tierra del Fuego to chart the wilderness, to list it and catalogue it, that it might be tamed and civilized, to bring the primordial darkness under control. But what of the darkness inside him?[104]

Watching him for years with his cold naturalist's eye Darwin arrives at his own diagnosis of FitzRoy's condition. In his opinion this aristocrat belongs to a "species [that] has reached the end of its natural lifespan"[105] and must therefore become extinct and be swept aside "like the great beasts of old". Modernity abhors idealists of royal blood who want the world to succumb to their spiritual vision of it and it is scientists and businessmen who are going to "inherit the earth".

---

104   Harry Thompson, *This Thing of Darkness* (London: Review, 2005), 72.
105   Thompson, *This Thing*, 428.

FitzRoy's conduct in Tierra del Fuego is symptomatic of this. Quite accidentally getting involved in a conflict and ending up with unclaimed native children on-board, he genuinely believed his duty was to bring light to Tierra del Fuego by educating the natives in England and then sending them back to spread knowledge and religion and "raise [their countrymen] from the brutal condition"[106]. This was bound to happen because God wanted this: all people were equally his prime creation, though because of sin some of them degenerated into pagans. FitzRoy's duty was to act as God's emissary and, moreover, he planned to use the success of the mission in scientific discussions with the advocates of an evolutionary view on the descent of men: "If I could go further and form them in the ways of polite society, then it would prove to the world that all men are created equal in the eyes of God"[107]. These two beliefs – in the literal truth of the Bible, and in the equality of all men – were so important for FitzRoy that he expected the whole of nature to behave accordingly and was genuinely shocked whenever such was not the case.

The failure of his Fuegian experiment made FitzRoy feel guilty for having involved the unwilling natives in it. He showed some a better life and then snatched it away from them, leaving them in the wilderness to suffer social ostracism. He realized that it was vain of him to impose his plan on other people's lives: "I have taken away their innocence, something I had no right to do. I wanted to bring them closer to God, but at the end it was I who played God with lives of other men"[108]. The feeling of guilt stayed with him to his last days, adding to his neurosis and in the final moment just before cutting his throat with a razor over thirty years later he thought about Jemmy Button, the brightest and most friendly of the natives, hoping they would meet in the other world: "Would Jemmy be there too... Or was Darwin right? Was he just another monkey, too highly developed for his own good?"[109]. It is FitzRoy who was the true humanist, and not Darwin – the latter saw all nature in hierarchical order, a continuum stretching from amoebae to men:

[T]he gap between the Caucasians and the lower apes... is greater than the gap between, say, the Negro and the gorilla... Come FitzRoy, look at the orangutan, its affections, its passion, its rage, its despair. Then look at the savage: naked, artless, roasting its parents.

106   Thompson, *This Thing*, 77.
107   Thompson, *This Thing*, 77.
108   Thompson, *This Thing*, 316.
109   Thompson, *This Thing*, 558.

Your Fuegians remind me of nothing so much as an orangutan taking tea at the zoological gardens.[110]

Darwin goes on to prove that Western man is in competition with both the natives of the New World and the great apes because he is expanding his empire and needs the lower creatures' niches. As a much more advanced life-form Caucasians must win – and Fuegians and orangutans are fated to perish. FitzRoy is very emotional in his arguments with Darwin and also desperate to find evidence for his theories. He interprets the Chilean native legends of the deluge as proof of the accuracy of the Book of Genesis and the universality of the Bible. Darwin's opinion is that Christianity better meets the demands of life than heathenism does, and that is why the Bible-worshipping West will conquer the rest of the world, and the question of whether God exists or not is here irrelevant.

The two men represent opposite ideologies and their conflict is a life-long affair. In later years, FitzRoy is devastated by the death of his beloved wife, Mary, but still believes there "had to be a reason... For without a reason, what point was there to her life, to his life, to anyone's life?"[111]. In the same period Darwin also suffered a great loss as his daughter, Annie, died at ten, but his emotional reaction was quite different. In Thompson's book at this moment he refused to "have anything to do with a God who delights in such cruelties"[112], though he was aware that he was causing still more pain for his family. In *This Thing of Darkness* FitzRoy makes Darwin give his word that he was never going to publish his heretical book, and in return he helps him in his research. After the return of the Beagle home Darwin pays a visit to FitzRoy to ask him for his samples of finches collected in the Galapagos Islands; Darwin did collect the birds himself, but he failed to label the specimen according to the particular island they come from, and so once he noticed the differences in their beaks he had no way of checking whether the different sub-genres came from different islands. It is FitzRoy's carefully inscribed boxes of finches that save the day. Besides, Darwin also fails to take an interest in the differing patterns on the tortoise carapaces, though the ship's surgeon did, at least in Thompson's version, tell him about it. Darwin in this novel is thus very far from ideal, scientifically as well as ethically. He is very passionate about nature, he does his research round the clock, for example playing with his children he makes notes concerning behavioral analogies between babies and piglets, and yet he loves his family very much and is loved in return.

---

110 Thompson, *This Thing*, 242.
111 Thompson, *This Thing*, 496.
112 Thompson, *This Thing*, 502.

Nonetheless, his success is partly sheer luck and partly the ability to use other people: FitzRoy, the surgeon, his own skillful servant, Covington, and finally, Alfred Russell Wallace. It is after Darwin receives Wallace's letter with the paper containing Wallace's independent discovery of the major assumptions of the theory of evolution that Darwin begs FitzRoy to forget the old promise and let him publish *On the Origin of Species*. FitzRoy agrees as a man of honor, but says he is going to ardently criticize the book in reviews, and Darwin's friends arrange for the joint presentation of Darwin's and Wallace's papers at a scientific society meeting. Yet FitzRoy is well aware that "in practice the name Wallace shall never escape the society's four walls"[113], because Darwin is an insider of upper-middle class London scientific milieux and Wallace is young, poor, and unknown.

Overall, *This Thing of Darkness* depicts FitzRoy as a tragic hero, noble and well-meaning, but a victim to melancholy and ill luck, and Darwin as a coward hiding behind the backs of his friends and his psychosomatic ailments, who uses other people and is afraid to face the consequences of his own discoveries. And yet ultimately he does face them, and this one brave thing he does in his life is remembered in history. Such a Darwin differs very much from the iconic figure we see in nature films, mainly because *This Thing of Darkness* is written from the point of view of FitzRoy, with whom the reader is made to identify and whose opinions he or she shares. All the same, Thompson remains faithful to the facts one can find in Darwinian archives, such as letters, diaries, and reports.

This is not the case with John Darnton's *The Darwin Conspiracy*, a book belonging to the recently arisen genre of neo-Victorian novel, the kind of fiction that flourished around the centenary of Queen Victoria's Golden Jubilee. It is concerned with the often nostalgic presentation of the Victorian past from a contemporary perspective and with post-modern narrative strategies, the main one of which is parallel narration: two storylines intertwine – one set in the mid-nineteenth century, the other today. The masterpiece of the genre and Darnton's main inspiration is Antonia Byatt's *Possession*, a novel whose late 20th-century protagonists, a pair of literary scholars, investigate the mystery surrounding the biographies of two well-known Victorian poets, whose scandalous love-affair was covered up. Darnton's novel repeats this scheme: a pair of scholars interested in Darwin and looking for a new research angle in the over-exploited field of evolutionary studies stumble upon a mystery and after many adventures realize that the official version of their idol's life is but an artifact, a skillfully manufactured lie. There was a Darwin conspiracy among the best naturalists of his days

---

113   Thompson, *This Thing*, 531.

who, for the sake of the theory of evolution, decided to keep a few things secret forever.

Darnton applies Byatt's narrative techniques: parallel to the contemporary plane we get to know the story from the past, the pivot of which is Darwin's youngest daughter Lizzie, who is always mentioned but in passing in official history as her spinster life is deemed bland. The truth is quite different; Lizzie, the mother of an illegitimate daughter (and an ancestor of one of the scholars, again like in Byatt) was the first to uncover her father's shameful secret. She conducts her own investigation thirty years after the Beagle voyage and writes everything down in her diary. This diary, period letters written by people so episodic in the Beagle story that no historian had bothered with them before, Darwin's confession in the lost epilogue of his *Autobiography*, the letters Lizzie wrote to her daughter and left with the lawyers who had arranged for the girl's adoption in America, and similar mysterious texts recovered in various manners (found in old family archives, hidden in secretive places in the Down house) are the sources from which the 'true' story is deduced by the scholars who in the end sit down to write the whole story in their book. Darnton refers throughout to Darwin's standard biography, but only in order to show that some of the oft'repeated facts are the falsified results of conspiracy.

At the beginning of the novel Darnton's protagonists are ardent lovers of Darwin and all the standard Darwiniana:

> He admired so many things about Darwin – his methodological exactitude, his boyish enthusiasm for experiments (imagine, playing the bassoon to see if earthworms could hear!), his demand for facts, nothing but facts and his willingness to follow them wherever they led wading knee-deep into lakes of hell-fire if need be… How extraordinary to possess sight that could stretch so far backward that the infinitesimal wheels of change and chance become apparent in their movement, like Galileo examining heavenly revolutions through the telescope. And how brave to measure yourself against the eons of all that time and recognize you live in a Godless universe and admit your nothingness.[114]

This description is focalized by a young scholar whose ideas about Darwin clearly come from the 'official' history dispersed by the media. His girlfriend, an American descendant of the scientist's illegitimate granddaughter, also cherishes illusions about her noble ancestor. Asked what she likes about him most she answers that she is moved by the fact that once on the Galapagos Darwin took with him inland but a single book:

---

114   Darnton, *The Darwin Conspiracy*, 12.

Paradise Lost. He read it here and then he thought about what he saw here and somehow he put the two together… He found Eden, he ate from the tree of knowledge, and the world hasn't been the same since.[115]

The above is the poetic vision of a genius who by the sheer power of reasoning pushes civilization in new directions, and sees what no one has seen before: a Galileo figure. The research starts with an attempt to find out why Darwin waited twenty-two years to publish the theory he had conceived in the Galapagos. The standard answers are that he did not want to upset his religious wife, and that his fellow naturalists who believed that God made-species are fixed and never-changing by studying their representatives were in their own opinion document-ing God's work, which was a very comfortable position. Christianity had been around more than eighteen hundred years and in comparison to this, according to one of the characters, the 22 years Darwin took to overturn it is not much, "a ratio of ninety to one"[116]. The characters visit the Down House museum with an exhibit depicting the Tree of Life with the drawings of animals in the balloons at the branches, just like bulbs on a Christmas tree. They imagine Darwin on the Galapagos when the naturalist observed the beauty and savagery of nature "and he felt he understood it as if he had observed a moment of creation"[117] (243). This was the moment of illumination "like St Paul on the road to Damascus and Archimedes in the bath"[118].

Gradually in the course of their research this too-good-to-be-true idyllic vision of Darwin the Sage becomes marred. The diary his inquisitive daugh-ter recovered from obscurity contains accounts of strange incidents happening towards the end of Darwin's life which cannot be explained within the tradi-tional frame of his biography. Thomas Huxley is overheard to talk about some inexplicable danger to Darwin's good name, Darwin is devastated having heard some otherwise innocuous words of Wallace, his reaction to Lizzie's question on whether he lost his faith after the death of Annie is inexplicable. Lizzie grows aware of some dark secrets in her father's young years and she decides to dis-cover them. She pays a visit to FitzRoy just a few days before his suicide, but the former captain is so deranged that she does not learn much from him, only that he believed Darwin was "the Devil's own Pied Piper"[119], guilty of violating the Commandments.

---

115   Darnton, *The Darwin Conspiracy*, 12.
116   Darnton, *The Darwin Conspiracy*, 111.
117   Darnton, *The Darwin Conspiracy*, 243.
118   Darnton, *The Darwin Conspiracy*, 271.
119   Darnton, *The Darwin Conspiracy*, 141.

Slowly, step by step, by comparing unedited and edited versions of Darwin's diaries and notes from the days he sailed on the HMS Beagle, and by reading the letters home of less prominent members of the crew, Lizzie realizes that there is a conspiracy of the famous naturalists who are determined to hide their guru's shameful past. The key to the riddle is the gradual change in the young Darwin's beliefs concerning nature reflected in his scribblings. In the beginning the major metaphor he uses is a ladder with some creatures occupying its bottom rung, others at the top. In later years this Linnaeus model is replaced by the Tree of Life scheme and the change did indeed occur during Darwin's voyages, but not in the way history has it. While editing his journals for publication Darwin for some reason got rid of all the entries mentioning the other naturalist onboard, McCormick. Thus, in the official version of the journey, McCormick left at the first port having quarreled with the crew. In fact McCormick participated in the journey for years – until his death in an accident – and accompanied Darwin in all his research. On the way home both scientists were exploring a small active volcano: they both had left camp in the morning and yet only Darwin came back, claiming that McCormick had fallen in. FitzRoy who had known of the mutual hatred of the two men was suspicious and his dislike of Darwin grew, which fact explains his behavior in the last years of life. Moreover, Lizzie discovers that the young Darwin purposefully mixed up his finch specimens from the Galapagos and then made much show of going to the FitzRoys' and begging for the captain's well-labeled collection[120], which resulted in the common belief that he got his great idea only after leaving the Galapagos archipelago and not, as in reality, before.

The final resolution of the enigma can be found in the letters of Matthew (the unfortunate missionary-to-be from Tierra del Fuego) to his mother. He describes there an excursion he, McCormick, and Darwin made in Tierra del Fuego when Jemmy Button nagged them to go along with him to his native village in the mountains and meet the local king, his father. His people are a very old tribe from the South American hinterland who had to flee and now are dying out in the inhospitable peninsula. The king organizes a feast in honor of his son's white friends and, kindly asked, he recounts to them his people's history and beliefs and in turn asks the Europeans their ideas concerning the commencement

---

120 The story of Darwin's finch collection and the mixed-up specimens from different islands is rather famous and the authors of Darwin's biographical and semi-biographical novels often refer to it, altering it to fit their own artistic agendas. Thus the discrepancies: in McDonald's account it is Syms Covington who saves the day, in Darnton's book – FitzRoy.

of life on Earth. Darwin, well-acquainted with the Bible (he was to have become a country parson) and his beloved *Paradise Lost*, tells the king the Christian story of the Garden of Eden, the Serpent, Adam and Eve, Noah, and then Christ and his Sacrifice. The native king is mistrustful: he says that it is far better to explain things in nature without introducing unseen powers in which one must believe though having no proofs and no way of falsifying the myths. The tribe has a completely different mythology: "there was the single little thing and everything grew out of it… when time is so long many little changes can occur and put together they add up to a big change[121].

The tribe worships no god other than nature and maintains that all elements of it are closely related as they have common roots. To explain this, the king points to a big tree. Its branches, leaves and twigs, he says, are like animals, plants and humans, separate but interconnected. In order to see how they change into new species it is enough, according to the king, to notice that worse things are always pushed away by better ones, and, given time, the better ones alone are going to stay on Earth. Both Darwin and McCormick listen to this, they both have been thinking and arguing about the origin of species and the mechanisms of their transmutation for years and they both realize simultaneously that the ancient lore of the tribe is the answer. They are also aware that the one who will publish the book on evolution elaborating the Tree of Life metaphor is going to get the fame and the money and the other one must be gotten rid of. This is how their mistrust and hostility toward each other was born. As it turns out, the third European present, Matthews, failed to appreciate the meaning of the tribesman's story, but remembered it as a nice conversation piece on native American folklore and, some years later, he shared it with a white explorer he accidentally met in the colonial wilderness on another continent, Alfred Russell Wallace. Thus, not only did Darwin help McCormick fall into a volcano, but much later he also had to deal with Wallace, who wanted to publish the Tree of Life book and pass it off as his own invention. Once he realized Darwin knew the folkloric story before writing *On the Origin of Species* he started to threaten him with exposing the whole truth and had to be appeased with blackmail money.

All of Darwin's life after his return home from the Beagle: his nervous and health problems, religious crisis, hesitation over whether to publish the book or not, tense relationships with FitzRoy and Wallace (he paid both to keep quiet) thus finds explanation. Lizzie, his brightest and wildest daughter, is the only person aware of the truth, but as an unmarried mother and a fallen woman she is

---

121   Darnton, *The Darwin Conspiracy*, 293.

a social outcast. She is sent to the continent to deliver her child, and after the subsequent adoption is sentenced to lifelong spinsterhood in her father's house, deprived of her right to speak her own mind and participate in social life. She dies without letting anyone know the secret, the only exception being her letter addressed to her daughter and left with the family's lawyers.

Thus, Nichols, Thompson, and Darnton give us their own versions of the 'true' life of Darwin. They vary in the degree of license they take: Nichols' version is in agreement with standard history, but the author adds his own interpretation of the facts; Thompson's novelty is in switching the focus from Darwin to FitzRoy and in this new perspective to make the villain of the previous biographies seem sympathetic and Darwin himself egoistic; Darnton's version is the most radical, his account being a neo-Victorian novel per se and replacing the non-fiction story of Darwin with a conspiracy theory. All three books are recently written, all reflect contemporary sensitivities and intellectual passions. Most attention is paid to long-extinct exotic tribes and their forgotten lore, ecology is considered important, colonization as such is mistrusted, and so is imperialism and the British middle class. The life of Darwin, a contemporary myth, proves to be malleable enough to serve diverse needs: one can easily re-write it according to his or her ideological agenda and it remains fascinating.

References to the Theory of Evolution in the Novels of John Fowles, A.S. Byatt, and Hilary Mantel

In the essay "The New Body of Writing: Darwin and Recent British fiction" A.S. Byatt claims that in the 1970s and 80s there is an "almost obsessive recurrence of Darwin"[122] in the British novel. Her examples include Graham Swift's *Waterland*, Peter Carey's *Oscar and Lucinda*, Julian Barnes' *A History of the World in 10 and ½ Chapters*, along with a number of other books. Byatt connects this Darwinian obsession of contemporary novelists to their interest in history, natural history included, and to their tendency to explain human destiny by references to modern science – the epitome of which in popular imagination is Darwin.

The aim of this chapter is to challenge the critical cliché that in recent fiction Darwinism replaces religion and that the scientific worldview is always in opposition to Christian belief. A close reading of three British novels written between the late 1960s and the early 1990s – namely, John Fowles' *The French Lieutenant's Woman*, A.S. Byatt's *Morpho Eugenia*[123], and Hilary Mantel's *A Change of Climate* – will demonstrate how Darwinian references themselves evolve over time. In order to do this I shall juxtapose three aspects of the novels: primarily, the way they depict Darwin and his fellow naturalists (both historical personages and fictive characters) and thus create the myth of how modern science was born in Victorian England. Moreover, I will establish what the novelists in question understand by the word 'science' and whether for them natural science is or is not science proper; and lastly, what is their attitude to the alleged conflict between Christian belief and the theory of evolution. In the quarter century dividing Fowles' novel from Mantel's much changes in the way each of these problems is handled.

In John Fowles' famous *The French Lieutenant's Woman*, a metafictive love story interspersed with authorial lectures on the Victorians and our erroneous understanding of their culture, naturalists are described as a laughingstock, though quite soon the narrator makes us aware that they apparently were far more intelligent and laborious than we presume. Most of the novel's action is set

---

122 Antonia S. Byatt, 'A New Body of Writing. Darwin and Recent British Fiction', in Antonia S. Byatt and Alan Hollinghurst, eds, *New Writing 4* (London: Vintage, 1995), 443.

123 A short novel published together with her other novelette *Conjugal Angel* in the volume entitled *Angels and Insects*.

in Lyme Regis, "a Mecca for a British paleontologist"[124], where rocks abound in interesting fossil specimens, and a certain Mary Anning, a natural-born geologist, who keeps an Old Fossil Shop where she trades her finds. The gentlemen who came to Lyme Regis were usually "Scholarly collectors of everything under the sun"[125], country squires who instead of drinking and womanizing indulged in natural science. Attracted by the promise of finding (or purchasing) often yet to be classified fossils and engaging in scientific discussions in the town's parlors they made Lyme Regis their pilgrimage center. As the narrator ironically puts it: "these last hundred years or more the commonest animal in the shore has been man – wielding a geologist's hammer"[126].

The novel's protagonist specializes in sea-urchins, so-called tests, which are beautiful but difficult to find, which fact makes them attractive enough for a gentleman to collect and moreover lets him fill the leisure time he has in abundance. The gentleman-geologist is well-equipped and his apparel is enormously uncomfortable, including stout nailed boots "as suitable as ice skates"[127] and a spacious rucksack filled with heavy hammers, wrappings, notebooks, and pillboxes. Being a naturalist is thus a matter of fashion: you dress up for the role and act it out. Yet Fowles' amateur naturalists are surprisingly competent. They follow the day's scientific debates and know a lot from different fields of natural science: geology, botany, ornithology, and so on. As they "can afford to dabble everywhere"[128] and are anything but narrow specialists. The novel's characters discuss the theories of Charles Lyell described in his seminal *Principles of Geology*, the book which opened up the way for yet more ridiculous theories of Earth's history by proving that our planet is millions not thousands of years old. They also laugh at the creationist ideas of people such as Gosse who tried to eliminate certain anomalies between science and the Bible by proposing that God created fossils of earlier forms together with modern species of animals. The narrator provides the reader with footnotes explaining who was who in the scientific life of the 1860s along with authorial comments summarizing the books the characters read and discuss.

Fowles' naturalists feel the nation's elite to be but "two grains of salt in a vast tureen of insipid broth"[129], and when they happen to recognize one another at a chance social gathering they strike one another like Crusoe and Man Friday,

124   John Fowles, *The French Lieutenant's Woman* (London: Penguin,1969), 43.
125   Fowles, *The French Lieutenant's Woman*, 16.
126   Fowles, *The French Lieutenant's Woman*, 43.
127   Fowles, *The French Lieutenant's Woman*, 45.
128   Fowles, *The French Lieutenant's Woman*, 46.
129   Fowles, *The French Lieutenant's Woman*, 141.

and greet each other with: "A Darwinianin?' 'Passionately'"[130]. Thus, overall, naturalists in *The French Lieutenant's Woman* are presented as affluent, decorative gentlemen-scientists and their hobby is a half-funny, half-serious quasi-scientific endeavor. As contemporary readers probably do not know the important scientific works of the period, with the possible exception of *On the Origin of Species*, necessary contexts and summaries are provided.

A quarter of a century later A.S. Byatt published her novel, which is set in the 1860s in a country manor. *Morpho Eugenia* casually refers to famous naturalists of the period and their publications and the narrator feels no need to explain who they were. The popularity of neo-Victorian themes in the 1970s and 80s made such explanations redundant, as a Darwinist gentleman reading his famous contemporaries had become a stock figure in this type of a novel.

Byatt's protagonist, an intellectually adept butcher's son skilled at skinning animals, made his way in the world by becoming a naturalist. He collected everything, mainly flowers and insects, categorized them with the help of available wildlife encyclopedias, and ultimately made this hobby his profession. He spent ten years in the Amazon collecting and sending back to England specimens which were then sold to affluent gentlemen collectors. After his return he pays a visit to one of them, a lord who boasts one of the most precious nature collections in the country. He finds most of the specimens rotting away in a stable, but some have been made by the lord's artistically-inclined daughter into beautiful collages. The lord of the manor cuts a stereotypical figure: he is a collector and a connoisseur of natural science. However, as the second son, before having inherited his fortune he had taken Holy Orders, and he is very much committed to both science and religion.

The protagonist, excepting his lowly pedigree, is a quite typical naturalist too, with "a ruling passion, the social insects. He peered into the regular cells of beehives, he observed trails of ants (…) Here was the clue to the world"[131]. His research into ants, both in the Amazon and in the native English countryside, becomes in the course of the narrative symbolic: ants reflect human societies and the microcosm of an ant heap is the model of the macrocosm of Nature. In his conversation with the scientifically-minded squire the protagonist refers to Henry Walter Bates' articles in *Zoology* and to the works of Bates' friend, Alfred Wallace (later famous for his discovery of natural selection, simultaneously with Darwin). These references are a little like name-dropping, quite irrelevant to the

---

130   Fowles, *The French Lieutenant's Woman*, 141.
131   Antonia S. Byatt, *Angels* (London: Chatto and Windus, 1992), 10.

action of the novel, yet they show that the author researched the natural science of the period.

The protagonist read Humboldt and W.H. Edwards, whose accounts of voyages influenced Lyell and Darwin, among others. His own research is described in reference to the real-life scientists of the epoch and their publications. The protagonist claims:

> I have come to be particularly interested in ants and termites. I should like to make a prolonged study of certain aspects of their life (…) I may have a better explanation than that put forward by Mr. Bates… and this would reinforce the observations of Mr. Darwin. Certain ants… appear to have affected the form of the plants over the millennium.[132]

The above is a two-folded allusion: apart from putting the protagonist's studies in the context of scientific and journalistic records of the 1860s – the real articles published in the really existing periodicals – it also anticipates what is going to happen in the next century, our times. As Byatt's late twentieth-century readers know, it was the research concerned with the behavior of social insects – ants, bees, and termites – that led E.O. Wilson to found a new branch of natural science: sociobiology. Sociobiology in turn (at least, according to Byatt's own essays) is of special importance to modern culture, as it proves that very many patterns of behavior which used to be considered exclusively human (such as altruism) have in fact evolved in the animal kingdom as they increase the species' chances for survival. Sociobiology is for Byatt, as she claims in the essay I cited above, a lay equivalent of Christian ethics: "We look for our morality in works like Richard Dawkins' *The Selfish Gene* or E.O. Wilson's *On Human Nature*."[133]

The decade of the 1860s is presented in *Morpho Eugenia* as the turning point in the way people see their place in nature. The theory of evolution infiltrates popular imagination and itself becomes a cultural myth, like in Lord Tennyson's famous poem, whose fragment Byatt quotes in the novel[134]:

Who trusted God was love indeed
And love Creation's final law –
Though Nature red in tooth and claw
With ravine, shrieked against his creed –[135]

---

132   Byatt, *Angels*, 16.
133   Byatt, 'A New Body', 443.
134   The original is: Who trusted God was love indeed/And love Creation's final law/Tho' Nature, red in tooth and claw/With ravine, shriek'd against his creed.
135   Qtd in Byatt, *Angels*, 87.

Attempts at reconciling the new myth of survival of the fittest (i.e., creation which works by "Nature, red in tooth and claw") with the older myth of God the Creator who is always good and loving were made numerous times by Victorian academics, and these often bizarre treatises are comically recalled in neo-Victorian novels. In John Fowles' book, as already mentioned, we read about the grotesque *Omphalos* by Mr. Gosse, and the conservative country squire in *Morpho Eugenia* plans to write a similar apologia himself.

Each of the three novels analyzed here is much concerned with definitions of science. The theory of evolution is their focus because Darwinism serves in the post 19th-century world as a handy metaphor of nearly everything – from social relations to the history of the Universe. In *The French Lieutenant's Woman* Darwin's *On the Origin of Species* is referred to in the epigraphs to chapters in such a way as to emphasize its symbolic nature, sometimes running the risk of becoming a parody of crude social Darwinism. For example, a chapter devoted to the description of an aristocrat whose education and habits are anachronistic in an early capitalist society is introduced by the following quote:

> The chief part of the organization of every living creature is due to inheritance and consequently, though each being assuredly is well-fitted for its place in nature, many structures have now no very close and direct relations to present habits of life.[136]

The quoted passages are often among the best known, so that even readers who know next to nothing about Darwinism would recognize definitions of phenomena such as the survival of the fittest or natural selection. The narrator comments on the importance of the theory of evolution by saying that Darwinism has infiltrated our way of thinking to such an extent that when we think about the historical Darwin we cannot help but realize that this flesh-and-blood Victorian scientist did not "understand Darwin himself"[137]. Yet even though contemporary readers are unconscious Darwinists, they tend to look down on Victorian natural science: "natural history had not then the pejorative sense it has today of a flight from reality"[138], complains the narrator, who then refers to Darwin's *The Voyage of the Beagle* and *On the Origin of Species* as to "triumphs of generalization not specialization"[139]. Darwin's genius, according to the novel's narrator, is only seen in retrospect and consists in upsetting the immobile vision

---

136   Qtd in Fowles, *The French Lieutenant's Woman*, 15.
137   Fowles, *The French Lieutenant's Woman*, 47.
138   Fowles, *The French Lieutenant's Woman*, 47.
139   Fowles, *The French Lieutenant's Woman*, 47.

of nature as never-changing and liable to classifying, pigeon-holing, and in general "fossilizing the existent"[140] in the way Linnaeus did.

The moment the protagonist finds a beautiful fossil in Lyme Regis is thus one of aesthetic satisfaction, not of discovery. The protagonist admires: "microcosm of macrocosm, whirled galaxies that Catherine-wheeled their way across ten inches"[141]. He appreciates the decorative character of the fossil, but fails to realize that he himself is very much like it, a beautiful relic of a no longer existent social order.

By way of contrast, Byatt's protagonist seems to be equipped with a better understanding of the ways of people and wildlife, and this allows him to apply his naturalist's instinct for observation to all aspects of life. Back from the Amazon he looks at an affluent English country manor in the same way he looked at savage villages. A provincial ball with its highly ritualized dancing patterns is for him but a variation of the same social institution he observed during an orgiastic palm-wine ritual in the Amazon jungle.

Being a naturalist is for Byatt synonymous with possessing an ability to pigeon-hole every phenomenon around: architecture, anthropology, manners, botany – all is accessible on the same plane. The basic question in each case is thus to find the ruling principle turning the mass of neatly grouped phenomena into a meaningful whole. Symbolically, the lord of the manner asks the protagonist to look through a vast collection of specimens he bought over the years from researchers working in the colonies and: "make sense of it, lay it all in some order or other"[142]. The failure of the protagonist – who for hours carries out examinations, establishes categories, makes labels, but finally abandons his task – shows his inability to build any meaningful system and create a model of nature. It is only when he starts to systematically observe the behavior of one small colony of English ants, make notes, and finally to write a book on the social history of this very anthill that he is successful. His valuable book proves to be "a supremely moving example of the inexorable secret work of natural selection"[143] and its readers are "struck by how completely Mr. Darwin's ideas might seem to explain [the life of ants]"[144]. It is thus with the small scale we should start – only then are we going to see how nature works, Byatt seems to suggest. However, the protagonist is unable to devise a general principle organizing the diverse

---

140 Fowles, *The French Lieutenant's Woman*, 47.
141 Fowles, *The French Lieutenant's Woman*, 48.
142 Byatt, *Angels*, 25.
143 Byatt, *Angels*, 102.
144 Byatt, *Angels*, 102.

specimens of wildlife into an orderly model of Nature. Nonetheless he eagerly explores the microcosm of an anthill and generalizes the results of his research cutting a proto- E.O. Wilson figure.

The most pessimistic discussion of what science is can be found in Mantel's *A Change of Climate*, the motto for which also comes from Darwin's writing and reads: "We are not here concerned with hopes and fears, only with truth as far as our reason allows us to discover it. I have given evidence to the best of my ability....". This passage comes not from *On the Origin of Species* but from *The Descent of Man*, a later book in which Darwin shares with his readers his observations concerning not wildlife, but the human race. Aware of the criticism the book is going to stir, he makes his reservations thus implying that the limits to knowledge are within the human mind: we research into truth as far as our intellectual abilities allow us.

Hilary Mantel's *A Change of Climate*, whose action, though concerned with the Victorian heritage of Darwinism and religion, takes place in the twentieth century (simultaneously on the two temporal planes – in the 1950s and the 1980s) refrains from entertaining the reader with portraits of period gentlemen-scientists. Instead the book tells the story of a young scientifically inclined boy who in the mid-twentieth century finds a beautiful fossil on an English beach during his holiday stay near Lyme Regis (an echo of a parallel scene in *The French Lieutenant's Woman*). Determined to become a geologist he studies natural science on his own and gets acquainted with the history of Darwinism and the diverse writings of Victorian illuminati. His decision to go to university is opposed by his fundamentalist father who blackmails him into submission and final resignation from geology, which he considers unbecoming for a Christian soul.

The protagonist gets married instead and with his young wife they go to South Africa to do charity work at a mission. Involuntarily drawn into political upheaval they are sent to prison and then to a very remote village near the Kalahari. It is there that their twin children are born and, some months later, kidnapped by run-away servants. The parents manage to retrieve the baby girl, abandoned by the criminals in a ditch, but the boy, whose blood and tissues are much more valuable in native rituals, is never found. The main action of the novel takes place twenty years later. The couple lives in provincial England where they work for a social charity fund. They are bringing up four children and provide temporary shelter for people deemed by the social services unable to cope on their own. The eldest girl, already a student, is not even aware she used to have a twin brother, but nevertheless she is somehow driven to volunteer to go to Africa with some humanitarian organization. The book is a prolonged study of how an unacknowledged tragedy, the death they never discuss but always think of, ruins the couple's life.

The Darwinian motto is bitterly ironic here: had the protagonist become an evolution-minded geologist dazzled by the beauty of fossils he would never have gone to Africa and his eldest son would not have been slain by a witch doctor. Science is beautiful yet quite useless when facing human atrocities. The moment the protagonist found his fossil is but the wonderful memory of a child: "a sharp pang of delight took hold of him, a feeling that was for a moment undistinguishable from fear. He had picked up a fossil: a ridged, gray-green curl, glassy and damp like a descending wave"[145]. All he learns about the geological past of our planet in general and the fossil in particular seems like a romantic vision: the warm seas of 150 million years previous, a small animal who died and whose shell was filled with sand and then compacted into a rock. Geology is a fairy tale: now that the ocean is reclaiming the east coast of England fossils are returning to the sea of their origin and only beach-combers stop them on their way. Reading natural history, mainly the history of the rocks and the soil, he performs a private trick:

> to look at the landscape and strip away the effects of man. England transforms itself under the geologist's eye, the scavenger sheep are herded away into the future and a forest grows in a peat bog, each tree seeded by imagination. Where others saw the lie of the land, Ralph saw the path of the glacier; he saw the desert beneath copse and stream: and the glories of Europe stewing beneath a warm clear shallow sea.[146]

He imagines time as a road on which he can walk back and forth and along which all geological epochs are situated and each form of life has its own place; sea-urchins and magnolias alike: "it was as clear in his mind as it might be in a child's picture book… the Irish elk, the woolly mammoth, then man, stooped hairy furrow-browed. It's a success story."[147] Bygone eras seem to him a kind of idyll with much sun and no people to enjoy it. Such a vision is drastically changed in the next thirty years of the protagonist's life. The tidy, clear vision he cherished in the 1950s becomes by the 1980s marred by experience and the knowledge of life and death, which turns out to be nothing but randomness. The Darwinian frame of mind explains nothing at all:

---

145 Hilary Mantel, *A Change of Climate* (Anstey, Leicestershire: F.A. Thorpe Publishing,1995), 64.

146 Mantel, *A Change*, 72.

147 Mantel, *A Change*, 75.

Every action contained its opposite... nothing was fixed, nothing in creation; cells made choices all the time. If we could rewind the tape of the Universe and play it over again we might find ourselves to be different: six-legged, intelligent creatures crawling on the seabed.[148]

He looks at his heirloom, the old fossil, which swells into a very complex symbol denoting his childhood, the romance of natural history, and the positivist dream of science in order to retrieve his long lost confidence in the logic of the world and find comfort: "the past doesn't change course: it lies behind you, petrified, immutable. What changes is the way you see it. Perception is everything".[149] This final conclusion belongs not to evolutionism, but to the more recent post-Heisenbergian vision of science: nature is unknowable, the uncertainty principle rules.

Fowles, Byatt, and Mantel feel equally obliged to include in their discussions of Darwinism the religious implications of the theory, but the way they do so is very different: from Fowles' simple statement that Darwinism was a mock-equivalent of religion for agnostically-minded gentlemen-scientists, through Byatt's claim that natural history may help the human race free itself from religious fallacies necessary at lower levels of civilizational development, to Mantel's pessimistic view that neither religion nor science can give any definite answers concerning humanity's place in the Universe. In *The French Lieutenant's Woman* Victorian illuminati proud of having grown out of religious needs playfully replace standard rituals such as vow-taking with their "Darwinist" new versions. In an emotionally charged moment the protagonist's friend: "turned and went to the bookshelves by his desk and then came back with... Darwin's great work... he laid his hand, as if swearing on the Bible, on *The Origin of Species*"[150].

In *Morpho Eugenia* two worldviews: religious and scientific stand in opposition and the only person who tries to reconcile both, the lord of the house, is doomed to fail. His study where he tries in vain to write a book proving that: "the extraordinary beauty of every creature is in itself the evidence of the work of a Creator"[151] is next to the manor's chapel where he preaches daily sermons to the family and servants. His lessons are kind, and yet to the protagonist (who was brought up in an orthodox Calvinist tradition) they seem to lack true religious zeal. The lord is aware that the vision of Christianity he had as a boy – with the stories of the First Parents in Eden, the Birth of Christ in a snow-covered stable

---

148   Mantel, *A Change*, 550.
149   Mantel, *A Change*, 660.
150   Fowles, *The French Lieutenant's Woman*, 192.
151   Byatt, *Angels*, 19.

with the Angels and the Magi – is now antiquated. In the post-Darwinian age he feels: "in a pit of despair itself"[152] because intellectually he accepts that "we are what we are because of mutations of soft jelly."[153] He persistently struggles to prove that "the world is the work of a Creator, a Designer"[154] in such a way as to confirm "the intricacy of the argument of Mr. Darwin."[155] at the same time. He knows Darwin is right, but he lacks the courage to accept the implications of the theory of evolution (ironically, in an anachronistic manner Byatt anticipates current debates: the intelligent design controversy and the discussion of "mutated soft jelly" genetics which the lord could not possibly have heard of in his own time). Such conflicting aims doom his book to failure: if ever written, it would have been a second *Omphalos*, a laughing stock for generations to come. The lord knows this somehow, and destroys all he writes, at the same time allowing his exceptional natural history collection to rot away as if to show that one cannot be both a Christian and a Darwinist.

His young guest, the protagonist, represents the next generation and a higher level of scientific awareness. Often in the Amazon among "passionate Portuguese friars"[156] and drug-taking Indian priests he feels suspicion toward all rituals: "not only the Amazon ceremonies but the English sermons seem strange, unreal, of uncertain nature"[157]. The protagonist, probably because of his Calvinist upbringing, tends to replace the doctrine of predestination he was once made to believe in with Darwinist "instinct". Consequently, though he is a full-fledged naturalist who looking at an ant community sees parallels with human social life and is capable of viewing both humans and ants as outcome of billions of years evolution of the Universe, he is still enslaved by the Calvinist belief that all organisms are predestined. Evolution, and the God of Calvin, seem powers external to life on Earth that control it entirely.

Only one character in the book is capable of imagining a fully materialist utopia: the ambitious though subdued governess who does follow the theological and scientific polemics of the day, and yet is very much interested in ants and men. She asks the protagonist at the end of the novel: "do you think it conceivable that there are finite beings with no afterlife – or that their natures may be

152   Byatt, *Angels*, 59.
153   Byatt, *Angels*, 59.
154   Byatt, *Angels*, 33.
155   Byatt, *Angels*, 33.
156   Byatt, *Angels*, 24.
157   Byatt, *Angels*, 24.

fully satisfied by the part they play in the life of the whole community?"[158]. This question, anticipating modern sociobiology, is yet left unanswered.

In *A Change of Climate* the discussions conducted by the characters of *The French Lieutenant's Woman* and *Morpho Eugenia* are already one hundred years old, but they still influence the way people think, especially in the evangelical east of England. The novel, apart from the already mentioned motto from *The Descent of Man*, also has a second epigraph, this time from the Book of Job: "Consider what innocent ever perished, or where have the righteous been destroyed?". The choice of this particular book of the Bible connects Mantel's text to the very old discussion on the reasons behind the obvious injustice so abundant in the world created by God who is said to be good and omnipotent. Yet Mantel's novel is definitely not a modern attempt at theodicy: according to his book there is no apparent aim of innocent suffering. It is rather people's naiveté that makes them think that if they are good no evil will be done to them. Young British missionaries going to Africa on church-financed charity missions are "familiar with the Psalms and (among other books) the Book of Job"[159]. Yet the narrator hastily adds: "they do not expect the Book of Job to have any practical applications"[160].

Mantel's protagonist in his youth wants to believe both in God and in evolution: "they weren't contradictory… Nobody thinks there's God on one side and Darwin on the other"[161], he tells his fundamentalist father who ironically anticipates Dawkins claiming that "Darwinism is atheism"[162]. He vainly tries to persuade his father that reading geology is not sinful; and that theirs is "an old debate, it's stale, it was never necessary in the first place"[163]. In the end, however, he succumbs to parental emotional blackmail and does not become a naturalist, and yet he goes to Africa firmly believing in a good God and the Human who is both the final product of Darwinian evolution and "has a unique place in creation"[164]. He agrees neither with his sister, who referring to their father quotes Freud's definition that "religion is a universal obsessional neurosis"[165], nor with people who call fossils "devil's toenails" and "maintained all fossils were planted in the rocks by Satan to tempt scholars into scientific hypotheses which led them

---

158   Byatt, *Angels*, 117.
159   Mantel, *A Change*, 453.
160   Mantel, *A Change*, 450.
161   Mantel, *A Change*, 79.
162   Mantel, *A Change*, 80.
163   Mantel, *A Change*, 81.
164   Mantel, *A Change*, 88.
165   Mantel, *A Change*, 99.

from the knowledge of God"[166]. Thus in the fifties his own geological finds seem to him trophies taken "in the battle for reason"[167].

Thirty years later and having experienced the loss of his child in a bestial ritual he considers his youthful infatuation with Darwin naïve and his simultaneous belief in the Christian God a matter of unconscious yet logical choice: "I thought it was more constructive to do so. I thought that not to believe was a vote for chaos"[168]. The world ruled by a benign God who decides for evolution to commence makes sense, it is orderly and patterned. Yet in such a world there is no place for African shamans buying white male children to use their body parts in cultish procedures and leaving female children behind as useless. The son's death is macabre and random. "But where is the pattern now?... Our lives have been ruined by malign chance. I do not see any pattern here, any reason why this had to happen"[169], he writes in a letter from Africa to the British clergyman he knows. The randomness of this tragedy and of other deaths, other tragedies, and the massive destruction so common in Earth's history make him remember his kid sister's question about the reasons for the dinosaurs' extinction and his own youthful and careless answer that "their habitat altered. A change of climate."[170]

The eponymous change of climate, denoting the randomness of the Universe, symbolically stands for the failure of theodicy. In Africa the protagonist felt that "it is God that took his child and cut him into pieces, dissected his child alive"[171]. God did not prevent this, thus there is no good and omnipotent God and survival seems a matter of pure chance: his baby daughter lived while his baby son, like many other white male infants, disappeared without a trace, leaving behind only "substances in bottles and jars"[172] at the captured and arrested witchdoctor's places. His friend, having received the letter with the tragic news concludes: "If it is a chance, can it be malign? If it is malign, can it be a chance?"[173].

Nor do death and survival follow a Darwinian pattern. Not only is the baby brother's death random, but so is the baby sister's survival:

It seems a strange impulse of grace to lay a baby down in a ditch, with a storm raging. She could have drowned in that ditch, or have died of cold before we found her, or have

---

166   Mantel, *A Change*, 259.
167   Mantel, *A Change*, 260.
168   Mantel, *A Change*, 404.
169   Mantel, *A Change*, 441.
170   Mantel, *A Change*, 999.
171   Mantel, *A Change*, 442.
172   Mantel, *A Change*, 439.
173   Mantel, *A Change*, 441.

been savaged by an animal. It seems to me that she has been selected for life and her brother for death.[174]

The unanswerable question is who did the selecting: both children were very healthy and physically strong, capable of survival in Darwinian terms. To say that God saved the girl would be to imply that God killed the boy. The protagonist feels abandoned by God and science, alone in the Universe, left for "eternity in the cold and the dark"[175] with no order and no pattern of whatever origin.

To conclude, the novels of Fowles, Byatt, and Mantel confirm the claim that the mid-nineteenth-century Darwinian naturalists and the intellectual debates provoked by their discoveries are still a very important subject for contemporary British novelists. Though the three novels discussed in this paper are very different stylistically and range from neo-Victorian metafiction to realism, their artistic effect springs from the juxtaposition of pre-Darwinian religiosity with the scientific worldview centered on the theory of evolution. Though the presentation of stereotypical gentlemen-Darwinists in the work of Fowles (and to a lesser degree the works of Byatt) is far from complex and rather entertaining, the introduction of Darwinian themes does serve serious ends – namely, the discussion of the changing understanding of what science is and whether evolutionism is or is not synonymous with atheism. The latter issue is definitely not a simple replacement of God by Natural History, but a very complex problem, especially inasmuch as, contrary to what some critics claim, Darwinism itself is not an impervious monolith. Indeed, as a frame of mind it now seems to be crumbling.

174 Mantel, *A Change*, 439.
175 Mantel, *A Change*, 443.

Echoes of the Mid-19th-Century Spiritual Crisis in Selected Contemporary Texts Referencing Charles Darwin

Our turn-of-the-millennium culture seems to attach great importance to Charles Darwin's works and biography. For instance, the debut of the 'Darwin online' project in October 2006 proved a gigantic media event. On the day of its inauguration this website, which offers free access to the enormous Darwinian archives, was headline news: it is estimated that some 400 million people learned of the debut, and several million tried to log on in the first forty-eight hours. Next, over the last forty years more than thirty actors have impersonated Darwin in a number of popular-science TV series and feature films, to say nothing of those who played Darwin on stage, most prominently in the numerous productions of the *Time Will Tell* music hall. What is more, new biographies and biographical novels devoted to the Darwins continue to be written and published, and so are books presenting Darwin's theories (and those of his followers) to non-scientific readers. Finally, reference to Darwin is a must in the recent histories of the English novel and of 19th-century British society, as we see with the recurrent thesis that it is through the spread of Darwin's ideas in the mid-19th century that Britain gradually lost faith in the literal truth of the Bible.

My aim in this chapter is to discuss the alleged relationship of Darwin to the mid-19th-century spiritual crisis as it is depicted in recent fiction and non-fiction. After a short historical introduction based on the books by Michael Ruse, Asa Gray, and Edward Caudill I am going to discuss three contemporary British books of fiction and semi-fiction by Randal Keynes, Jenny Diski, and Graham Swift in order to examine a number of standard associations the name of Darwin evokes in popular imagination. The most prominent of these are: the loss of faith, the search for forebears symbolized by inheriting certain 19th-century heirlooms, the feeling of acute nostalgia and suicidal depression, and the act of re-creating the past based on family documents, papers, or books.

Randal Keynes' *Annie's Box* is an account of the married life of his great-great-grandfather, Charles Darwin, and might be called, to use a fashionable term, 'faction', a blend of fact and fiction. Referring to the old papers of the Darwin family, the naturalist's notebooks and letters, Keynes tells the story of the last forty years of Darwin's life focusing on his relations with his children and on the spiritual crisis he went through when working on his theory. The two focal issues come together in the years following his eldest daughter's death, a tragedy which added to the trauma Darwin felt over his Bible-defying discoveries. At that time Darwin was experiencing a very dark period of psychosomatic ailments, pangs of guilt, and disappointment with the Anglican church.

In Graham Swift's *Ever After* the narrator, Bill Unwin, is recovering from a suicide attempt caused by the untimely death of his beloved wife, followed by the demise of his mother and step-father. In his misery Unwin is trying to come to terms with his childhood: his father's suicide and the too rapid re-marriage of his mother, which was (as he realizes only now) the subliminal reason why as a teenager he became fascinated with "Hamlet" and decided to become a Shakespeare scholar. Researching his family's history he reads the notebooks left by his great-great-grandfather, Matthew Pearce, a 19th-century engineer who lost his faith having seen the skeleton of an ichthyosaurus and having read *On the Origin of Species* when his infant son died.

Jenny Diski in *Monkey's Uncle* describes Charlotte FitzRoy's nervous breakdown following the death of her daughter Miranda and the collapse of her utopian social ideals prompted by the fall of the Berlin Wall. Charlotte is a microbiologist and her field of research, genetics, suggests a subliminal urge to study the problems of paternity and descent which have haunted her since childhood. Charlotte suffers from temporary insanity, which leads to an unsuccessful suicide attempt. In her madness she is persuaded that she is a descendant of Captain FitzRoy who sailed Darwin over the world, unintentionally providing the then young scientist with opportunities to research wildlife, ones which years later enabled him to publish his theory of evolution. Having read *On the Origin of Species* FitzRoy, who was a fundamentalist Christian, committed suicide and Charlotte feels genetically bound to do likewise. She hallucinates and finds herself in an Alice-in-Wonderland-like hell, where she converses with Charles Darwin, Sigmund Freud, and Karl Marx in a re-assessment of the social and scientific history of the past 150 years.

"The more you understand the significance of evolution, the more you are pushed away from the agnostic position and towards atheism. Complex, statistically improbable things are by their nature more difficult to explain than simple, statistically probable things,"[176] claims Richard Dawkins. The ideas he repeats in his numerous interviews and books may be summarized as follows: not everybody believes that Darwinian theory explains everything about life and the

---

176   http://old.richarddawkins.net/articles/95-the-likelihood-of-god, accessed on January 18, 2014.

universe, but this is only because not everybody knows enough about evolution. Once they are properly educated, all people are sure to replace God with Darwin. Therefore, a proper scientific education will result in the unequivocal acceptance of the Darwinian paradigm as crucial to the contemporary understanding of the workings of the Universe and everybody will be "pushed towards atheism". The authors of the most influential American textbook on Darwinism for students of the humanities seem to have a very similar view:

> The intellect evolved to crack the defenses of things in the natural and social world. It is made up of modules for reasoning about how objects, artifacts, living things, animals, and other human minds work. There are problems in the universe other than those: where the universe came from, how physical flesh can give rise to sentient minds, why bad things happen to good people, what happens to our thoughts and feelings when we die. The mind can pose such questions but may not be equipped to answer them... Almost all our attempts at deeper explanations are likely to be flawed and skewed, as the hundred thousand religious explanations of the world suggest.[177]

Today's prominent Darwinians advocate atheism, claiming that early humans devised religion because they needed it at a certain stage of homo sapiens' development. As a species we are now leaving that stage behind. Frequent repetitions of the above claims in the media and popular science have resulted in a series of oversimplifications. In the popular imagination an association has been made between Charles Darwin – a Victorian naturalist – and the absence of God in post-Victorian British public discourse. The novelist Antonia Byatt, who is also an eminent scholar of the Victorian era and its cultural heritage, writes in her essay entitled "A New Body of Writing: Darwin and the Recent British Fiction" that the recurrent references to Victorian natural history in the contemporary novel are directly connected to the fact that with Darwin's oeuvre the millennia-long Christian tradition has started to fade away. In our post-Darwinian society we live in a unique socio-cultural situation, as: "Man is a religious creature, says the socio-biologist E.O. Wilson. No human society has not had a religion. Until now."[178] Thus on the one hand we are evolutionally conditioned to believe and, on the other hand, thanks to evolutionary science we know this and thus cannot believe anymore. Moreover, as most of the grand narratives we inherited from the past were created from the Christian perspective, we are now left without credible histories and it is the contemporary novel that should fill the gap and

---

177   Brian Boyd, Joseph Carroll, and Jonathan Gottschall, eds, *Evolution, Literature and Film* (New York: Columbia University Press, 2010), 129.

178   Byatt, 'A New Body', 444.

work on "a project of reassessing the past, our own ancestry, without the old framing certainties of Christianity."[179]

In each of these three books Darwin epitomizes the loss of faith and the 'Darwinism is atheism' thesis is neither subverted nor even discussed, as it is an undeniable truth too obvious to analyze. Darwin was traumatized and lost his faith, the people who read Darwin in the 19th century were traumatized and lost their faith, the faithless people of today who are traumatized think about Darwin and the loss of faith. Yet, as historians of religion claim, Darwin's religious odyssey: "went through several stages. The orthodox Christianity of the first third of his life gradually shifted into the unorthodox theism for the next forty years. In his last decade that theism mingled with agnosticism."[180] The opinion that he drifted from Christian orthodoxy to aesthetic materialism is simply wrong, and since the 1987 publication of the transcripts of his notebooks from the period when he worked on his theory (1836–44) everyone can read about his restless attempts to reconcile his views on evolution with Christian theology. The naturalist asked himself very many questions concerning God, Creation, and how the Universe works – but to his constant distress he could not find satisfactory answers. In a letter written during the last days of his life we read: "whether the existence of a conscious God can be proved from the existence of the so-called laws of nature is a perplexing subject on which I have often thought but cannot see my way clearly."[181] Yet not many people examine source material, and most of the books about Darwin are based on the earlier books about him. The special status of Darwin in contemporary British culture as a national icon and the greatest Englishman since Newton makes it difficult to challenge even a patent misunderstanding concerning his too-well-known biography. As Phipps explains:

> Those calling Darwin irreligious are guilty of the illogic of equating disbelief in one form of Christianity with the rejection of all religion. By narrowly assuming that being a religious person means being a Christian, a scriptural literalist, or a pietistic churchman, these interpreters claim that he was not religious for most of his life… But the scope of religion is much wider.[182]

In an attempt to recreate Darwin's doubts and apprehensions from the time when he was working on his theory one should look at the epigraphs[183] the naturalist

---

179 Byatt, 'A New Body', 445.
180 Phipps, *Darwin's Religious Odyssey*, 106.
181 Qtd in Phipps, *Darwin's Religious Odyssey*, 153.
182 Phipps, *Darwin's Religious Odyssey*, 164.
183 Compare William E. Phipps, *Darwin's Religious Odyssey*, pp. 61–2.

chose for the frontispiece of the book he was then preparing. There are three of them, and the first, a quote from William Whewell, sums up Isaac Newton's attitude towards science: "But with regard to the material world events can be brought about not by insulated interposition of Divine power exerted in each particular case, but by the establishment of general laws." It is by studying the rules governing nature and not by cataloguing each and every phenomenon that we may approach and admire the Divine power; the Universe is not a collection of separate things created in separate acts of God, but rather one ingeniously interconnected whole made of myriads of phenomena governed by knowable laws.

The second epigraph is taken from the works of the 18th-century philosopher and the bishop of Durham, Joseph Butler, for whom reason and revelation are intertwined: "the only distinct meaning of the word 'natural' is stated, fixed, or settled; since what is natural as much requires and presupposes an intellectual agent to render it so, i.e. to effect it continually or at stated times, as what is supernatural or miraculous does to effect it for once." The final and the most prominent motto is a quote from the writings of Francis Bacon, who proposes that "let no man upon a weak conceit of sobriety or an ill-applied moderation, think or maintain that a man can search too far or be too well studied in the book of God's word or in the book of God's works, divinity or philosophy; but rather let men endeavor on endless progress or proficiency in both." Thus theology and natural history may and can go together and getting to know nature, which is itself the work of God, enables you to learn about Him, just as studying of the Holy Scripture does. In fact, we should do both.

Choosing his epigraphs Darwin tried to inscribe *On the Origin of Species* into the centuries-old tradition of God-inspired scientific treatises where the study of the natural world leads to a greater understanding of the world's Maker. Yet, simultaneously, the need of Reason and rationality is emphasized. Darwin's book was finally published in 1859, years after the tragic death of his daughter and his subsequent crisis of disillusionment with Christianity. Thus the opinion that his sailing away from religion was completed during this crisis once and forever and he became an atheist for the rest of his days can easily be debunked. And yet such a belief endures. As early as in the 1860s the American readers of his book ignored the epigraphs and the veiled references to the Maker in the text and pronounced the work atheistic. Darwin's American friend, the naturalist Asa Gray, wrote: "to deny that anything was specially designed to be what it is is one preposition, while to deny the Designer supernaturally, or immediately made it so, is another: though the reviewers appear not to recognize the distinction."[184] Gray goes on to compare Darwin to Newton and maintains that the two scientists approach nature in a similar way, and yet no one accuses Newton of "the atheism of fortuity"[185] in the way they accuse Darwin. Already a century and a half ago Darwin was for fundamentalist Christians (especially in America) an epitome of vile, unholy science: his devil-orchestrated project symbolized a threat to the good old vision of the Universe as a safe and godly place.

In the US, where creationist sentiments have always been acutely felt, just a few years after Darwin's death the story of his alleged conversion surfaced. Despite its patent nonsense, the story proved enduring and is part of the creationist subculture's folklore to this day. The mental image of Darwin in his dying hour renouncing the theory of evolution and summoning the minister to seek safety in the blood of the Lamb is strangely reassuring. Edward Caudill in *Darwinian Myths. The Legends and the Misuses of a Theory* calls this recantation story a myth, because myths are always devised to bring order out of chaos and to comfort anxious humans. If Darwin first had traded God for knowledge and then, realizing his bargain meant damnation, regained his God by recanting his knowledge, we all can break our Faustian contract and be saved:

> [T]he world was returned to order, with faith again reigning supreme over mere knowledge. The myth was an attempt to return the old system of preeminence. It was a restoration of divine, universal unity over cold, fragmentary calculations about the material world...The story involves conflict at two levels: religion versus science and Darwin versus his conscience. There is, for the creationist subculture, the satisfying finality of the story's conclusion: the triumph of Christ, the conversion of a man who was a symbol of 'godless science' and a reassertion of the power of faith."[186]

Both in America and Britain Darwin has always been 'a symbol of godless science' and not a person who lived in his own time, suffered his own anxieties, and arrived at certain compromises. It is to this disembodied symbol that contemporary British novelists really refer.

In the case of Keynes a more complex process is involved; himself a descendant of Darwin and the keeper of family records and memorabilia, he attempts to describe in his 'faction' Darwin the man. Yet at the same time he does not want to resign from perpetuating the 'godless science' cliché. As a result, he chooses the

---

184   Asa Gray, *Darwiniana. Essays and Reviews Pertaining to Darwinism* (Cambridge, Melbourne, New York: Cambridge University Press, 2009), 138.

185   Gray, *Darwiniana*, 95.

186   Caudill, *Darwinian Myths*, 57–59.

most tragic personal event of Darwin's life, the death of his daughter (which did bring about a religious crisis) and describes it as a final straw that caused Darwin to depart from all religion. In Keynes' opinion this moment was very painful but also liberating, and only after such a spiritual shock could the scientist write *On the Origin of Species*. He is both very human and an embodiment of modern science, not 'godless' but 'god-free'. Keynes discusses the entire spiritual life of his famous forebear prior to the traumatic period as a mere prelude or preparation for the true breach with religion. Thus, the young and naïve Darwin read William Paley's *Natural Theology* where all the world is presented as evidence of God's creative power and almighty goodness.[187] Keynes argues that Darwin grew up in a culture which considered every aspect of life to be proof of God's power and wisdom; even in cookery books you would come across exclamations rooted in natural theology. While describing how to cook a fish the famous Mrs. Beaton makes a long digression about the anatomy and purpose of a fish's air bladder and ends in the praise of God: "how simply, yet how wonderfully, has the Supreme Being adopted certain means to the attainment of certain ends."[188]

Young Darwin initially agreed with such opinions, but when over the course of his research he realized that humans were a species of mammals like all other species and had evolved from some ancestral apes he felt anxious not knowing how to reconcile this discovery with his belief that God governs our lives according to special moral purposes. Later still, the scientist studied the organic view of mind and body and the way we have inherited mental capacities and instinctive behavior from our forebears. He strove to reconcile the organic brain hypothesis with the possibility that we do have immortal souls which are rewarded or punished when we die. Next, reading Lyell and travelling to the Americas he grasped the vastness of geological time and the impossibility of the literal truth of Genesis. Once disappointed with the Hebrew Bible, he started to doubt the Christian Revelation: "he could not believe that God expected us to accept Christ's message on the authority of the Old Testament because the Gospels placed such emphasis on the fulfillment of the Old Testament prophesies. If God had meant

---

187 It was Paley who coined the famous comparison of God – the clockmaker: "if you find a watch on the ground and you do not know how it got there, you are able to say it did have a maker by noticing its purposeful design. Some artificer must have intentionally construct it, otherwise it would not tell time. By analogy the same is true in the case of the complex and wonderfully efficient nature, and the whole universe".

188 Randal Keynes, *Annie's Box; Charles Darwin, His Daughter, and Human Evolution* (London: Fourth Estate, 2001), 18.

us to accept Christ's message, he would surely have given it a more credible and persuasive foundation."[189] The discrepancies between his learning and his religion and the illogic of Anglican doctrine worried Darwin, which fact Keynes proves by describing the dreams Darwin had in this period of his life about letters and manuscripts of the great ancient philosophers found at Pompeii confirming everything that was said in the Gospels.[190] Disbelief was then "creeping over Darwin"[191] and, in the 1840s when paternal grief struck he was intellectually ready to say no to religion:

> After Anne's death Charles set the Christian faith firmly behind him. He did not attend Church services with the family, he walked with them to the church door, but left them to enter on their own and stood talking with the village constable or walked along the lanes around the parish.[192] (Keynes, 222)

Darwin, in his great-great-grandson's opinion, not only abandoned the church, but also ceased to believe in the infinite goodness of the Maker of the Universe, who now seemed to him: "a shadowy, inscrutable and ruthless figure". It was at this time that Darwin wrote in his notes the oft-quoted observation: "what a book a Devil's chaplain might write on the clumsy, wasteful, blundering low and horridly cruel works of nature"[193].

> Untimely deaths, struggle and pain in the natural world (and the world of humans) – this all became evident to him. With his daughter's demise he firmly opposed the standard view that suffering has a moral sense and is a part of God's design. He noticed that all of nature suffers while the number of people in comparison to the myriads of other beings is but a pittance. What kind of God would therefore make millions of lower animals suffer greatly without any moral improvement throughout "almost endless time"?[194]

Keynes goes on to offer a psychoanalytical interpretation of Darwin's acute trauma after the bereavement. Depicting his psychosomatic ailments and clinical depression Keynes suggests a displacement of guilt. Subliminally horrified by the repressed thought that the tragedy was God's punishment for his rejection of religion and reluctant to accept such a superstitious idea, Darwin started to blame

---

189   Keynes, *Annie's Box*, 43.
190   Keynes, *Annie's Box*, 119.
191   Keynes, *Annie's Box*, 121.
192   Keynes, *Annie's Box*, 222.
193   Keynes, *Annie's Box*, 222.
194   Keynes, *Annie's Box*, 278.

himself for weakening his children "by a natural process"[195]. Mr. and Mrs. Darwin were first cousins whose families had intermarried for three generations – the husband and wife had all their four grandparents in common. During years of research the scientist grew aware of the dangers of in-breeding and the health hazard it has for progeny. His own health was failing and he suspected he had passed his weak blood to his children; his obsessive worry about the condition of the remaining eight of them was – according to Keynes – a symptom of the psychological conflict he suffered having departed from Christianity. Once he had left faith in God's Providence behind he felt lonely and left to his own devices in the hostile universe. Keynes believes such was the price his forebear had to pay for a genius that forever changed not only our science, but our (at least British) culture as well.

Swift's and Diski's protagonists live in the next century after Darwin and do not even think about religion and God when they ponder their own lives, though they do when they consider the lives of their 19th-century ancestors. The subject of ancestry is in general very prominent in all the three books, fiction as well as faction. Keynes sketches family trees, reproduces old daguerreotypes, depicts heirlooms he inherited not only from Charles and Emma, but also from their son, George, and from Gwen, George's American bride who looked at British turn-of-the-century life with an unbiased self-confidence. One of these heirlooms is the eponymous Annie's box, where in the 1850s Charles and Emma put memoranda of their daughter: her drawings, letters, notes, Darwin's account of her illness, and a few personal items they wanted to keep. It is the discovery of the box which, we are told, prompted Keynes to write the book and share his obsessive interest in his ancestry. Swift's protagonist, in recovering from his trauma, also grows obsessed with ancestry. He recounts all the details of his father's suicide, which officially was caused by his mother's infidelity, but now it is disclosed that his father was involved in classified negotiations between the allies during the Second World War and probably had something to do with the Manhattan Project. His death following Hiroshima and Nagasaki might therefore have resulted from guilt or might not have been suicidal at all. For Unwin, a Shakespearean scholar who cherished his Hamletesque feelings for decades, and even thought about his step-father as 'Claudius', this comes as a shock. Moreover, he also learns he was conceived during a brief affair his mother had with an engine-driver who subsequently was killed in World War II. He cannot tell if his official father was aware of this or not.

---

195   Keynes, *Annie's Box*, 208.

Unwin, who used to think about his life in theatrical metaphors (his deceased wife was an actress), feels bereft in many ways. Not only did he lose love and companionship, but also an emotional frame of reference as well; he feels like Hamlet, who during the famous bedroom scene learns from Gertrude that he is illegitimate and in fact not related to King Hamlet at all, so he should not begrudge Claudius. Unwin also inherits heirlooms: a Victorian mantelpiece clock whose accuracy is emphasized in phrases echoing Paley's praising of the Divine Clockmaker and a suitcase of notebooks recording the gradual loss of faith of his atheist Victorian forebear, Pearce. Predictably, it was the loss of his baby son, Felix, that brought about Pearce's spiritual crisis following the years of concealed doubt. As a young engineer Pearce used to relish his mother's Bible, which gave him relief after the rigors of science: "a wondrous thing, this central fact, a wonderful clarifier, encourager and liberator."[196] Yet on the way home from his university Pearce by chance saw an ichthyosaurus on the Lyme Regis beach, "the long toothed jaw, the massive eye that stared through millions of years,"[197] and which contradicted the history he had learned from his Bible. Decades later he called this moment "the beginning of my disbelief"[198]. Yet for years he pretended he believed; and only after Felix's death did the long-suppressed questions re-emerge: "Is the Creator to be viewed as mere Experimenter?"[199]; "If the world existed so long without Man upon it, why should we suppose that futurity holds for us any guaranteed estate and that we occupy any special and permanent place in Creation?"[200]. It is precisely when such anxieties pester him that Pearce reads *On the Origin of Species*, which educes his disbelief and causes him to sacrifice his family life and professional career. His father-in-law blames Darwin: "not for his assault on religion… but for timing the publication of his dangerous work as to clinch his son-in-law's apostasy, bringing scandal on his family and parish."[201]

Charles Darwin grows to symbolize the spiritual transformation of mid-19th-century Britain, which for some is "the work of Darwin's master, the devil" (Swift, 194), while for others the victory of the new sobriety over the old folly. Pearce feels that *On the Origin of Species* merely prompted a paradigm shift which had long been in the air: "did people have souls before 1859, when Darwin published his momentous work, then, suddenly cease to have them? (Swift, 197) asks

---

196   Graham Swift, *Ever After* (London: Picador, 1992), 100.
197   Swift, *Ever After,* 111.
198   Swift, *Ever After,* 110.
199   Swift, *Ever After,* 144.
200   Swift, *Ever After,* 146.
201   Swift, *Ever After,* 155.

Unwin, who under the influence of his ancestor, Pearce, starts reading *The Origin* himself and finds it dull, predictable, and not revolutionary at all. He concludes that this means Darwin was right: species adapt and we have adapted to the godless universe, and we miss the mid-19th century because then it was still possible to feel agitated and to fight for one's intellectual and religious ideas: "in some medieval vision full of demons and terror… before the walls of some beleaguered city in which the pious cower for safety and … their champion is beating back, with the intrepid zeal… the ravening beast, Darwin."[202]

Diski's protagonist, Charlotte, turns to Charles Darwin looking for a correlative for the repressed emotions she does not fully comprehend. She is the illegitimate child of a hotel maid and an otherwise childless upper-class married man who loved her very much and took care of her in her early life, and yet renounced her and her mother when his wife discovered this long-standing extra-marital relationship. As a parting gift he gave his daughter a pearl set in gold and then disappeared from her life forever, soon afterwards taking his own life.

Now in her late forties Charlotte is obsessed by the idea of one's own uniqueness, heritage, and ancestry which the pearl heirloom symbolizes and which she professionally researches in the genetic laboratory. Her heirloom and her set of DNA both stand for her identity and her destiny written down in her genotype, and a part of this heritage might be proneness to suicide. After the death of her grownup daughter, a model whom Charlotte has always despised, and the demise of the socialist ideas and militant feminism Charlotte has devoted all her life and energy to, the woman feels at a loss. Never religious and always single she has nothing and no one to rely on, and, quite accidently she comes across a book with her family name on it, FitzRoy. The name is also an heirloom, her father's name which she took by a deed poll, and a link with her unclaimed heritage, the history of her father's line. She buys the book, which is a biography of Captain FitzRoy, and reads it obsessively as her mental state deteriorates. She is sure she has found her true ancestor, and in his life story she looks for clues suggesting her own genetically-determined destiny. The book recounts the fervent discussions of FitzRoy and Darwin, stereotypically and not too accurately presented as a conflict between fundamentalist Christianity and self-assured atheism over the flood, Creation, geology, and the issues of science and religion. Even when Charlotte is placed in a mental institution, and all through her unsuccessful suicide attempt and subsequent recovery, she rereads the biography and finds her own story there: the agonizing inner struggle of FitzRoy (as described in this

---

202   Swift, *Ever After*, 202.

biography, again not accurate, historically speaking) is what she should have felt in her own life, but never let herself feel.

FitzRoy suffers psychotically because while defending Almighty God against Darwin's skepticism he cannot help suspecting that Darwin might be right: "however much he might try to hide it from himself he had doubted, and once doubting could never feel perfect security again."[203] To conquer his lurking uncertainty FitzRoy grew devout and yet suffered from the suppressed guilt which surfaced over twenty years after the return of the Beagle, when *On the Origin of Species* was published. He felt like an accomplice and, tormented by his anguish, first he tried to destroy the book by writing devastating reviews and frequenting scholarly meetings devoted to natural history. Outraged by a discussion concerning the apish descent of Man he interrupted everyone:

> He picked up the large Bible... God would not be silenced. He held the book aloft with one hand now and bellowed over the collective derision of the crowd: 'The Book! The Book! The Book!' to counterbalance the derisive cry of monkeys... His voice was drowned by the crowd aping him and he could not stop it happening. Finally defeated, he let the Bible fall to his side and dropped back onto his seat in a terrible confusion of shame.[204]

The scene when a few months later he cuts his throat with a razor blade in a spectacular and messy suicide is for Charlotte a gruesome reminder of her own gene-determined destiny. Fantasizing about life after death Charlotte descends in a fit of madness into a bizarre godless hell where Jenny, an orangutan Darwin described in his notebooks, is her Virgil showing her around and introducing her to the long-dead sages. The Darwin she converses with is himself a little surprised by the total absence of a Supreme Being, yet misses not Him but Emma:

> Emma with her rock-solid love of religion and her husband had been sure that they would be together through all eternity, in spite of Charles's skepticism and downright ungodly ideas. And Charles, hardly believing, yet hoping she might be right had held to the image of being reunited with Emma in an afterlife he had no longer credited. But she had been wrong.[205]

There is no heaven, no God, and no doubts either; hell is empty. This picture of the afterlife is created by the mind of Charlotte, a contemporary mind certain that religion is a thing of the past which used to be important for people of Emma's generation. In her imagination she and Darwin confirm to one another

---

203   Jenny Diski, *Monkey's Uncle* (London: Phoenix, 1994), 78.
204   Diski, *Monkey's Uncle,* 166.
205   Diski, *Monkey's Uncle,* 181.

that, "the human race was the result of an aimless process."[206] There is no discussion and indeed nothing to discuss really, everybody agrees on everything. The talks Charlotte has with Freud and Marx are equally dull: each of the three sages whose oeuvres seem to have influenced 20th-century Western culture the most turns out to be boring and ignorant. Marx and Freud accuse Darwin of paving the way for the contemporary nihilism by depriving Europe of her illusions. "And what had Charles Darwin done? He had stolen faith from the world. He had deprived the world of the loving care of an infinitely good Creator and put a damned monkey in his place."[207] Marx and Freud's own theories only followed suit and now, with no illusions left, Europe is an emotional desert full of people who vaguely feel they have lost something. That is why they are obsessed with ancestors, inheritance, heirlooms, and destiny.

Charlotte clings to her surname, FitzRoy, because she wants to be like the famous FitzRoy, a passionate, spiritually over-active Christian, dying for a lost cause. This is why Charlotte altered the spelling of her father's name, Fitzroy, capitalizing 'R' to create an imaginary link with the tragic Captain who was everything she is not. She often thinks and talks about this surname, yet never discusses her first name, Charlotte, a feminine version of FitzRoy's greatest enemy's name, Charles. But in sober moments she knows FitzRoy was not her forebear and that she in fact takes after Darwin, the epitome of "godless science". Obsessive echoes of the naturalist's name are yet another feature Swift and Diski have in common: one can argue that Unwin is a pun on 'Darwin' and suggests 'a loser'.

Interestingly, in each of the three books described above the protagonists attempt to re-create the past by reading about it in retrieved 19th-century texts: Pearce's notebooks, the papers from the inherited box, a biography of an alleged ancestor. They are all tormented by the thought of ancestry: Keynes focuses his attention on the past and spends years recreating what his family did day by day in the mid-19th century; Unwin and Charlotte suffer because they were illegitimate children and are traumatized by the suicides of their fathers, believing they have inherited an inclination for self-destruction. The past is embodied in emotionally charged artifacts: a jewel, a box, a clock. Moreover, following the popular myth of Darwin losing faith after his daughter died, the motif of the death of a child accompanies atheism, as in the case of Felix and Miranda's deaths. Suffering from mental breakdown at the end of the 20th century, the protagonists compulsively come back to the mid-19th-century spiritual crisis. Their trauma narratives

---

206   Diski, *Monkey's Uncle*, 163.
207   Diski, *Monkey's Uncle*, 190.

resemble talking cures: recurrent, repetitive accounts of similar situations using similar motifs, objects, and settings. The stories read like variations on a theme, and the theme is, to refer back to Caudill's term, the myth of Darwin's loss of faith treated as signifying the beginning of the spiritual crisis of the West.

Darwin's Problem with Human Ancestry as Reflected in Recent Fiction

Darwin: A Life in Science by Michael White and John Gribbin, one of the best recent biographies of Charles Darwin, aims at presenting Darwin's theories and ideas rather than his private life and personality. The book begins with a paragraph explaining the never-ending popularity of Darwinism and the huge amount of new books and films devoted to Darwinian themes that appear every year. Gribbin and White's answer to the question *why does the twenty-first century so adore Darwin?* is that in our age of mass culture Darwin alone offers us a comprehensive, universal theory that can be easily summarized and made into a story. Its simplified versions may be applied to explain a vast range of phenomena concerning nature, natural history, and the workings of the Universe to non-specialists and even to children. Darwinism is a paradigm everybody refers to when they are explaining the ways of nature: from how the biosphere came into existence to how humans evolved; from the rivalry between better and worse adapted species, to the rivalry between better and worse adapted cells. White and Gribbin claim that other paradigms of science available today – say, Newtonian and Einsteinian – are too difficult to grasp, too abstract, and too exclusive for contemporary egalitarian culture.

White and Gribbin wrote about Darwin in the late 90s; fifty years earlier William Irvine had addressed the issue of Darwinism as contrasted to more abstract theories in his then influential joint biography of Charles Darwin and Thomas Huxley entitled *Apes, Angels, and the Victorians*, but his point is slightly different:

> Darwin's theory is less esthetically satisfying, less beautiful, than Newton's. It lacks the elegant precision of a mathematical solution. It is the prose realism, not the poetic truth of science... In any case, natural selection has not been supplanted by a more graceful substitute. Darwin has proved only superficially confused and, on the whole, astonishingly right.[208]

The juxtaposition of these two opinions divided by half a century tells much about how science is defined and referred to in popular books: 50 years ago Darwinism seemed less beautiful than physics and yet good for practical reasons; now it is less complicated than physics and thus good. Apart from some

---

208 William Irvine, *Apes, Angels, and Victorians* (London: Readers Union Weidenfeld and Nicolson, 1956), 77.

implications concerning the intellectual horizons of the target readers of popular science in the 1950s and the 2000s, the above change in the perception of Darwinism may suggest something more: the notion of evolution has entered popular imagination and is now a part of our frame of mind. The universe in general and life on Earth in particular (humankind included) seem to us to be governed not by a set of mathematical equations, but by the rules of natural selection. We see nature in Darwinian terms. As Thomas Kuhn claims in *The Structure of Scientific Revolutions*, theoretical paradigms we grow up to accept always influence the way we see facts:

> No natural history can be interpreted in the absence of at least some implicit body of intertwined theoretical and methodological belief that permits selection, evaluation, and criticism. If that body of belief is not already implicit in the collection of facts – in which case more than "mere facts" are at hand – it must be externally supplied, perhaps by a current metaphysics, by another science, or by a personal and historical accident.[209]

Nowadays it is Darwinism that supplies popular culture with "the body of intertwined theoretical and methodological belief" Kuhn is writing about. Darwin in popular imagination is the most eminent sage ever, one who gave us the paradigm we live by and use to explain nearly everything, our own past included. The latter issue has always been the most controversial in the Darwinist debate and numerous cliché simplifications of human descent have been made from the proposition that people and apes derive from the same ancestral species. Indeed, the famous conflict concerned with monkeys and men erupted as early as the 1860s, just after the publication of *On the Origin of Species*, and thus well before *The Descent of Man* was written.

The aim of this chapter is to explore how contemporary popular culture deals with the implications of Darwinism – namely, that humanity, especially human babies and savages, display features and behaviors also to be seen in primates, especially young ones. The unabashed political incorrectness of describing non-Europeans as ape-like and the provocative treatment of infants as little monkeys, which goes against the sentimentalization of babies prevalent in mass culture, are challenges to the writers and film-makers creatively at work within the dominant paradigm. After all, they not only have to preserve the vision of Darwin as a paragon of the devoted and loving parent and an abolitionist outraged by slavery, which popular culture traditionally reinforces, but also to discuss his ideologically provocative ideas. Darwin is a genius who cannot be wrong, and

---

209 Thomas Kuhn, *The Structure of Scientific Revolutions* (Chicago and London: University of Chicago Press, 1996), 17.

yet his theories resulted from observations which today are clearly politically incorrect.

In order to show the diverse ways in which this contradiction is resolved I shall first prove that Darwinism does provide contemporary culture with a Kuhnian frame of reference, an instant theory of everything. Secondly, I shall focus on one particular aspect of the Darwinian debate: the alleged shared ancestry of apes and humans which reveals itself via similarities between, on the one hand, orangutans and savages, and, on the other hand, human and orangutan babies. This issue will be shown in historical perspective, initially by referring to non-fictive accounts of the Victorian evolutionary debate. Thereafter I will take up its literary echoes: Peter Nichols' and Harry Thompson's recent novels, followed by Randal Keynes' quasi-biography *Annie's Box: Charles Darwin, His Daughter, and Human Evolution* together with its film adaptation, the biopic *Creation*. In my conclusion I hope to show through the orangutans-savages-babies chain of associations the ability of popular culture to incorporate provocative issues into the dominant paradigm and to render them uncontroversial.

In his *The Structure of Scientific Revolutions* Kuhn challenges the traditional principle that science is always objective and rational. As one can see in the two-way traffic between natural history and literature, the post-Enlightenment belief in science as progressive, empirical, verifiable, and falsifiable had to be revised. The ruling paradigm that allowed the Victorians, Darwin among them, to make sense of facts and observations was more a product of Western thought than mere facts of nature. Darwin's theory came right at the moment the world was ready for a new paradigm to emerge. His books in turn influenced the thinking of his contemporaries: Charles Dickens, George Eliot (Mary Anne Evans), and G.H. Lewes among them. The vision of nature Darwin offers is apparently first and foremost both abounding and chaotic: there is no design – whether in the myriad diverse plants, or in the hosts of insects and animals multiplying, competing for food and space, and dying in the millions. Yet if one looks very closely, there is an element of order in this chaos: pure chance produces variations within each species and, by natural selection, the fittest of them survive and adapt. Life is abundance and possibility, nature is full of 'grandeur' and, moreover, all organisms are interrelated in a network of dependencies. There is no vacuum, every new niche immediately gets filled by adaptive organisms. As they adapt, some groups within a given genus change, while others stay the same. Thus, the boundaries between species and varieties forever blur, the human species being no exception here.

As George Levine writes in *Darwin and the Patterns of Science in Victorian Fiction*, what Darwin challenges is not that "phenomena are adapted to one another but that the adaptation is to man"[210]. Although in *On the Origin of Species* Darwin omits mentioning human ancestry, he does include an allusion to the topic claiming that thanks to his theory "light will be thrown on the origin of man and his history", a promise he fulfills over ten years later in *The Descent of Man*. According to Levine it is easy to assimilate Darwin's theory to the anthropocentric and Europocentric view of nature[211], as the story of evolution does have a happy ending: the seemingly aimless evolution did produce white civilized man. Moreover, not only is his theory a happy-ending cause-and-effect narrative convergent with the European vision of history, but it is also very much influenced by literary tradition: it springs from fascination with books describing journeys to exotic places. Darwin himself often recalls the emotions he felt while reading Alexander von Humboldt as a young boy in his father's hothouse when he swore to himself he was going to travel to the tropics when he was grown up[212]. It is partly because of such personal touches and the ability to engage the reader emotionally that Darwin's books were so influential in non-scientific circles:

> Darwin's cautious, charming, self-depreciatory style constructs a large trope of modesty which is not merely personal. That is, modesty in self-presentation, refusal to advertise self, caution and detail in argument… Darwin's language helped his ideas subversively enter the culture. The apparently simple language quietly adopts personal anecdote, tentative speculation, hard presentation of data, careful logical argument, broad generalization, metaphysical expansiveness, rigorous scholarship.[213]

Darwin is a character in his own books, the standard Victorian protagonist-narrator remembering his youth, referring to his travels, modestly diminishing his achievements, and emphasizing his shortcomings as a thinker and a theorist. What he does throughout is to unperceptively persuade his reader that his subject – organic life, one part of which is human life – is entirely a matter of scientific observation and interpretation, and that the facts one can notice in nature fit certain general laws. The power of science over nature extends over human beings as well: "the human subject becomes equivalent to the planetary

---

210   George Levine, *Darwin and the Patterns of Science in Victorian Fiction* (Chicago and London: University of Chicago Press, 1991), 26.

211   Levine, *Darwin and the Patterns*, 27.

212   Levine, *Darwin and the Patterns*, 85.

213   Levine, *Darwin and the Patterns*, 87–95.

or the geological… as the observer gains the power over the observed, he or she becomes the observed and is more particularly vulnerable"[214].

Darwin tells his readers stories, the most prominent of which are ones about themselves, and he does so from a very well-defined vantage point. The narrator himself is the subject of numerous embedded digressions: his travels, his memories, his ideas, his children, and his observations are narrated. He writes about us, and not only himself. Thus, even before the publication of *The Descent of Man* in 1871, people understood the theory of evolution to aim at explaining the origin of humankind: the implication of his ideas for humans was evident to both his enthusiasts and his critics.

In June 1860, six months after the appearance of *On the Origin of Species*, a famous confrontation between Darwinists and anti-Darwinists took place at Oxford, when the British Association met to discuss a number of papers on natural history. Although Darwin was absent everyone knew that the lines of battle were between those who believed in the Darwinian single parent-stock of apes and humans and those who contested the hypothesis. The leader of the anti-Darwinians, Bishop Samuel Wilberforce, is said to have turned at one moment to the leader of the Darwinians, Thomas Huxley, and asked, "was it through his grandfather or his grandmother that [Huxley] claimed his descent from a monkey"[215]. In his reply, Huxley, having demonstrated the bishop's obvious ignorance of the theory, explained its leading ideas and "said that he would not be ashamed to have a monkey for his ancestor, but he would be ashamed to be connected with a man who used great gifts to obscure the truth"[216]. His answer immediately was turned into the anecdote that Huxley had said he would rather be a descendant of a monkey than of a bishop, which helped the Oxford meeting go down in history. In retrospect, that meeting marked the beginning of the renegotiation of humanity's place in nature as presented in popular science and literature. If apes and savages have as much in common as Darwin is said to have noticed while in Tierra del Fuego, and the 'civilized Fuegians' onboard the Beagle are so much like other civilized people, then there must be some continuity in nature: slow, steady progress and constant struggle must allow for improvement in the species, and the features we consider human in their nascent form must be found in lower regions of the animal kingdom. If this is true, Huxley was right: vain attempts

---

214   Levine, *Darwin and the Patterns*, 210.
215   Irvine, *Apes*, 5.
216   Irvine, *Apes*, 5.

"to obscure it" are indeed pathetic and any ape is better than a creationist. Peter Nichols refers to the Oxford meeting as the moment the paradigm shifted:

> ... it was the great confrontation between science and religion, between God and ape, the moment when that ceased to be the concern only of philosophers, scientists, academics and clerics, but passed into the public consciousness and became the question of the age. History has for once determined the victors rather than the other way round, but the battle raged furiously on sixty-five years later, in the famous "Monkey Trial".[217]

The belief in the unbridgeable gap between the realm of humans and the animal kingdom made obsolete by Darwinism and how this happened is one of the recurrent motifs in the works of fiction and semi-fiction concerned with the historical figure of Darwin. Their authors generally agree that the idea that there is continuity in nature entered Darwin's mind onboard the Beagle. Having noticed that the difference between a Caucasian and a baboon is much greater than that between a Negro and a gorilla, Darwin – the protagonist of Harry Thompson's *This Thing of Darkness* – goes on to suggest in a discussion with his fundamentalist Christian captain that savages are but evolved apes.

> Look at the orangutan – its affection, its passion, its rage, its sulkiness, its despair. Then look at the savage – naked, artless, roasting its parents... Compare the Fuegians and the orangutan and dare say the difference is so great.[218]

Such Bible-defying statements point to the gradual idea of humanness as a feature which can be found in varying intensity in higher ranks of animals. Moreover, the fictional Darwin in the above conversation, taking place during his trip round the world, compares humans to orangutans and not to, say, chimps for some very special though anachronistic reasons (Thompson interposes this statement from Darwin's diaries written a couple of years later). Orangutans, especially the young female named Jenny, are of much importance in Darwinian studies because of what happened after Darwin's return to London. There, at the Zoo in 1838 Darwin met Jenny. On the same day he noted down in his notebook:

> Let man visit Orangutan in domestication, hear its expressive whine, see its intelligence when spoken to: as if it understands every word said – see its affection to those he knew – see its passion & rage, sulkiness & very actions of despair;... and then boast of his proud pre-eminence... Man in his arrogance thinks himself a great work, worthy the interposition of a deity. More humble and I believe true to consider him created from animals.[219]

---

217 Peter Nichols, *Evolution's Captain* (New York: HarperCollins, 2003), 317.
218 Henry Thompson, *This Thing of Darkness* (London: Review, 2005), 242.
219 Charles Darwin, notes, <http://www.pbs.org/wgbh/evolution/darwin/diary/1838.html> accessed 16 December 2014.

The chain of associations – monkeys-apes-savages-Europeans – Darwin makes in his diaries in the late 1830s is elaborated upon in his mature works. In *The Descent of Man* observations of the human-like behavior of lower animals and animalistic features of American natives are scattered all over the book and are juxtaposed with the opinion that all the apparently diverse human races are in fact very closely related. In conclusion we see nature as united and continuous, the frontiers between the species and varieties are blurred, progress is gradual, and humanness is a question of quantity, not quality.

Darwin quotes stories about the baboons who waged a well-prepared war with the geladas; about the orangutans who planned to attack the vehicles of people they hated, and in order to do so had collected and stored pebbles in a handy place[220]; and about a male baboon who heroically and cunningly saved from being torn to pieces by dogs "a young one about six months old, who, loudly calling for aid climbed on a block of rock and was surrounded"[221]. Such human-like behavior seems to prove, at least for the author, that apes do possess self-awareness and are capable of feelings such as longing for revenge, pain caused by injustice, and loyalty to the group they belong in. Thus probably "some ancient member of the anthropomorphous sub-group (of the apes) gave birth to man"[222].

It is implausible that the same mental propensities evolved twice: in apes and in ancient people: thus, the shared ancestry hypothesis is much more promising. Darwin enumerates arguments in favor of it: psychological similarities between people and other primates, similarities in the embryological growth of apish and human fetuses, and the presence of vestigial organs humans have. Thus, all the human races must have descended from some ape-like ancestor, and must be closely related; only differences in the environment make them seem superficially different: "I was incessantly struck while living with the Fuegians onboard of the *Beagle* with the many little traits of character showing how similar their minds were to ours, and so it was with a full-blood negro with whom I happened once to be intimate"[223].

Darwin goes on to enlist proofs of "the close similarity between men of all races in tastes, dispositions and habits"[224]; he also notices that fetuses and infants

220   Charles Darwin, *The Descent of Man And the Selection in Relation to* (New York: A.L. Bert Company Publishers, 1874), 92.
221   Charles Darwin, *The Descent*, 115.
222   Charles Darwin, *The Descent*, 175.
223   Charles Darwin, *The Descent*, 203.
224   Charles Darwin, *The Descent*, 203.

of primates and humans are much more alike than the mature form, as it is in the immature organisms that the ancestral heritage is easier to discern. This is true as far as anatomy is concerned and in behavior, as well. In Nichols' *Evolution Captain* a fictive Darwin compares the Fuegians to Adam and Eve, but for him the first parents are not virtuous prelapsarian creatures, but half-apes who often behave in infantile ways. In April 1838 Charles Darwin wrote in a private letter that their resemblance to Jenny the young orangutan is conspicuous:

> The keeper showed her [Jenny the orangutan] an apple but would not give it her where-upon she threw herself on her back, kicked and cried precisely like a naughty child. She then looked very sulky and after two or three fits of passion the keeper said: 'Jenny if you will stop bawling and be a good girl, I will give you the apple'. She certainly understood every word of this and though like a child she had a great work to stop whining, she at last succeeded, and then got the apple with which she jumped into an armchair and began eating it with the most contented countenance imaginable.[225]

Darwin is overtly emotional about Jenny and her human features; he looks at her with a scientific eye and yet his observations are not at all naturalistic but sentimental, very similar in tone to the way he later described his own children. Most Darwinian fictions gladly reinforce the image of the naturalist as the perfect and benign father who emerges from the memories of his children, most of whom published their memories from childhood. Yet in Darwin's own letters and notebooks there are patently bizarre notes he made when his children were small concerning the inborn patterns of behavior they displayed and how they fit into his theories. These are also referred to by contemporary novelists and biographers and the perfect father cliché is paired with the half-crazed naturalist cliché – the scientist of a Pickwick or Dr. Doolittle kind who studies nature everywhere and all is for him equally fascinating and mysterious. In Thompson's *This Thing of Darkness* Darwin is playing with his children in the garden and at one moment he says to himself while looking at them: "'that is behaviorally analogous to young pigs hiding themselves', thought Darwin and made a mental note to investigate the subject. 'It surely represents the hereditary remains of our savage stock'"[226].

When the children were already mature Darwin wrote a letter to Alfred Wallace, begging his younger colleague to carefully observe his newly born son and note down when the baby starts to secrete tears, as he needs this piece of information for his book *The Expression of the Emotions in Man and Animals*. The book,

---

225 Nichols, *Evolution's Captain*, 216.
226 Thompson, *This Thing*, 485.

eventually published in 1872, is a prolonged anthropological study comparing facial muscles and the facial expressions they produce in humans and primates. Darwin proves that contrary to the belief among the scientists of his day, facial expressiveness is not exclusively human: the similarities and differences between people and their animal cousins may be explained by the gradual evolution of these muscles and, consequently, by the gradual evolution of complexities in social life which made it profitable to be able to express and to interpret emotions among the members of one's pride or tribe.

The dotting father cum the comically devoted naturalist makes a promising figure for popular writing: biographers repeat how Darwin made his children play the bassoon[227] to earthworms to check whether they sense sounds coming from above the ground[228] and often quote the anecdote about how one of his children tried to bribe him with pocket money to stay in the children's room and play during hours customarily devoted to writing. William Irvine's *Apes, Angels, and the Victorians*, written in the heyday of Neo-Darwinism when the evolutionary paradigm entered the popular imagination, introduces the sentimental take on Darwin's family into culture;

> Himself a charming combination of man and boy, and a fabulous store-house of knowledge and accomplishments, Charles was inevitably a hero to his children. He read them Scott's novels, explained them engines, taught them how to covet rare beetles and stamps, and shared their youthful experiences with easy quality and at the same time a persistent suggestion of the mature view.[229]

By consistent emphases on the youthfulness of his spirit, Darwin's biographers seem to suggest, again in accordance with sentimental clichés, that it is only with a child's eyes that one can see the world as it really is – with all the subtleties that escape jaded adults. Thus, in noticing patterns of evolution in nature Darwin saw what everyone else could see, granted they are not blinded by convention and prejudice. Stereotypically, genius is able to see truths so simple and elemental that nobody else notices them. In yet another stock Darwinian anecdote, Thomas Huxley, having finished reading *On the Origin of Species*, shouted that it had been very stupid of him not to have discerned natural selection first.

---

227 Here we see another example of discrepancies between diverse novelistic accounts of Darwin's life. In some versions the earthworms are made to listen to the piano, in other versions – to a bassoon.

228 Irvine, *Apes*, 96.

229 Irvine, *Apes*, 91.

Writing from within a culture which believes in the Darwinian paradigm, authors tend to follow Huxley's attitudes and consider the theory of natural selection self-evident. Irvine quotes Darwin boasting about one of his son's evolutionary intuitions:

> Horace said to me yesterday: 'If everyone would kill adders they would come to sting less.' I answered: 'Of course they would, for there would be fewer.' He replied indignantly: 'I did not mean that, but the timid adders which run away would be saved, and in time they would never sting at all.' Natural selection of cowards! [230]

Some recent books try to subvert the saccharine sweet picture of Darwin and his numerous children living happily together, playing in the garden, and theorizing about wildlife. In John Darton's *Darwin's Conspiracy* one of Darwin's daughters writes a secret diary in which she paints a much bleaker picture of her father, claiming that only in official accounts (these letters and notebooks the family chooses to preserve) was he always noble, loving, and selfless. Nichols' *Evolution's Captain* also subverts the iconic biography of Darwin by focusing on his archenemy Captain FitzRoy, who sailed him round the world and went down in history as a hysterical religious maniac. FitzRoy, in Nichols' quasi-historical account, sees many shortcomings of the *Beagle*'s naturalist – as a father, as a gentleman, and as a scientist – shortcomings which got erased in historical records. Similarly, in Thompson's *This Thing of Darkness*, the fictional FitzRoy is critical about Darwin and yet knows that his voice is unreliable and will not be remembered.

Nonetheless, contemporary fiction overall reinforces the myth of Darwin as the family-loving genius who watches his babies grow, remembers savages in America, and plays with apes in the Zoo, eventually arriving at the Theory – the most important universal frame of reference for structuring our knowledge of nature – and giving us the Paradigms. In this interpretation the Darwinian 'myth' denotes both the narrative enabling people to understand the ways of the universe and the human stance in it and, at the same time, a canonical story about how Darwin got his Theory, a story which circulates in many versions and is widely known.

---

230 Irvine, *Apes*, 93. The theory of evolution by natural selection nicknamed 'the natural history of babies' is depicted in *Creation* – following the already mentioned stereotype – as a truth so elemental children grasp it intuitively. Annie Darwin (just like Horace Darwin in the letter quoted above) is able to think in terms of ecology. In a flashback we see the Darwin children hide in the forest with their father to watch wild rabbits. Unexpectedly a fox appears and kills a rabbit which makes Henrietta, Annie's younger sister, hysterical. Only Annie manages to calm her down saying: "it has to be like this. Otherwise fox's babies will die. It's the balance of things".

The perfect example of such reinforcement may be found in Randal Keynes' *Annie's Box; Charles Darwin, His Daughter, and Human Evolution* and in Jon Amiel's film adaptation of the book, *Creation*. To begin with, Keynes is the great grandson of Darwin, one of many descendants of Charles and Emma's numerous children, which gives him a certain authority as far as the family life of the Darwins is concerned. Combining facts with imagination *Annie's Box* does not try to encompass the entire life of its subject the way standard biographies do, but focuses only on the years Darwin spent with his growing family in the rural surroundings of their country home. Keynes refers to the diaries of the children, heirloom photographs, and stories he heard at home in order to show how the loss of his eldest daughter affected Darwin and his writings about human evolution. In this book Darwin struggles with himself over whether to publish his theory and whether to study human nature, which necessarily involves an investigation of the natural history of humankind: parent-species, savagery, the vast chain of beings people evolved from, and the features we still have in common with apes. The book gives much prominence to the episode with Jenny the orangutan and the Fuegians, as Darwin remembers them while observing his own infant children. Keynes presents a sentimental picture of his forebear's research into the Theory and depicts the emergent Darwinian paradigm as politically correct and an easily acceptable point of reference embracing humans and apes, slaves and slave-drivers, the savage and the civilized – all of which, despite their apparent differences, are basically the same: they all are products of millennia-long progress.

The book has the shape of a sentimental return to the Victorian era when the most important Theory ever was formulated by Charles Darwin. Keynes refers to the family heirloom, the eponymous box whence all the memories come. Emma Darwin put her daughter's things there: letters, the sewing kit, envelopes, quills, Darwin's notes on Annie's illness. The recovery of the box in the real world compelled Keynes to write his book, while in the fictive world Darwin – the character of Keynes' book – kept the box in his study and his fear to write down his theory was symbolized by his reluctance to open the box and relive his anguish after the loss of his beloved daughter.

Such symbolic meaning of the box is retained in Amiel's biopic *Creation*, based on Keynes' book. The box, 'Annie's box', is visually conspicuous in the movie: when near it, Darwin keeps imagining long discussions with his dead daughter concerning, as she calls it, 'the natural history of babies', which is *On the Origin of Species* in its nascent form. The orangutans-savages-babies chain of associations is presented as the key to the story: Annie wants Darwin to make use of it, he is reluctant, afraid, clinically depressed. It is in the box that we see an old issue of *Penny Magazine* with an orangutan on the cover – this is Jenny from the short period when she was the star of the London Zoo.

The Annie-Jenny analogy is crucial in the movie: the two young female primates Darwin loves and loses uncannily resemble each other. Annie makes Darwin tell her half-imagined stories about Jenny: how she was captured on Borneo, sold for three hundred guineas to the Zoo, and dressed up in the long robes of Victorian babies and "made presentable for polite society". These are Annie's very favorite stories, she loves them better than the stories about the Fuegians, the Beagle voyage, the earthquake in Argentina, or the gigantic sloth. Darwin has to repeat how he came to see Jenny, how he played with her, observed her communication skills, and amused her with dangling his keys in front of her face. The scene with the keys is visually analogous to another flashback in the film in which Darwin in the same manner amuses the six-week-old Annie in the cradle while making notes on the expressiveness of human infants.

In the Jenny episode we see in close-up the human-like hand of the young orangutan touching Darwin's fingers. When Darwin tells her goodbye and she does not want to let go of his hand, for a second the two index fingers – the human and the apish – are frozen in the Sistine Chapel fresco image of God and Man touching fingers. The most sentimental analogy is between the death of Annie, who in her terminal illness suffers patiently, is silent and lovable (Darwin noted "her sensitiveness lest she should displease those she loved, and her tender love was never ever weary of displaying itself"[231]) and the death of Jenny in the London Zoo. The dying Annie begs Darwin to tell her for the hundredth time how Jenny contracted pneumonia, and how her heart-broken keeper tried to feed her with a spoon. In this story Jenny is patient, sad, and sweet; "shakes her hand as if to say you cannot help me, puts her arms around the keeper's neck, looks into his eyes in the most human fashion" and dies. The close-up of Jenny's immobile dead hand is followed by one showing Annie's dead face. The orangutan and the girl were both in Victorian night-gowns with laced collars and both died quietly with somebody they loved by their side. The sketch of Jenny on the magazine cover is in Annie's box just like Annie's own photograph. When Darwin looks at it he remembers how the daguerreotype was made. With Annie posing for the camera, Darwin told her the story of the Fuegians who onboard the Beagle dressed and behaved just like the British, but once in Tierra del Fuego they immediately shed

---

231   Darwin, notes, April, 30, 1851 < http://www.darwinproject.ac.uk/death-of-anne-darwin>
      accessed 16 December 2014.

their 1830s clothes and reverted to savagery, proving that people are capable of altering their ways depending on their environment, although deep inside they are basically the same.

Darwin the proto-ecologist, Darwin the greatest genius of humankind, Darwin the loving father, and Darwin the reluctant rebel against the Biblical paradigm of creation is a stock-figure in contemporary culture. Popular texts reinforce such an image by making stories about the origin of the Darwinian paradigm of science we are accustomed to accept. The orangutans-savages-babies chain of associations which one hundred-fifty years ago provoked so much controversy during the Oxford meeting of the Victorian naturalists is now 'tamed' by mass culture, which has rendered it sentimental by producing numerous versions of Darwin's struggle with the implications of his own discovery.

Darwinism and the Humanities

In 1996 Alan Sokal, a French physicist outraged by the careless and pompous way in which French humanist illuminati refer in their essays to scientific terms and notions they plainly do not understand, decided to test the gullibility of Academia by preparing what is now known as Sokal's hoax. Compiling fragments of articles written by postmodern philosophers (Gilles Delleuze, Jacques Derrida, Luce Irigaray, Jacques Lacan, Jean-Francois Lyotard, and others) he produced an apparently very 'learned' and patently nonsensical piece of writing he entitled "Transgressing the Boundaries. Towards a Transformative Hermeneutics of Quantum Gravity". His 'essay' was absurd: its paragraphs taken from diverse sources went together neither thematically nor logically, and the conclusion was not at all related to the argument. As far as ideologically-charged terms and phrases go, "Transgressing the Boundaries...." aped post-structural philosophy in its effort to prove that Enlightenment dogmas are dated. It argued that there is no outside world independent of the human perception of it, that what we call reality is a culture-dependent socio-linguistic concept, and that both Euclid and Newton are wrong. Sokal demonstrated that last matter using a number of very twisted references to the newest theories of space-time and 'post-Heisenbergian physics'. Generally, Sokal's article claims that empiricism is always tainted by cultural relativism and that virtually nothing can be said to be universal – even gravity itself. What Newton erroneously described as a universal "g" factor is in fact relative and embedded in the bygone historical situation of 17th-century England.

Sokal sent his article to the fashionable American social studies periodical, *Social Text*. Not only was his paper accepted, but it was also published in the *Social Text Special Issue* devoted to polemics with those who criticize postmodernism. Sokal immediately made everything public and disclosed his provocation, explaining in a number of texts[232] that his aim had been to prove that postmodernism had degenerated into some kind of pseudoscientific discourse devoid of meaning. What followed can only be described as an international scandal, with French, American, and British academics publishing polemics, letters of protest, and articles in scholarly periodicals and such daily newspapers as *The Guardian* and *The New York Times*.

---

232   Notably in Alan Sokal, 'Transgressing the Boundaries. An Afterword', *Descent* 43 (4) (1996), 93–94; and Alan Sokal 'A plea for reason, evidence and logic', *New Politics* 7 (2) (1997),126–129.

There is a number of reasons why it is worth remembering Sokal's hoax two decades afterward, as with the perspective of time this story looms ever more starkly indicative of the mid-1990s crisis of literary and social studies. It also shows how humanist scholars envied their colleagues from the science departments and wanted their own texts to sound equally hermetic and learned. Thus, in the twilight of postmodernism there is no mutual understanding between science and the arts and the two realms continue to drift in opposite directions. Literature, film, and social studies departments (then, as well as today) are increasingly marginalized, and the effected scholars are aware of this. Their predilection for 'fashionable nonsense'[233], i.e., proving their theses by referencing optics, topology, or cybernetics, seems to point to an innate inferiority complex among practitioners of disciplines from within the social sciences.

It is definitely worth examining what has changed in relations between the arts and sciences in the early 21st century, as certain new and interesting attitudes have emerged within the humanities. The aim of this chapter is to examine one such perspective that has been introduced to literary, film, and social studies – namely, the Darwinian vantage point – and to show that by references to evolutionary biology literary scholars are re-introducing to their papers the human universals which in the late 1960s were deemed false and the mere products of Western culture. The most important of them is the concept of some universal human nature independent of the particular cultures people are brought up in. The return of human nature to literary criticism by the Darwinian backdoor is the single most important event in the last few years, as I shall show here. In order to do so, the late twentieth century's postmodern cultural relativism will first be presented, followed by a summary of the theses of its early critics. Then, controversies concerning the adaptation of a biological perspective to literary criticism will be discussed. Here the new set of universals put forward by neo-Darwinian anthropologists will be presented, and the controversial matter of Edward O. Wilson's consilience introduced. In my concluding thoughts the titular notion of the return of human nature will be explained.

The feeling that literature studies are in crisis is shared by very many scholars of the discipline – falling enrollments and funding, along with the eroding prestige of the humanist departments outside Academia, cannot be denied. But worst of all is the sense of exhaustion and the repetition of dogmatic truths dating from

---

233 *Fashionable Nonsense. Postmodern Intellectuals' Abuse of Science* (1998) is Alan Sokal and Jean Bricmont's book describing the Sokal hoax and proving its points by analyzing science metaphors in recent criticism written by various leading humanists.

the heyday of post-structuralism in the 1970s. As Brian Boyd, Joseph Carroll, and Jonathan Gottschall prove[234], all that is new in the discipline in the last thirty years – namely, women's studies, queer theory, ethnic literature studies, cultural studies, postcolonial studies, and ecocriticism (which indeed do offer different angles for looking at literature) – goes on applying the same old research methods. Boyd's essay "Getting it All Wrong" is even more aggressive in driving home the point Darwinism-oriented scholars make:

> Until literature departments take into account that humans are not just cultural or textual phenomena but something more complex, English and related disciplines will continue to be the laughingstock of the academic world that they have been for years because of their obscurantist dogmatism... Until they listen to searching criticism of their doctrine, rather than dismissing it as the language of the devil, literature will continue to be betrayed in academe and academic literary departments will continue to lose students and to isolate themselves from the intellectual advances of our time.[235]

It is precisely the occupation with words alone, along with the belief that all meaning is generated by differences between various kinds of texts, that make postmodernism unacceptable to contemporary Darwinism-inspired scholars. The standard set of questions that post-structuralists ask when they analyze discourses – what is said, what is left unsaid, and how, depending on who is speaking (and who is silenced), meanings change – seems insufficient to interpret literature as stories about human beings and written by human beings.

The post-structural approach to literature is thus abstract: we study texts without referring to any extra-textual reality (and sometimes claiming there is no extra-textual reality), and as a result literature studies are alienated within Academia. Moreover, the gap between the two broad magisteria – the natural sciences and the humanities – is more prominent than ever. As Edward Slingerland claims in the book entitled *What Science Offers the Humanities: Integrating Body and Culture*, not only are methodologies and background theoretical assumptions radically different, even funding agencies serving the departments are separate. One may recall here *The Two Cultures*, an influential lecture delivered by the British scientist and novelist C.P. Snow way back in 1959. Its thesis was that the intellectual life of the whole of Western society was split into the titular two cultures – namely, the sciences and the humanities – and that this was

---

234   Brian Boyd, Joseph Carroll, and Jonathan Gottschall, 'Literary and Film Studies Now: Death or Rebirth', in Brian Boyd, Joseph Carroll, and Jonathan Gottschall, eds, *Evolution, Literature and Film. A Reader* (New York: Columbia University Press, 2010).

235   Boyd, Carroll and Gottschall, *Evolution*, 198.

a major hindrance to solving the world's problems. Apparently, natural science concerns itself with the deterministic laws governing matter, whilst the humanities study the unrestricted products of the ever free mind (literature, art, beliefs). Thus the former must value empiricism and the latter cannot. The underlying assumption is that matter and mind are two utterly distinct substances which cannot be studied using a common scientific approach.

Post-structural critics pushed this assumption to the extreme, the most provocative examples of which are the works of Jean Baudrillard, one of the most famous philosophers of the last decades of the 20th century. Baudrillard, applying the fashionable style of a science-inspired essayist, claims that in late capitalism there is no longer any material reality, as it has been replaced by hyperreal textuality. Referring to the theory of "Einsteinian relativity", which postulates "the absorption of the distinct poles of space and time",[236] he claims that "substance" and "information" are just empty concepts nowadays:

> Nothing separates one pole from another, the beginning from the end, there is a kind of contraction of one over the other, a fantastic telescoping, a collapse of the two traditional poles into each other: implosion – an absorption of the radiating mode of causality, of the differential mode of determination, with the positive and negative charge – an implosion of meaning.[237]

Instead of the world of matter we live in and refer to using a structured language which reflects our experiences of this world, we are embedded in something which only simulates reality. Such a point is argued by a number of examples hailing from different kinds of scientific discourses. The first ever-present instance of simulation is our DNA, which for Baudrillard (in the hyperreal world of today when cloning is possible) stands for the whole of a being condensed into "information", i.e., neither matter nor spirit but "an abstract matrix, from which will be able to emerge, not even through reproduction (…) the identical beings".[238] Secondly, what we call nature is also an instance of simulation, as there is no "free, virgin nature" – nor has there ever been[239]. Instead, we project "nature, desire, animality" onto some vague territory which we place somewhere else, outside our realm of experience, and which is in fact immaterial. It is not

---

236 Jean Baudrillard, *Simulacra and Simulation,* trans. Sheila Faria Glaser (Ann Arbor: The University of Michigan Press, 2003), 31.
237 Baudrillard, *Simulacra*, 31.
238 Baudrillard, *Simulacra*, 99.
239 Baudrillard, *Simulacra*, 140.

nature we talk about and miss, it is "the very schema of deterritorialization"[240] we indulge in. In *Simulacra and Simulation* Baudrillard arrives at the final conclusion that in the poststructural perspective there is no meaning, and he considers this circumstance "a good thing"[241] because we can now examine "that on which [meaning] has imposed its ephemeral reign, what it hoped to liquidate in order to impose the reign of the Enlightenment, that is, appearances."[242]

Baudrillard's core point that, firstly, there is no material reality; secondly, we live embedded in textuality; and thirdly, what we discuss when we think we discuss the tangible world of phenomena are but appearances, is the dominant assumption of post-structural humanist critics. This would seem to offer little hope for fruitful discussion between literary critics and their colleagues from the science departments. Some contemporary critics, among them Francis Wheen, consider the close of the twentieth century a period of "the sleep of reason", when philosophers and other humanist illuminati produced incomprehensible abstract theories in their own dilapidated ivory tower, while the people outside it were left on their own to search for meanings in bizarre ways. Wheen reflects on the late twentieth-century flourishing of astrology, the art of crystals, the interest in UFOs, or in healing vibrations and other new age-esque methods of striving for illumination and spirituality in his book *How Mumbo-Jumbo Conquered the World*.[243] In his opinion, in the 1990s, the heyday of post-structural theories, strange bestselling self-help manuals flooded the market, all of them promising to teach you how to find enlightenment just because enlightenment is what we all now need. Their authors use the term enlightenment not of course in reference to the eighteenth-century intellectual movement based on rational thought deemed by postmodernist obsolete, but some new age spiritual illumination. Today Amazon's search engine still finds 22,785 books with the word "Enlightenment" in their titles currently on sale. And yet their titles are a far cry from eighteenth-century rationalism, as the examples indicate: *Three Steps to Enlightenment*; *The Simplicity of Spiritual Enlightenment*; *A Spiritual Awakening. How the Enlightenment Code Can Change Your Life*; and *Enlightenment Ain't What It's Cracked Up to Be. A Journey of Discovery*.

Thus, more or less consciously, people do strive to be enlightened because, one may hazard, the need to know and understand is a part of human nature.

---

240 Baudrillard, *Simulacra*, 140.
241 Baudrillard, *Simulacra*, 164.
242 Baudrillard, *Simulacra*, 164.
243 Francis Wheen, *How Mumbo-Jumbo Conquered the World* (New York, London, Toronto, and Sydney: Harper Perennial, 2004).

And yet, the very concept of human nature as universal and inherent is precisely what post-structural critics deny. According to cultural relativists people are just centers of consciousness and each of us creates their own vision of reality by accepting or rejecting endlessly shifting linguistic signs. We are conditioned to do so, to accept some and reject the others, by the particular culture we happen to grow up in and be shaped by. Indeed, what we call "nature" is just a social construct. The literary scholars who believe in the above creed tend to interpret characters in literary works as allegoric embodiments of the terms they happen to be looking for, as terms defined by the source theories that produce the standard postmodern blend – most importantly deconstruction, feminism, psychoanalysis, and Marxism:

> In their postmodern form, all these component theories emphasize the exclusively cultural character of symbolic constructs, "Nature" and "Human Nature" in this conception are themselves cultural artifacts. Because they are contained in and produced by culture, they can exercise no constraining force on culture.[244]

Recent criticism written from the Darwinian perspective challenges the above opinion with two basic counterarguments: primarily, the idea that there is no universal truth is in itself a universal truth (and as such it disqualifies itself) and, secondarily:

> Not everything in human lives is culture. There is also biology. Human senses, emotions and thought existed before language, and as a consequence of biological evolution. Though deeply inflected by language, they are not the product of language.[245]

Such a belief, which purposefully negates the dominant post-structural ideology and which comes from the 2006 essay "Getting it All Wrong" by Brian Boyd, is in fact quite similar to ideas expressed as early as in the 1940s by Karl Popper. In the mid-20th century Popper claimed that instead of pursuing the essence of words and terms, philosophers should examine theories which refer to the natural phenomena of the tangible world and try to establish how proficient they are in both explaining why what happened did happen and in predicting what will happen. It is worth noting that Popper considered Darwinian natural selection to be one of the most proficient theories. He referred to it in *The Logic of Scientific*

---

244 Joseph Carroll, "An Evolutionary Paradigm for Literary Study" <http://www.umsl.edu/~carrolljc/Documents%20linked%20to%20indiex/Target%20Pieces/1_Target_article.htm> accessed on 23 November 2012.

245 Brian Boyd "Getting It All Wrong" *The American Scholar* Autumn 2006<http://theamericanscholar.org/getting-it-all-wrong> accessed 23 November 2012.

Discovery (1935)[246] to explain how theories compete with one another. The winner is the theory which is most often "reproduced", i.e., scholars refer to it most often in their lectures and academic papers, students learn about it and are then encouraged to apply it in their assignments. This theory survives and breeds new theories and the ones which are rarely "reproduced" plunge into oblivion for the historians of science to unearth sometime in the future and describe as curiosa. One may venture to say that Popper anticipates Richard Dawkins' memes by attempting to apply Darwinian paradigms to discuss the way science (which is a part of culture) works.

The New Synthesis, the twentieth-century blend of the theory of natural selection and modern genetics "reproduces" easily, as it is now used to explain a wide range of phenomena, human psychology included. Growing ever more popular, it has bred sub-theories such as sociobiology and evolutionary psychology, which from university departments made their way to popular science bestsellers and nature films and are a part of our cultural competence. Thus, for example, it is now generally accepted that, according to modern Darwinism, every organism is said to bear marks from its evolutionary past which make it behave in a way that was optimal for survival and efficiently reproducing in the primary environment where the given species evolved.

Yet even general awareness of nature-culture evolutionary linking has next to no influence on the post-structural humanities. Since the turn of the twentieth century Emile Durkheim's postulate that what we call "human nature" is produced by culture scholars has imagined people to be containers filled by what is going on around them. In *The Rules of Sociological Method* (1895) Durkheim deemed culture-dependent even such strong emotions as paternal and maternal love, sexual jealously, hatred, and longing for revenge.[247] Sociologists, anthropologists, and even psychologists in the twentieth century believed in the *tabula rasa* conception of the human mind and tried not to refer to human universals. It was the emergence of sociobiology, a new branch of natural science, that marked the beginning of some renegotiations as far as the mutual dependence of biology and cultural studies go: in 1970, E.O. Wilson published *Sociobiology. The New Synthesis*, calling for much closer cooperation between scholars doing science and those working within the humanities. Yet it was mainly biologists who tried to incorporate elements of culture studies into their research than *vice versa*: in

---

246 Karl Popper *The Logic of Scientific Discovery* (London, New York: Routledge Classics, 2002).
247 Emile Durkheim *The Rules of Sociological Method*, trans. W.D. Halls (New York: The Free Press, 1982).

the last decades of the century proto-societies of apes, wolves, elephants, and crows were studied and their "cultures", i.e., non-genetic transmission of data from generation to generation, were described (such as patterns of behavior, knowledge of food resources, and how to get to them, war, and warning calls).[248] Literary scholars for their part were still reluctant to look at humans as beings shaped by evolution.

Robert Wright in *The Moral Animal*[249] asks why Wilson's theories were rejected in liberal arts departments and suggests that perhaps the history of misusing Darwinism by fascist and racist ideologues to back up their agendas in the early twentieth century is to blame – some scholars did not even read *Sociobiology*, but immediately deemed it as "social Darwinism" and thus unacceptable. Moreover, it is difficult for any of us to study human nature as a single and universal product of evolution because we humans are not universally equipped by the logic of this very same evolution with introspective talents. The self is not easily studied, something which Sigmund Freud noticed as early as in the 1890s – our conscious minds are not aware of our real motivations in wanting something, being afraid of something, or feeling attracted to something. Of course, Darwinism is not psychoanalysis and does not follow Freudian principles. For instance, the Oedipus complex is a pattern of behavior that could not have evolved, as it reduces one's chances of producing lots of well-adopted healthy off-spring. Nonetheless, the two schools of thought believe that people do what they do for mostly unconscious motives and that every human being is to a certain extent volitionless in their lives. Darwinian scholars consider evolution to have neither aims nor foresights, and simply to have equipped humans with the propensity to behave in a way which, in the African plains hundreds of thousand years ago, helped them survive and reproduce better than other members of the tribe. Humans did not and do not know why such kinds of behavior are "natural" for them, they simply feel an urge to follow instincts – social, family, and moral instincts included.

Some decisions simply feel morally right and taking them makes us feel noble and good despite all the troubles and hardships they may cause, and thus we consider morality to be exclusively human and to derive from the realm of culture, not nature. Yet moral instinct itself must have evolved and altruistic behavior must be in the long run profitable. In *The Descent of Man* Darwin claims:

248   Konrad Lorenz, Vitus B. Droscher
249   Robert Wright, *The Moral Animal: Why We Are, the Way We Are: The New Science of Evolutionary Psychology* (New York: Routledge, 1995).

[P]rimeval man, at the very remote period was influenced by the praise and blame of his fellows. It is obvious that the members of the same tribe would approve of conduct which appeared to them for the general good, and would reprobate that which appeared evil. To do good unto others – to do onto others as ye would they should do unto you – is the foundation stone of morality.[250]

Darwin goes on to explain that the primeval man who sacrificed himself for the good of others and was praised in the tribe would influence other tribesmen in two ways: exciting the same wish for glory in his fellows and begetting off-spring with a tendency to inherit his own noble character. Thus, though unaware what genes are, Darwin defines both cultural and genetic paths for transmitting patterns of behavior, which is precisely what contemporary sociobiology does. Darwin knows that the "lowly origin"[251] of noble features is difficult to accept and in the famous last paragraph of *The Descent of Man* he feels obliged to defend why he feels the mechanisms of pre-human apes becoming "savages", and then "savages" turning into humans, should be exposed at all: "we are here not concerned with hopes or fears, only with the truth as far as our reason permits us to discover it, and I have given the evidence to the best of my ability."[252]

By these reservations Darwin anticipates criticism of those who feel offended by the thesis that human nature is produced by natural selection. Interestingly, precisely a hundred years later, with the publication of *Sociobiology*, Wilson stirred the very same objections. Aiming at integrating the findings of numerous branches of natural science into a single theoretical framework of cause-and-effect explaining the social behavior of many species from ants to men, he too did offend. Though the book was highly praised for using Darwinian paradigms to explain the seemingly bizarre conduct of animals, it was strongly criticized for its last chapter devoted to homo sapiens. Readers were scandalized by the thought that human social norms may be reducible to biology and that human nature is stored in our genes and thus, firstly, we are genetically predetermined and not so free in our choices, and, secondly, much in humanness is universal, which goes against the contemporary cult of individualism, as well as the principles of postmodern philosophy.

In the last chapter of *Sociobiology* and in his next book *On Human Nature*, Wilson elaborates upon his vision of human civilization as the ultimate product of biological evolution. He argues that the humanities and social studies might

---

250   Charles Darwin, *The Descent of Man And Selection in Relation to Sex* (New York: A.L. Bert Company Publishers, 1874),149.
251   Darwin, *The Descent*, 708.
252   Darwin, *The Descent*, 708.

be considered branches of biology: "history, biogeography, and fiction are the research protocols of human ethnology; and anthropology and sociology together constitute the sociobiology of a single primate species".[253] Wilson also argues for a fundamental reform in both the sciences and the humanities and for the emergence of a new paradigm based on consilience[254] that seeks to narrow the gap between them. For example, neuroscience may explain how animal brains work, genetics, how heredity works, and evolutionary biology "why brain works, or more precisely, in the light of the natural selection theory, what adaptations if any led to the assembly of their respective parts and processes".[255]

Sociobiology-inspired zoologists did very much research on the social life of animals, mainly primates, in the 1970s and 1980s, and now the term "culture" has broader meaning, at least within the natural sciences, denoting not the opposition of nature but its product. Yet the furtherance of consilience among branches of humanist learning is yet to transpire. Wilson wrote his most important books in the same decade that Jean Baudrillard did, and their impact on, respectively, the natural sciences and the humanities is comparable. Yet the juxtaposition of their essays shows how huge the gap between their approaches to human civilization is: Wilson calls for the objective, multi-perspective study of human nature, Baudrillard claims that there is no more nature, nor reality either, and that we live surrounded by simulacra.

Yet many contemporary novelists (Ian McEwan, Antonia Byatt, and Julian Barnes among them) are fascinated by Wilson just because his theses go against the grain of post-structural dogmas and allow them to talk about such old-fashioned concepts as objectivity, truth, and humanism:

> The objective meaning of human nature is attainable in the borderland of disciplines. We have come to understand that human nature is not the genes that prescribe it. Nor is it the cultural universals such as incest taboos and rites of passage, which are its products. Rather, human nature is the epigenetic rules, the inherited regularities of mental development. These rules are the genetic biases in the way our senses perceive the world, the symbolic coding by which our brains represent the world, the options we open to ourselves and the responses we find easiest and most rewarding to make.[256]

---

253 Edward O. Wilson *Sociobiology: The New Synthesis* (Cambridge Mass.: The Belknap Press of Harvard University Press, 2000), 547.

254 "Literally a 'jumping together' of knowledge by the linking of facts and fact-based theory across disciplines to create a common groundwork of explanation", Edward O. Wilson, *Consilience: The Unity of Knowledge* (New York: Alfred A. Knopf, 1998), 7.

255 Edward O. Wilson *Sociobiology*, vii.

256 Wilson, *Sociobiology*, vii.

Once scholars felt it was alright again to talk about human nature understood as a set of inclinations and propensities we are born with independently of the time and the place we are born in, anthropological studies started to slowly change. Instead of denying the existence of human universals, some cultural studies specialists try to see underlying universal patterns in the multiplicity of human social systems. Donald E. Brown in *Human Universals*[257] ask the simple question, what precisely do all people, all cultures, and all their languages have in common? In order to give an answer and explain his findings better he describes what he calls the UP, the Universal People, a non-existent tribe of archetypal humans whose culture – patterns of doing things and thinking about the world – are passed from generation to generation, and who are self-aware enough to know that their way of doing things, their culture is not the way of nature.

The UP have a language with which they can describe both their internal states and the external world: things abstract, things concrete, and things which are not. The proficiency in using this language gives prestige to speakers, the language allows the UP to, among other things, gossip, which is an important part of their social life. Their language is symbolic and built of basic speech units – phonemes – and it has a grammar. The UP language can be used for special purposes, e.g., poetry making, and thus it can give esthetic pleasure to its speakers. It also allows for naming people, objects, and places with proper names – and it has a set of kinship terms reflecting family structures and sex terminology which is fundamentally dualistic.

Apart from the linguistic means of communication, the UP use gestures and very complex facial expressions and each of them is aware of her or his unique personality, they are capable of empathy, they understand that other people also have feelings, they distinguish between a private inner life (memories, dreams, hopes) and public behavior. The UP are moved by sexual attraction, are attached to their children, fear snakes, spiders and rodents, but are able to overcome their fears. They recognize individuals by their faces, they are toolmakers skillful enough to use tools to make tools. Some of these tools are weapons. They may not know how to make fire, but they definitely know how to use it and they cook. Thanks to tools and fire they feel secure. They can also take substances which alter their moods: stimulants, narcotics, or intoxicants. They shelter themselves from the weather, they prepare for birth, and have patterns of postnatal care. They do not live alone: their basic social unit is the family, but there are bigger

---

257 Donald E. Brown, *Human Universals* (New York: McGraw-Hill, 1991). Also at <http://www.udel.edu/anthro/.../universal_people.pdf) accessed 20 November 2012.

groups too. They feel they are the owners of some kind of territory even when they are nomads.

The core of a UP family is a mother with her children and she is usually in a more or less stable relationship with a man. Children are socialized according to traditional patterns, mating with close kin is prohibited, prestige and status are very important. There is some distinction of labor based on sex and age, some tasks are done collectively, goods, services and labor might be reciprocally exchanged, people engage in trade, and there are some affairs which are considered public: "in the sense that decisions binding on a collectivity are made".[258] There are leaders, though they do not have to rule for long – and some of them are situational. The UP value generosity, they know what rights and obligations are and they have traditional ways of dealing with conflicts between each other. There are certain standards of sexual modesty, religious or supernatural beliefs, music, dance, and burials. They adorn their bodies and have patterns of hygienic care.

Brown goes on to enumerate many more universals and he ends by appealing to scholars to undertake further research aimed at establishing the relationships between these universals and at offering explanation on why they have evolved. Once a new field of studies is opened for research it needs to be charted with the help of the New Synthesis, especially its sub-branch, evolutionary psychology. To quote Brown: "Anthropology has scarcely begun to illuminate the architecture of human universals. It is time to get on with the task."[259]

Adopting a new perspective in anthropological studies and the acceptance of some biological unity beneath the apparent diversity of human cultures is refreshing and allows for some new insights into the movies and literary texts we study. For example, Murray Smith in "Darwin and the Directors" refers to an early evolution studies classic, Charles Darwin's *The Expression of the Emotions in Man and Animals* (1872) in his discussion of the way emotions are presented in films by gestures, postures, and faces. He claims that we respond to some underlying emotive patterns in a film, and the generic classification of films very often reflects these patterns rather than the film's setting or style (thrillers, weepies, horrors). Moreover, the above-mentioned patterns are universal – at least Smith finds them in Japanese, British, and American films alike.

Steven Pinker in "How the Mind Works" analyzes a huge number of canonical works (by Nathaniel Hawthorne, William Shakespeare, Tennessee Williams, and many others) in respect to interpersonal relationships between the characters

---

258   Brown, *Human Universals.*
259   Brown, *Human Universals.*

and our intuitive understanding of them. Pinker tries to establish the basic structures of meaning which are connected to the strategies and tactics humans instinctively use in social situations to pursue their subliminal Darwinian goals: to increase by social interactions their chances to survive and multiply. Pinker also argues that literature, just like gossip, is interesting for us because through getting to know other people's stories we see for ourselves the outcomes of the decisions one takes in diverse interpersonal situations. In this way, by reading books we exercise our own decision-making strategies and get prepared to one day face similar challenges in life. Thus, on some deep level, in literature we are dealing with very basic conflicts older than civilization: mostly connected to love or sex, revenge, and threats to the safety of the protagonist and his kin. Pinker's examples include "the mistaken jealously" motif, "vengeance taken for kindred upon kindred" and "the discovery of the dishonor of the loved one"[260]. Literature is not reducible to such patterns, but they can be found in very diverse texts: from reports on gory murders in the tabloid press to Fyodor Dostoyevsky.

The taste for literature is thus a kind of adaptation: in real life humans face social situations so very complex that it is impossible to have innate ways of dealing with them. The mechanisms of natural selection which might have prepared us for challenges require thousands of generations to produce new genes which would make us behave in an appropriate manner. The lifespan of an individual is too short to collect enough experience by means of generalized learning:

> There was not enough time for human heredity to cope with the vastness of new contingent possibilities revealed by high intelligence… The arts filled the gap. Early humans invented them in an attempt to express and control through magic the abundance of the environment, the power of solidarity, and other forces in their lives that mattered most to survival and reproduction… The arts still perform this primal function, and in much the same ancient way. Their quality is measured by their humanness, by the precision of their adherence to human nature.[261]

Surprisingly, the above quote comes from a book written in 1998. The very terms applied here such as "humanness" and "human nature" and the concept that arts can magically control the environment (thus, there must be some tangible extratextual reality to control) are unacceptable within the difference-oriented schools of postmodern criticism that was hegemonic in the 1980s. The Darwinist take on literature is of course limiting too: one has to prove that the propensity

---

260   Steven Pinker, *How the Mind Works* (New York: Norton, 1997), reprinted in Boyd, Carroll, and Gottschall *Evolution*, 132.

261   Wilson, *Consilience*, 225.

for telling stories and listening to stories had an adaptive function in the days of our cave-dwelling great grandparents and that the taste for literature increased their chances for survival. Yet it is a genuinely new approach and it frees literary studies from post-structural dogmatism. Moreover, it recalls certain pre-postmodern literary theories, if only to name Northorp Frye's *Anatomy of Criticism*, the book where literary works are classified on the basis of social relations and the corresponding emotions (happiness, sorrow, and hostility in comedy, tragedy and satire).

Postmodernism claims culture is everything we have access to, Darwinian critics want to prove that it is a product of nature which in turn produces a vision of nature we can grasp and study. To sum up, the return of human nature into the humanities is a hard fact which, moreover, may bring about a reconciliation of the two discourses: that of the arts and that of sciences. Such a reconciliation is not a purely linguistic "borrowing" of terms and notions of the kind Alan Sokal ridiculed, but it may happen on some more profound level. To quote Edward O. Wilson again:

> Can the opposed Apollonian and Dionysian impulses, cool reason against passionate abandonment which drive the mood swings of the arts and criticism, be reconciled? This is I believe an empirical question, its answer depends on the existence or nonexistence of an inborn human nature. The evidence accumulated to date leaves little room for doubt. Human nature exists and it is both deep and highly structured.[262]

---

262 Northrop Frye, *Anatomy of Criticism: Four Essays* (Princeton, N.J.: Princeton University Press, 1957) 140.

The Motif of Human Evolution in Selected Fiction and Non-Fiction

Popular culture nowadays is very much fascinated by 'pre-history': with the help of advanced computer techniques which allow us to make the non-existent visible on screen, life-like simulacra of pre-human apes or charging T-Rexes are produced and distributed in millions of copies. Recently, producers of educational TV shows, especially nature films devoted to evolution, often decide to apply similar special effects. Thus, a new film genre has been created which apparently belongs to non-fiction because of the way it refers to real-life scientific discoveries and yet freely makes use of numerous visual clichés drawn from works of fiction.

Two films produced by France 3 and the Discovery Channel and directed by Jacques Malaterre, *A Species Odyssey* (2003) and *Homo Sapiens* (2005), blend fiction and non-fiction, thereby creating a picture of human evolution which is both in accordance with textbooks of anthropology and simple to grasp by its similarity to the scenarios we already know from mass culture. Based on the example of Malaterre's films, the aim of this chapter is to show how science and entertainment coexist on screen and, additionally, how anthropologists inspire fiction writers who are in turn themselves inspired by novels and films. In order to achieve this I am going to precede my analysis of the films by providing some suitable context: on the one hand literary texts concerned with human evolution by H.G. Wells and William Golding, on the other scientific essays by Charles Darwin, Misia Landau, Roger Lewin, and Geoffrey Miller. Moreover, some references will be made to the epic tradition in literature as this convention seems to influence Malaterre the most.

Darwin's *The Descent of Man And Selection in Relation to Sex* in its first part, which is devoted to paleoanthropology, strives to indicate the most probable paths of human evolution as based on the theories Darwin presented some years earlier in *On the Origin of Species*. Though *The Descent of Man* is a scientific work, the narrative voice systematically stresses the grandeur of the story of our brave ancestors who by their toils made our reality come true. Their story is our story, the most important in the history of our planet.

> The world… appears as if it had long been preparing for the advent of man, and this in one sense is strictly true, for he owes his birth to a long line of progenitors. If any single link in this chain had never existed, man would not have been exactly as he is now.[263]

Darwin writes for late 19th-century readers who have already become acquainted with his grand theory of evolution and who generally are interested in the progress of natural science and, for example, have been thrilled by the quite recent discovery of the gorilla, which for a long time had remained a half-legendary creature haunting European imagination. Fascination with the beastly yet nearly human apes living in the colonial wilderness and the growing awareness that humanity's ancestors must have been similar prevail in the text. Darwin's picture of the early progenitors of modern humans "covered with hair, both sexes having beards,… ears were probably pointed,… with a tail"[264] emphasizes the human-animal affinities. It is disputable whether these creatures may be called humans; in the series of forms graduating "from some ape-like creature to man as he now exists it would be impossible to fix on any definite point when the term "man" ought to be used"[265], and yet this half-animal commenced a millennia-long process of intellectual advancement. First it descended from the trees to the firm ground, then it stood upright on its two hind legs, thereby freeing its hands and arms and, by exposing itself to challenges and solving problems it gradually developed a large brain and ultimately became fully human and created civilization.

> He has discovered the art of making fire, probably the greatest ever made by man excepting language… These several inventions by which man in the rudest state has become so pro-eminent, are the direct results of the development of his powers of observation, memory, curiosity, imagination and reason.[266]

Thus, mental virtues made up for the physical weakness of early humans and we today, no matter how distant we are from our apish forefathers, boast the same abilities.

The impact of *The Descent of Man*, along with the turn of the century discoveries of pre-human fossils and pre-historic cave paintings, made the subject of human evolution attractive to the reading public. H.G. Wells' story "The Grisly Folk" (1921) reflects the attitudes people at that time had towards their alleged

---

263   Charles Darwin, *The Descent of Man And the Selection in Relation to Sex* (New York: A.L. Bert Company Publishers, 1874), 188.

264   Darwin, *The Descent*, 182.

265   Darwin, *The Descent*, 205.

266   Darwin, *The Descent*, 54.

progenitors. The text starts in press-article fashion by describing a visit to the natural history museum and the exposition of the "Neanderthaler's" bones. The narrator imagines this creature as a hairy or grisly half-animal with an apish face walking head forward not head up like a baboon. He calls the story of the rise and fall of this hominid species "the most fascinating riddle of all these riddles of ice and hardship"[267]. "Neanderthalers" in his opinion must have been far from human: living alone or in twos (food was too scarce for any bigger groups), deprived of language, with no notion of what family is. In Wells' opinion when "Neanderthalers" found their own young annoying they ate them up unless the children managed to run away in time. In this latter case the child would come back when grown and eat the father. After such initial remarks the narrator asks the reader to imagine a story: a small group of early Homo sapiens are heading north at the close of the last Ice Age. On their way they encounter a terrible inhuman beast, a hungry "Neanderthaler" crouching in a middle of a rocky field, eager to charge at the tribe. They manage to escape thanks to cooperation. In the years to come brave Homo sapiens have to wage a continuous war against their much stronger hominid neighbor. They are superior in intelligence, they have strong communal feelings and they are willing to die in defense of their own folk. Thus in the long run they are sure to win the battle against the evolutionary older and more primitive race. Yet for generations the grisly folk are the bane of Homo sapiens' young, whom they kidnap at night while lurking outside the blaze of cave fires. The narrator suggests that the traditional nursery room horrors of ogres or trolls or wicked giants devouring young children are remnants of these prehistoric tragedies as well as archetypal nightmares and the fear of darkness our race suffers.

William Golding in his famous novel of 1953 entitled *The Inheritors* counters Wells by giving his own absolutely contrasting version of the Neanderthal-Homo sapiens encounter. He clearly writes in response to Wells' text where he describes the extinction of the last remaining tribe of Neanderthals at the hands of the more sophisticated newcomers. Before the merciless enemies come we see the very simple life of the Neanderthal tribe through the eyes of one of its members, a young male named Lok. The Neanderthals know burials, speak simple language partly helped by some inborn telepathic abilities, they experience strong emotions, feel quasi-religious awe of life and birth and have a basic spiritual system with a female-centric principle. What they worship is the power of nature

---

267   H.G. Wells, 'The Grisly Folk'< http://gutenberg.net.au/ebooks06/0602061.txt > accessed 8 January 2012.

embodied by a mother-like goddess in whose unseen presence they perform solemn rituals to commemorate their dead relatives. They also have an outstanding knowledge of food resources: roots, vegetables, and occasionally carrion stolen from hyenas. They eat meat when they find it, but they are biased against killing, believing that "blood is blame".

The skillful narration makes the reader automatically presume that Lok's tribe is "us", an early Homo sapiens clan who in a child-like manner still live in the eternal present broken only by the changing seasons which have not yet given rise to the concept of cyclicity. Only when superior folk come in their boats with arrows and weapons and kill all Lok's family one by one do we realize that we are the heirs of the killers, not the victims. Although the people of Homo sapiens are mortally dangerous, they are also fascinating for the last Neanderthal tribe whom they outlast not by their technology, but primarily by their superior understanding of violence. The guilt from having stolen the territory of the older race and having slain innocents is immediately transferred in the Homo sapiens mythical frame of thinking onto the victims – hence, the Neanderthals become "forest devils". Demonized and deprived of humanity they are scapegoats who have to be caught and ceremoniously murdered, as such a sacrifice purifies the newly conquered land and a new metaphysical order of things emerges: the tribe of Homo sapiens lives surrounded by a wilderness peopled by ghosts of half-human beasts. The novel's motto taken from Wells: "we know very little of the appearance of Neanderthal man"[268] sounds ironic in this context, as we never knew them yet we exterminated them. Indeed, all the hominid tribes seem peaceful to themselves, but in their metaphysics violence and the sacred go together, "blood is blame" and civilizations, just as Freud would have it, are founded on crime.

In all the numerous texts devoted to human evolution, both fiction and nonfiction alike, a dearth of material findings are capable of inspiring mostly imagined stories and, moreover, these stories generate new texts written in response to them, just as Wells inspired Golding. Thus, the bulk of writing on human evolution refers to itself rather than to any material reality and even anthropologists interpreting the scarce data use literary conventions in forming new theories. Every theory is a story with a beginning, a middle, and an end which should be chronologically and casuistically connected. According to Misia Landau, in working out these connections theorists of human evolution more or less consciously apply schemes they know from literature. In 'Human Evolution

---

268 William Golding, *The Inheritors* (New York, Harcourt: Brace & World, 1955), motto.

as Narrative'[269] Landau analyzes the way they do so: she points out four basic cause-and-effect chains anthropologists employ to present in a logical way the very same facts. The first of them is the already mentioned Darwinian scheme: an early hominid climbed down from a tree, thus stood on hind legs, thus developed bigger brains, and thus built civilization. The second reverses this logic: the ape learned to stand on two legs, thus it left the trees and built civilization with resulted in rapid growth of its brain. The remaining two offer a pair of other orderings for the four events which after all must have happened inasmuch as we now are, in contrast to simians, biped, big-brained, ground-dwelling civilization-builders.

Landau argues that, like all people, anthropologists are born story-tellers: they tend to present the history of the human race as a meaningful totality where past events result in the present situation and where, from time to time, a ground-breaking revolution occurs, a turning point after which history takes a new course. By arranging such important events into a whole, evolutionists have arrived at an epic-like version of human pre-history which they agree on, but which belongs to literature, not science. This history reads like a heroic tale: in the beginning, in the period of uninterrupted existence, the pre-human ape lived happily in the trees. It was smart yet weaker than, say, tigers and in most anthropological texts this ape resembles Cinderella or the youngest brother from a folk-tale: it is the least privileged and yet we know it is going to get the kingdom in the end. One day there comes a change and the ape leaves the trees. At this point specialists quarrel: some say an environmental change made the ape leave the trees, others maintain the ape wanted to leave the trees (e.g., it had developed a bigger brain and, hence, curiosity). The next stage is that of adventures in the new terrestrial realm. The ape faces hardships and tests, it has to develop new skills and, in order to make up for its weak physical frame, it needs help. In most stories at this point the ape's big brain acts like a donor of magical items in a folk tale: it furnishes the ape with intelligence, reason, moral sense – and thus the ape is transformed into a human. The end is happy: after the final test, for example the survival of the European Ice Age or the encounter with the Neanderthals, the former ape now boasting the name of Homo sapiens builds empires and is ready to go into outer space in search of new planets to populate. Landau also mentions more pessimistic scenarios: the ex-ape is so successful that he becomes guilty of hubris, destroys his own planet and now is in danger of extinction.

---

269   Misia Landau, 'Human Evolution as Narrative', *American Scientist* 72:262–268 (1984) <http://www.unc.edu/~akakalio/Landau.pdf> accessed 8 January 2012.

Roger Lewin in his influential textbook *Human Evolution: An Illustrated Introduction* agrees with Landau and embellishes her thesis with some more observations.[270] He notices that all the accounts of human evolution are written from the point of view of the present condition of the human race and thus they tend to interpret all history as leading to today's world. Therefore, every past event serves the sole purpose of making the present come about – that is, evolution has a goal, this goal is us, and the prehistoric people did nothing but strive to be turned into creatures very much like members of modern Western culture.

Similarly, Geoffrey Miller in *The Mating Mind*, a book devoted to the human brain's evolution in relation to sexual selection, follows the steps of Landau and Lewin: he discusses images of human prehistory in mass culture proving that they vary according to target audience. First, some children's movies and cartoons show ethnically diverse cavemen happily living with dinosaurs in the "primordial" jungle hunting mammoths. In these productions there is no sex, no violence, and suburban family values are promoted. The next group, which is geared at adolescent viewers, does show some violent scenes and does have some love story motifs, but generally they are quite decent and the stress falls on adventure, danger, and survival. Frequently a group of contemporary teenagers travels back in time. The last group is that of films for an adult audience where prehistoric scenery serves as a backdrop to very violent fantasies; on screen we see cannibalistic feasts, crude bloody fighting with clubs and stones, collective rapes, and animals devouring human flesh. Of course none of these conventions is concerned with any conceivable "truth" about the dawn of humanity.

Conversely, Jacque Malaterre's films, devoted to reconstructing human evolution and financed by educational and science networks, are meant to teach rather than entertain. They are made with the help of leading paleoanthropologists and their producers are very much concerned with the scientific veracity of the smallest detail. And yet, Malaterre is heavily dependent on the show business tradition of film pre-history and, as far as the organization and logic of the story he shows is concerned, he takes his ideas from the tradition of heroic epic tales. The very title of the first of his films, *A Species Odyssey* points to both sources of inspiration: primarily to Kubrick's *A Space Odyssey*, whose first mute scenes of pre-human apes learning to use bones in order to kill one another are deeply rooted in the imagination of contemporary moviegoers. Secondly, the word "odyssey" sends us back to the most famous epic ever created, while Malaterre's

---

270   Roger Lewin, *Human Evolution: An Illustrated Introduction* (Oxford: Blackwell Publishing, 1999).

protagonist, who in consecutive episodes becomes less hairy and more human, is a bit like Odysseus traveling across the world, fighting his own weaknesses and performing heroic deeds.

The film's structure is episodic just like in the case of primary epics, Homer's *Odyssey* included. It starts the moment the first primate stood up on its hind legs. The voice-off narrates the story of the origin of hominids stressing the grave difficulties apes had to face when Africa split into two parts, one of which became very dry. The luckier jungle apes who lived in the wet half of the continent went on eating fruit in the trees, but their poor dry-land cousins were dying in the newly emerged savannahs clinging to the last trees. The moment the first of them decides to jump down and, seeing nothing in the grass stands up and struggles to march on its two legs is given much prominence on screen. The narrator assures us that all of us, our ethnic background and geographical location notwithstanding, are descendants of this very ape. Its deed is epic and though it soon dies (when spotted by a wild cat) it is the greatest of human heroes.

The next episode of *A Species Odyssey* shows us a dramatization of the story of Lucy, the famous hominid whose bones were dug up by Donald Johanson in Africa in 1974. Malaterre chooses to adopt the tragic convention: Lucy is last of her tribe and, like all hominids, cannot stand loneliness. Leaving the trees she joins the carrion-eating tribe of apes and tries to imitate their ways and adapt to survive. She dies in a river crossing, but there is a suggestion that her genes survived as she had had children, while the narrator explains that eating meat helped early humans to develop brains.

In the following scenes Malaterre shows a number of "firsts" in human history, for instance the first time a human used a tool in order to cut through the skin of a dead crocodile: he used a sharp edge of a stone he had cut his own skin with just a moment earlier. We also see the first mourning after a lost mate when some very smart hominid realized what death was; the first time an early human died to rescue some other member of the clan and thus altruism was born; and the first time an elderly human became a teacher passing on to the younger generation his knowledge of how to work stones. The narrator explains the importance of each and every step: forming tribes, migrating to Asia and Europe, peopling the Earth. Between episodes we see the solar system, revolving planets, the Moon, and the Earth – with both prehistoric and today's coastlines. This creates the visual effect of grandeur, as we witness a story of planetary and cosmic importance.

The last episodes show the most recent and most advanced species of hominids: the Neanderthals and Homo sapiens. A Neanderthal's little drama takes place near the entrance to a bear cave and has the simplicity and gravity of an archetypal universal struggle. A few hunters, among them the clan chief and his envious comrade, go inside to chase the bear away, leaving the chief's adolescent son at the entrance, keeping watch. When they come out the chief is already dead, allegedly killed by the bear and the envious warrior takes over his power represented by the chief's skins which he then puts on. Later he also claims the dead leader's woman and daughter while the young heir can do nothing but escape. Yet he swears to come back when he is grown up and find the old traitor to exact his revenge. This episode owes much visually to *The Clan of the Cave Bear* by Michael Chapman, but as far as the subject matter is concerned it reaches back to much older texts: unspecified mythic stories of old gods replaced by new ones, and revenge tragedies of the *Hamlet* kind. The voice-off makes us aware that what we are watching is not a specific event but an allegory of how early social structures were formed.

This scene is followed by the final episode in which epic tradition goes together with the tragic unfulfilled love motif. In the same cave, thousands of years after the treacherous murder of the clan chief, one of the last remaining Neanderthal families dwell there coping with a low birth rate and diseases. The hunters intuitively sense their time is running out and when one day they see a tribe of Homo sapiens nomads camping nearby they decide to go to them out of loneliness and despair rather than to rob. As all the adult Homo sapiens males are out of the camp hunting, the Neanderthals retreat taking with them one of the alien women and leaving in return some game. The leader is fascinated by this woman, he tries to befriend her and communicates with her demonstrating his tribe's artifacts. The Neanderthals even learn from their hostage how to produce musical sounds using de-marrowed bones. Here Malaterre borrows some ideas from Jean Jacques Annaud's *The Quest for Fire*, the movie in which a hostage Homo sapiens woman teaches cavemen more advanced techniques, and both women look very similar with their black make-up stripe round their eyes. In Malaterre's version (already dated, scientifically speaking) no inter-racial love affair is possible, the voice-off tells us that the two species are too different genetically to produce off-spring and fresh Homo sapiens blood cannot strengthen the exhausted Neanderthal genotypes. The Neanderthals are somehow (the narrator does not know how) doomed to extinction, though he is sure that they passed on to our ancestors some of their cultural heritage. The Homo sapiens woman is claimed back by her kinsmen and the nomads depart, leaving the Neanderthals to die out alone. In the closing scenes we see the leader many years later. He is old and the last of his tribe, and before his death he goes to visit the place the Homo Sapiens woman's people once camped in.

Homo Sapiens, the sequel to A Species Odyssey, narrates the remaining part of our evolutionary past. This time Malaterre tells the story of humankind's intellectual progress, staging the probable "first times" when people performed certain tasks. The two films overlap historically speaking, as in the initial episode of Homo Sapiens we return to Africa in the days of Homo erectus. One very smart hominid notices that an animal skin wrapped around you shields against insect bites when you cut the flesh of freshly killed prey. He decides to take the skin as his trophy and a sign of leadership to the camp where he discovers that his woman is giving birth in the pouring rain. After much thinking he covers the newly born girl in the skin to shelter her against the cold, which is the first time a hominid ever got dressed. The episode ends showing the parents licking up blood and pieces of tissue from their daughter's face, who sleeps contently wrapped up in the hide and the voice-off tells us that this baby is the first representative of the new species, our own species, which is going to rule the world.

In the following scenes similar nameless geniuses for the first time tame fire, bury the dead, sail the ocean, perform a pre-religious ritual, paint a cave wall with their palms, enter a shamanic trance, domesticate a wolf, build a permanent village, and discover that wild wheat grows in the spot where you have dropped its seeds into the soil. Two episodes in the middle are again devoted to Neanderthal-Homo Sapiens encounters which from the days of Wells and Golding seems to be an obsessive recurrence in all the texts narrating human evolution in fictive or non-fictive mode. In the first episode we see a matriarchal Homo sapiens clan led by a sun-worshipping shaman woman across the Alps over which the sun is setting. Most of her people die in the crossing and the few survivors are rescued by the Neanderthals whom they consider atrocious, but from whom they learn how to live in snow-covered Ice Age Europe. The second story repeats the impossible interracial love motif: a starving Homo Sapiens clan enters Neanderthal territory, the chief goes hunting and there he encounters a strange looking attractive and strong-willed woman hunter of the local tribe. After initial mistrust the two peoples get together to kill and eat a mammoth. The cooperation is fruitful, the woman joins the Homo Sapiens chief's harem and all are for some time comparatively happy. Yet no child is born out of this union, the woman dies of some fever she contracted and the leader mourns her.

In both A Species Odyssey and Homo Sapiens Malaterre stresses the homogeneity of humankind, all the brave apes and early humans we see contributed to the making of our civilization – if not via genes, then via cultural heritage.

We all are united, all differences between people are just the superficial adaptations of only tens, not hundreds, of thousands of years. As in the above anthropologic texts that Misia Landau discusses, the narrator claims human evolution is a success story and now humankind should travel to the stars in search of new planets.

Such a frame: from the dawn of humanity to the departure for the stars (judging from the films it is absolutely certain that we shall depart very soon) gives Malaterre's story an epic dimension, as this is a mythic account of how Homo sapiens conquered the powers of nature. And just what is the epic, according to literary scholars? Here I refer to one of the most renowned definitions given by C.S. Lewis in his 'A Preface to *Paradise Lost*' – namely, that the epic is above all a grand scale story. Malaterre's films do show grand scale events: this is our planet, our heritage, and it is only we who decide what the future shall bring, as the voice-off assures us at the end of the film, when on the screen we see a stag-headed shaman dancing in an Ice Age cave. Secondly, epics narrate important events in (natural) history and often incorporate legends, folk tales, and myth. This is also true in *A Species Odyssey* and *Homo Sapiens*, as the "firsts" the films show are such important events, while the mythic and folk stylization is also present, for example in the bear cave episode. Next, according to C.S. Lewis, epics describe noble heroes, mostly warriors who in the distant past fight allegorical monsters. Malaterre's hominids certainly are nameless heroes; their ancient toils do have a symbolic dimension and, moreover, just like in the case of the secondary epics Lewis discusses, what they achieved in the past influences the present situation, i.e., we read about the origins of our world.

Lastly, epics are narrated in solemn and lofty language. In Malaterre not only is the style of the commentaries pompous, but a certain grandeur is also suggested visually. Though in the first episodes the apes are computer simulations and in the rest of the films played by heavily made-up actors, in all the scenes "acting" is equally exaggerated, the figures on screen display the most rudimentary emotions in their pure and strong form: grief, awe, curiosity, ire. As there is no comprehensible language used (apart from the voice-off), the scenes have the lurid quality of a mime show or *commedia del'arte* performance in their staging primordial dramas of jealously, tragic love, and cruel death.

Thus *A Species Odyssey* and *Homo Sapiens*, though technically more advanced than any earlier narratives of human evolution, perfectly inscribe themselves within the tradition commenced by Darwin and Wells. They are strange mixtures of fact and fiction combining anthropology with literary and cinematic inspiration. They present human evolution as a heroic endeavor and, in trying to depict history as a coherent whole, freely borrow from conventions we all know from the most diverse sources, including myths, cartoons, revenge tragedies, contemporary novels, and fantasy films.

Annie Dillard and Kurt Vonnegut on the Galapagos Archipelago as the Archetypal Darwinian Setting

The turn of the millennium debate concerning the relationship between the sciences and the humanities is one of the most exhilarating issues in the contemporary intellectual life of the West. Ever since Edward O. Wilson advocated 'consilience' among all the branches of learning in the last decades of the 20th century, neo-Darwinist scholars have dreamed of charting an integrated body of knowledge extending from the theories of narratology and aesthetics all the way to theories explaining how atomic particles and photons behave. The only way for researching such a vast territory is within the Darwinian paradigm of evolutionary studies. Darwin's theory fascinates numerous scholars and writers precisely because of its universality: it brings an enormously large range of phenomena (from the scope of psychology, geology, biology, anthropology, and many other branches of science) within the simple compass of casual explanation.

The theory of adaptation by means of natural selection is crucial for the contemporary worldview and yet it stirs a lot of controversies. In Britain, the homeland of both Charles Darwin and Richard Dawkins, novelists reference the theory of evolution and describe 19th-century Darwinian naturalists in order to discuss such issues as religion, rationalism, and human nature. Antonia Byatt in *Angels and Insects* depicts the mid-Victorian spiritual crisis evoked by the publication of *On the Origin of Species*; Graham Swift in *Ever After* focuses on the loss of faith of the first readers of Darwin's book; Julian Barnes in *Before She Met Me* applies evolutionary psychology to describe jealousy; Hilary Mantel in *A Change of Climate* poses questions concerning the reconciliation of Darwinism and Fundamentalist Christianity. All these authors, among many others, look back to previous epochs – the Victorian era or the distant past of the human race – in order to explain diverse aspects of the human nature we have inherited from our ancestors. Yet, as far as American culture goes, the public debate on Darwinism and the theories targeted at proving Darwin was wrong is definitely not a thing of the past. Thus, American writers who apply Darwinian metaphors or references in their fiction are at the same time making a sort of ideological, if not to say political statement – just as was the case in 19th-century Britain.

The aim of this chapter is to compare how Darwinian references are used allegorically in the writings of two late 20th-century American authors – namely,

Annie Dillard and Kurt Vonnegut. Although neither Dillard nor Vonnegut have a conspicuously political agenda, they both consider the theory of evolution a heavily ideological subject and both apply the Darwinian paradigm to describe nature and the human race within nature in a metaphoric manner. Interestingly enough, they also both choose the Galapagos archipelago as the focal setting of their symbolical narratives, as we see in Vonnegut's novel *Galápagos* and in Dillard's essay "Life on the Rock: the Galápagos". As far as Dillard's prose is concerned, she also depicts the archipelago in other short narratives – i.e., *Teaching a Stone to Talk* and *Pilgrim at Tinker Creek*.

Nevertheless, Vonnegut and Dillard's texts are generically very different. Vonnegut's novel is a work of science fiction and a bitter social satire which depicts a luxurious tourist cruise to the Galapagos and a simultaneous global crisis followed by the outbreak of a virulent plague which kills everybody on Earth except for a handful of tourists marooned on a deserted island in the archipelago. They live on raw iguanas and fish, they breed and their children do the same, as do their children's children until, finally, after a million years of evolution in the hardship of the Galapagos, the human genotype 'improves' – we change into big, friendly, seal-like marine mammals who have flippers and long toothy faces to catch fish with and who are morally good and kind. With no hands and very small brains they are literally unable to do any harm to themselves, other creatures, or the planet – which fact represents huge progress in comparison to what we are capable of doing (and what we are doing) now.

Dillard's *Teaching a Stone to Talk. Expeditions and Encounters* and *Pilgrim at Tinker Creek* are essay collections whose main subject is nature. In the former, a travel book, it is the nature of exotic places – in the latter, it is the natural life of a creek in Virginia near the narrator's home, as described in a number of snapshots in consecutive seasons of the year. Vonnegut's perspective is enormously vast, his narrative spans across the millennia showing how the mechanisms of natural selection work on an entire species which in its original shape is a dangerous misbegotten genus keen on ruining its members' lives and the global biosphere. Dillard's perspective is minute and she focuses on small creatures (muskrats, snails, snakes, and praying mantises) and on precise settings: one puddle, a small shrub, a hedgerow. Vonnegut writes a full-fledged allegory of human nature; Dillard by meticulous descriptions of tiny things depicts the ways of nature, human nature included.

Both Dillard and Vonnegut systematically and obsessively reference Charles Darwin and both would agree with the following statement made by Michael T.

Ghiselin, a Darwinian historian of science, where he praises the eminent Victorian as the founder of the modern scientific method.

> Darwin was a great scientist because he asked great questions. He was an influential scientist because he seized upon those problems which, at the time, could be exploited in further research. His works retain their interest for the working biologist because they continue to generate new and useful theories. His thoughts have been historically important because they illuminated the path of investigation, regardless of where that path may lead.[271]

The origins of this method may be found in the young Darwin's trip to the New World, and primarily in his stay in the Falklands and the Galapagos. In one of his diaries, dated 1837, he writes: "In July opened first note book on 'Transmutation of species' – had been greatly struck from about month of previous March on character of S. American fossils – and species on Galápagos Archipelago. These facts origin (especially latter) of all my views."[272]. Darwin's short visit to these islands is now a part of popular science folklore, numerous nature films mention the event, and the naturalist's name remains associated with the archipelago and its wildlife, particularly the rare animals with bizarre adaptations, the finches being the best example.

In Vonnegut's novel we see the first trip of a new passenger ship called the Bahia de Darwin to the Galapagos. It is publicized and advertised all over the world as 'the Nature Cruise of the Century'. Bahia de Darwin is to re-trace Darwin's route in order to celebrate the famous voyage during which *On the Origin of Species* was conceived. The narrator who is scandalized by the publicity of the cruise describes Darwin's 1835 visit in the islands in far less romantic terms. He calls the naturalist "a mere stripling of twenty-six"[273] who is "underspoken and gentlemanly, impersonal and asexual"[274] and who came to see boring, gray, disappointing, and rocky islands. Only the tremendous success of *On the Origin of Species* made people falsely maintain that the archipelago was interesting at all. The ship-wrecked passengers of Bahia de Darwin found them as they really were: dull, inhospitable, and chilly. The contrast of what things are in nature and how they are described in culture is very sharp, though admittedly, "there were no woodpeckers on the islands, but there was a finch which ate what woodpeckers

271 Michael T. Ghiselin, *The Triumph of the Darwinian Method* (Mineola, New York: Dover Publications, Inc., 2003), 241.
272 Quoted after Ghiselin, *The Triumph*, 33.
273 Kurt Vonnegut, *Galápagos* (New York: Delacorte Press, 1985), 12.
274 Vonnegut, *Galápagos*, 16.

would have eaten. It couldn't peck wood, and so it took a twig or a spine from a cactus in its blunt little beak and used that to dig insects out of their hiding places"[275]. Interesting as the finch is, it definitely does not make the archipelago worth visiting.

The picture of Darwin Dillard believes in is quite different and apparently derives from the standard text-books on the history of biology:

> Charles Darwin came to the Galapagos in 1835, on the Beagle, he was twenty-six. He threw the marine iguanas as far as he could into the water; he rode on tortoises and sampled their meat. He noticed that the tortoises' carapaces varied wildly from island to island, so also did the forms of various mockingbirds. He made collections. Nine years later he wrote in a letter: 'I am most convinced (quite contrary to the opinion I started with) that species are not (it is like confessing a murder) immutable…' it is fashionable now to disparage Darwin's originality; not even the surliest of his detractors however, faults his painstaking methods or denies his impact.[276]

And yet his discoveries made all the difference and altered the way we view the universe, ourselves, and God. Before Darwin came –

> We were all crouched in a small room against the comforting back wall awaiting the millennium which had been gathering impetus since Adam and Eve. Up there was a universe and down here would be a small strip of man come and gone, created, taught, redeemed and gathered up in a bright twinkling, like a sprinkling of confetti torn from colored papers tossed from windows, and swept from the streets by morning. The Darwinian revolution knocked out the back wall revealing eerie lighted landscapes as far back as we can see. Almost at once Albert Einstein and astronauts… knocked out the other walls and the ceiling, leaving us sunlit, exposed, and drifting.[277]

In the light of this statement the Galapagos are the first, primordial place, both metaphorically and literally. Dillard describes these islands as "just plain here"[278]. They are rocky plots of ground which blew up out of the ocean. Some animals drifted aboard, some plants were blown to them, and in the austere conditions these organisms evolved weird forms: "you can go there and watch it happen, and try to figure it out. The Galapagos are a kind of metaphysics laboratory, almost wholly uncluttered by human culture"[279]. For Dillard each of the islands

---

275 Vonnegut, *Galápagos*, 131.
276 Anne Dillard, *Teaching a Stone to Talk. Expeditions and Encounters* (New York, London Toronto, Sydney, New Delhi, Auckland: Harper Perennial, 2013), 117.
277 Dillard, *Teaching*, 121.
278 Dillard, *Teaching*, 91.
279 Dillard, *Teaching*, 91.

rises from the sea as "a chunk of chaos"[280] with rough and smooth parts and devoid of any life. It is empty and uninviting and yet stowaway creatures, shipwrecked creatures, and flotsam get there and evolve unmolested into "a Hieronymus Bosch assortment"[281].

Wildlife conquers all the space available, life abounds and yet is thrifty enough to make use of every particle. Such a statement, one which both Vonnegut and Dillard consider valid, is of course very old, it dates back to the very famous passage in *On the Origin of Species* describing the so-called 'entangled bank' vision of nature:

> It is interesting to contemplate an entangled bank, clothed with many plants of many kinds, with birds singing on the bushes, with various insects flitting about, and with worms crawling through the damp earth, and to reflect that these elaborately constructed forms, so different from each other, and dependent on each other in so complex a manner, have all been produced by laws acting around us... a Ratio of Increase so high as to lead to a Struggle for Life, and as a consequence to Natural Selection.[282]

Vonnegut's narrator is outraged that "Darwin's law of Natural Selection"[283] has worked ceaselessly for millennia filling the Earth with resilient yet senseless life of every imaginable kind. The best-adapted organisms are born and die in the myriads and the only goal of all this life is to produce yet more life. In the Galapagos lives a blue-foot booby which is but a big stupid bird famous for its very complicated and majestic courtship dance. Before the global disaster, Mary, the protagonist of Vonnegut's novel and a high school biology teacher, used to give her students extra credits if they wrote an essay on the courtship dance. Most of those who undertook the task claimed in their papers that boobies worship God. Only one insightful boy, subsequently killed in Vietnam, saw the dance for what it was: a manifestation of the mindless, never-ending drive to multiply. Instead of an essay he wrote a poem, the boobies' eternal love song:

> Of course I love you,
> So let's have a kid
> Who will say exactly
> What its parents did
> Of course I love you,
> So let's have a kid
> Who will say exactly

---

280   Dillard, *Teaching*, 109.
281   Dillard, *Teaching*, 110.
282   Dillard, *Teaching*, 54.
283   Vonnegut, *Galápagos*, 79.

What its parents did
Of course I love you...[284]

Mechanically repeated the song goes on and on, generation after generation, but there is no meaning in it beyond generating yet another repetition. Nature is plentiful and tolerant of the clearly ridiculous mistakes evolution has committed. Vonnegut's examples of horridly maladapted and yet long-surviving species are the Irish elk with antlers the size of a ballroom chandelier that make it highly difficult for the animal to feed at all, and humans with their poisonous, overgrown brain keen on destruction of every kind.

Dillard conversely adores the entangled banks in the world and the bounty of nature, and the pressure the environment has on every creature, propelling them to evolve into an unimaginable richness of shapes: "Extravagance! Nature will try everything once. No form is too gruesome, no behavior too grotesque. If you are dealing with organic compounds then let them combine!"[285], she exclaims in *Pilgrim at Tinker Creek* where one plot of ground is the world in miniature. Its narrator, an avid reader of Darwinian natural history looks at the grass and the insects and finds out that that, yes, everything is just as the biologists say and "that the insects have adapted is obvious"[286]. She ponders the top inch of soil and considers it to be the whole world squirming under her palm with an average of 1,356 larger organisms in every square foot and, probably "up to a billion" bacteria, fungi, and protozoa. All this richness is somehow connected to the narrator herself as they all belong to the gigantic living macrocosm. Thus, being capable of logical thinking, the narrator feels obliged to look for the meaning of nature: "If I did not know about the rotifers and paramecia... fine, but since I've seen it I must somehow deal with it, take it into account"[287].

Humans, thanks to their spiritual place in the Universe, have to speak for the rest of Creation and the Darwinian perspective allows people to see the grand design of the universe. For the narrator, who is a reader of Pierre Teilhard de Chardin, Darwinism and Christianity complement each other –

De Chardin, a paleontologist, examined the evolution of species itself, and discovered in that flow a surge towards complexity and consciousness, a free ascent capped with man and propelled from within and attracted from without by God the holy freedom

---

284   Vonnegut, *Galápagos*, 108.
285   Anne Dillard, *Pilgrim at Tinker Creek* (London: Canterbury Press, 2011), 66.
286   Dillard, *Pilgrim*, 66.
287   Dillard, *Pilgrim*, 95.

and awareness that is Creation's beginning and end. And so forth. Like flatworms, like languages ideas evolve... in the supple flux of an open mind.[288]

Darwin himself was aware that if the organic scale is topped by humanity it is so only because humankind fought to rise that high, which fact gives us all "hope for a still higher destiny in the distant future"[289]. Yet, as he claims in the very last sentence of *The Descent of Man* –

> We must however acknowledge, as it seems to me, that man with his all noble qualities, with sympathy that he feels for the most debased, with benevolence which extends not only to other men but to the humblest living creature, with his god-like intellect which has penetrated into the movements and constitution of the solar system – with all these exalted powers – Man still bears in his bodily frame the indelible stamp of his lowly origin.[290]

Human minds are thus what they are because they have evolved from earlier forms. 'Much to the distress of our planet', Vonnegut's narrator adds, because he firmly believes that the human brain with its lethal potential is the greatest mistake of nature. He rhetorically asks:

> So I raise this question, although there is nobody around to answer it: Can it be doubted that three-kilogram brains were once nearly fatal defects in the evolution of the human race?

> A second query: What source was there back then, save for our overelaborate nervous circuitry, for the evils we were seeing or hearing about simply everywhere?

> My answer: there was no other source. This was a very innocent planet, except for these great big brains.[291]

Yet for Dillard humans were created "from a clot and set in proud, free motion"[292] by the apparently merciless laws of nature. Evolution loves death and births equally and is "this whole business of reproducing and dying by the billion"[293]. Yet all of it happens "ad majorem dei gloriam" and "we little blobs of soft tissue crawling around on this planet's skin"[294] are entitled to ask the big question, to look at the universe, and to worship its Creator. People or finches, we all are "embellishments of random chromosomal mutations selected by natural

---

288 Dillard, *Pilgrim*, 120.
289 Darwin, *The Descent*, 78.
290 Darwin, *The Descent*, 78.
291 Vonnegut, *Galápagos*, 8–9.
292 Dillard, *Pilgrim*, 12.
293 Dillard, *Pilgrim*, 170.
294 Dillard, *Pilgrim*, 175.

selection and preserved in geographically isolated gene pools"[295] because all the organic matter participates in the gigantic Darwinian game:

> Ça va. It goes on everywhere tit for tat, action and reaction, triggers and inhibitors ascending in a spiral like spatting butterflies. Within life we are pushing each other around. How many animal forms have evolved just so because there are, for instance, trees? We pass the nitrogen around, and vital gases, we feed and nest, plucking this and that and planting seeds.[296]

Thus all the life on Earth is like a gigantic dance and a great race. Everybody is dependent on everybody else, and having a brain – i.e., being rational, being capable of seeing this dance and understanding its rules – is one of the greatest privileges imaginable. Once you have evolved and have acquired culture you start studying nature and you realize, thanks to, among other things, Darwinian biology, the intricacies of its design. We are the acme of Creation.

Vonnegut in his novel turns a similar idea of a perfectly adapted human race into a bitter irony. Over a million-year period the descendants of the Bahia de Darwin survivors evolve into perfect creatures. Thanks to the bottleneck effect their genetic pool is easily re-design so they will nevermore threaten the ecological balance of the Earth.

> As for human beings making a comeback, of starting to use tools and build houses and play musical instruments and so on again: They would have to do it with their beaks at the time. Their arms have become flippers in which the hand bones are almost entirely imprisoned and immobilized. Each flipper is studded with five purely ornamental nubbins, attractive to members of the opposite sex at mating time. These are in fact the tips of four suppressed fingers and a thumb. Those parts of people's brains which used to control their hands, moreover, simply don't exist anymore, and human skulls are now much more streamlined on that account. The more streamlined the skull, the more successful the fisher person.[297]

In light of the above passage the Darwinian bon mot quoted at the end of the novel reads very ironically: "progress has been much more general than retrogression"[298]. This is paradoxically true – the overdeveloped human brain was a dangerous mistake of nature, and nature working slowly but steadily set this right by altering the human species in such a way as to make it harmless. *Galápagos* is the record of this alteration done in Darwinian discourse.

---

295   Dillard, *Teaching*, 175.
296   Dillard, *Teaching*, 126.
297   Vonnegut, *Galápagos*, 185.
298   Vonnegut, *Galápagos*, 291.

Yet Dillard applies the very same Darwinian apparatus to emphasize the glory of Creation and the greatness of the Universe. She considers it tragic that "Fundamentalist Christians... feel they have to make a choice between the Bible and modern science"[299] because only with the help of modern science can you truly appreciate God's greatness and see beyond the apparent cruelty of death-loving evolution.

The Darwinian perspective thus allows both Dillard and Vonnegut to express their attitudes towards human civilization and its place within the natural environment of the planet, the human past and future, and the way culture and nature depend on each other. Both share a fascination with Darwin as well as the very profound expertise in the subject of his theory. For both authors the two most important issues Darwin discusses in his imposing oeuvre are 'the entangled bank' metaphor of wildlife depicted in *On the Origin of Species*, and the hypothesis concerning the evolution of the human brain and the human mind discussed in the final sections of *The Descent of Man*. And although their intimations provoked by the Galapagos islands are as ideologically far apart as possible, the above analysis of their texts inspired by this Archipelago clearly shows that they both are artists-cum-evolutionary theorists whose output is – as Wilson would have it – 'consilient'. Being evolutionary theorists they attempt to re-shape the paradigm within which the research in all possible fields of learning is conducted in order to achieve a consilient picture of how the universe works and how its nature can be studied. As artists they are neo-Darwinists because neo-Darwinism is the pivotal approach uniting the human sciences, the arts, and the hard sciences. Thus, using precisely such a perspective both Dillard and Vonnegut seek to achieve new insights into the very nature of human beings.

> An evolutionary perspective allows us to see ourselves both in the widest angle and with the most precise focus, as individuals solving particular problems within specific contexts, physical and social, using the cognitive equipment – including the predilection for culture – acquired through natural selection.[300]

---

299 Dillard, *Teaching*, 119.

300 Brian Boyd, Joseph Carroll, and Jonathan Gottschall, 'Literary and Film Studies Now: Death or Rebirth', in Brian Boyd, Joseph Carroll, and Jonathan Gottschall, eds, *Evolution, Literature and Film. A Reader* (New York: Columbia University Press, 2010), 3.

Human beings are therefore primarily creatures who have evolved and the theory of natural selection teaches us why and how this has happened. Despite all their differences, the two depictions of the Galapagos Archipelago, by Annie Dillard and by Kurt Vonnegut, are written with this very same idea in mind.

References

Ballard, J. G., The Drowned World (Harmondsworth, Ringwood: Penguin Books, 1974).

Barthes, Roland, Mythologies, trans. Anette Lavers (London: Vintage, 1993).

Baudrillard, Jean, Simulacra and Simulation, trans. Sheila Faria Glaser (Ann Arbor: University of Michigan Press, 1994).

Boyd, Brian, 'Getting It All Wrong' *The American Scholar* <http://theamericanscholar. org/getting-it-all-wrong> accessed 23 November 2012.

Boyd, Brian, Joseph Carroll, and Jonathan Gottschall, eds, *Evolution, Literature and Film* (New York: Columbia University Press, 2010).

Brown, Donald E, Human Universals (New York: McGraw-Hill, 1991).

Browne, Janet, Charles Darwin. Voyaging (Princeton: Princeton University Press, 1996).

Byatt, Antonia S., 'A New Body of Writing. Darwin and Recent British Fiction', in Antonia S. Byatt and Alan Hollinghurst, eds, *New Writing 4* (London: Vintage, 1995).

— *Angels and Insects* (London: Chatto and Windus, 1992).

Carey, Peter, Oscar and Lucinda (New York: Vintage International, 1988).

Carroll, Joseph, 'An Evolutionary Paradigm for Literary Study'<http://www. umsl.edu/~carrolljc/Documents%20linked%20to%20indiex/ Target%20Pieces/1_Target_article.htm> accessed 23 November 2012.

Carter, Angela, The Sadeian Woman. An Exercise in Cultural History (London: Virago, 1979).

Caudill, Edward, Darwinian Myths (Knoxville: The University of Tennessee Press, 1999).

Darnton, John, The Darwin Conspiracy (New York: Anchor Books, 2005).

Darwin, Charles, On the Origin of Species (New York: Bantam Dell, 1999).

— *The Voyage of the Beagle: Journal of Researches into the Natural History and Geology of the Countries Visited During the Voyage of H.M.S. Beagle Round the World* (Ware Hertfordshire: Wordsworth Editions, 1997).

— *The Descent of Man And the Selection in Relation to Sex* (New York: A.L. Bert Company Publishers, 1874).

Dillard, Annie, *Pilgrim at Tinker Creek* (London: Canterbury Press, 2011).

— *Teaching a Stone to Talk. Expeditions and Encounters* (New York, London Toronto, Sydney, New Delhi, Auckland: Harper Perennial, 2013).

Diski, Jenny, *Monkey's Uncle* (London: Phoenix, 1994).

Eliade, Mircea, *Myth and Reality*, trans. Williard R. Trask (New York: Harper Torchbooks, 1968).

Fowles, John, *The French Lieutenant's Woman* (London: Penguin,1969).

Frye, Northrop *Anatomy of Criticism: Four Essays* (Princeton, N.J.: Princeton University Press, 1957).

Ghiselin, Michael T., *The Triumph of the Darwinian Method* (Mineola, New York: Dover Publications, Inc., 2003).

Golding, William, *The Inheritors* (New York, Harcourt: Brace & World, 1955).

Gray, Asa, *Darwiniana. Essays and Reviews Pertaining to Darwinism* (Cambridge, Melbourne, New York: Cambridge University Press, 2009).

Gribbin, John and Michael White, *Darwin: A Life in Science* (London, New York, Sidney, Toronto: Pocket Books, 2009).

Healey, Edna, *Emma Darwin. The Inspirational Wife of a Genius* (London: Review, 2001).

Holland, Henry, *Travels in the Ionian Isles, Albania, Thessaly, Macedonia etc. during the Years 1812 and 1813* (Cambridge, New York, Melbourne, Madrid, Cape Town, Singapore, São Paolo, Delhi, Mexico City: Cambridge University Press, 2012).

Hutcheon, Linda, 1988, 'Historiographic Metafiction. Parody and the Intertextuality of History' <http://hdl.handle.net/1807/10252> accessed 12 November 2011.

Irvine, William, *Apes, Angels, and Victorians* (London: Readers Union Weidenfeld and Nicolson, 1956).

Keynes, Randal, *Annie's Box; Charles Darwin, His Daughter, and Human Evolution* (London: Fourth Estate, 2001).

King, Philip Parker, *Narrative of a Survey of the Intertropical and Western Coast of Australia*, 1826 http://gutenberg.net.au/ebooks/e00027.html accessed 24 November 2014.

Kuhn, Thomas, *The Structure of Scientific Revolutions* (Chicago and London: University of Chicago Press, 1996).

Landau, Misia, 'Human Evolution as Narrative', *American Scientist* 72:262–268 (1984) <http://www.unc.edu/~akakalio/Landau.pdf> accessed 8 January 2012.

Levine, George, *Darwin and the Patterns of Science in Victorian Fiction* (Chicago and London: University of Chicago Press, 1991).

Lewin, Roger, *Human Evolution: An Illustrated Introduction* (Oxford: Blackwell Publishing, 1999).

Lewis, C.S., *A Preface to Paradise Lost* New (York Oxford University Press, 1961).

Llewellyn, Mark, 'What Is Neo-Victorian Studies?', *Neo-Victorian Studies* 1.1 <http://www.neovictorianstudies.com/past_issues/Autumn2008/NVS%20 1-1%20M-Llewellyn> accessed 12 November 2011.

Lyotard, Jean-Francois, *The Postmodern Condition: A Report on Knowledge*, trans. Geoffrey Bennington and Brian Massumi (Manchester: Manchester University Press, 1984).

Mantel, Hilary, *A Change of Climate* (Anstey, Leicestershire: F.A. Thorpe Publishing,1995).

McDonald, Roger, *Mr. Darwin's Shooter* (Milsons Point and London: Anchor 1998).

— 'Evolution of a Novel: Mr. Darwin's Shooter', *Australian Humanities Review 12* <http://www.australianhumanitiesreview.org/archive/Issue-December-1998/ mcdonald.html> accessed 12 November 2011.

Miller, Geoffrey, *The Mating Mind: How Sexual Choice Shaped the Evolution of Human Nature* (New York: Anchor Books, 2001).

Moorehead, Alan, *Darwin and the Beagle* (London Hamish: Hamilton, 1969).

Nichols Peter, *Evolution's Captain* (New York: HarperCollins, 2003).

Oramus, Dominika, 'Darwinian obsessions: References to the Theory of Evolution in the Novels of John Fowles, A.S. Byatt, and Hilary Mantel', *Anglica. An International Journal of English Studies* 21/1 (2012), 5–17.

— 'The Voyage of Charles Darwin on the HMS Beagle in Recent Fiction and Non-Fiction', in Grażyna Bystydzieńska and Emma Harris, eds, *From Queen Anne to Queen Victoria. Readings in 18th and 19th century British literature and culture.* Vol. 3. (Warszawa: Ośrodek Studiów Brytyjskich, 2012), 379–388.

— 'The Theory of Evolution or a Conspiracy Theory? Recent Fiction about Charles Darwin', *Acta Philologica* 41 (2012), 164–171.

— 'In the Beginning was Darwin: History and Simulation in Thorvald Steen's *Don Carlos and Giovanni* and Roger McDonald's *Mr. Darwin's Shooter*' in Dorota Babilads and Lucyna Krawczyk, eds, *We the Neo-Victorians* (Warszawa: Instytut Anglistyki, Uniwersytet Warszawski, 2013).

— 'Orangutans, Savages, and Babies: Darwin's Problem with Human Ancestry as Reflected in Recent Fiction' in Grażyna Bystydzieńska and Emma Harris, eds, *From Queen Anne to Queen Victoria. Readings in 18th and 19th century British literature and culture.* Vol. 4 (Warszawa: Ośrodek Studiów Brytyjskich, 2014), 407–420.

— 'Echoes of the Mid-19th-Century Spiritual Crisis in Selected Contemporary Texts Referencing Charles Darwin', in Grażyna Bystydzieńska, ed., *The New Review. An International Journal of British Studies* (Glasgow: University of Glasgow, University of Warsaw, 2014), 101–117.

— 'Two Exercises in Consilience. Annie Dillard and Kurt Vonnegut on the Galapagos Archipelago as the Archetypal Darwinian Setting', *Anglica. An International Journal of English Studies* 23/1, 2014, 19–31.

Phipps, William E., *Darwin's Religious Odyssey* (Harrisburg: Trinity Press International, 2002).

Popper, Karl, *The Logic of Scientific Discovery* (London, New York: Routledge Classics, 2002).

Ruse, Michael, *The Darwinian Paradigm* (London and New York: Routledge, 1989).

Segal, Robert A., *Theorizing about Myth* (Amherst: University of Massachusetts Press, 1999).

Sokal, Alan, 'Transgressing the Boundaries. An Afterword', *Descent* 43/4 (1996), 93–94.

—'A plea for reason, evidence and logic', *New Politics* 7/2 (1997), 126–129.

Steen, Thorvald, *Don Carlos and Giovanni*, trans. James Anderson (København and Los Angeles: Green Integer, 2004).

Stone, Irving, *The Origin. A Biographical Novel of Charles Darwin* (London: Corgi Books, 1982).

Swift, Graham, *Ever After* (London: Picador, 1992).

Thompson, Henry, *This Thing of Darkness* (London: Review, 2005).

Vonnegut, Kurt, Galápagos (New York: Delacorte Press, 1985).

Wells, G.H., 'The Grisly Folk'< http://gutenberg.net.au/ebooks06/0602061.txt > accessed 8 January 2012.

Wheen, Francis, How Mumbo-Jumbo Conquered the World (New York, London, Toronto, and Sydney: Harper Perennial, 2004).

Wilson, Edward O., Sociobiology: The New Synthesis (Cambridge Mass.: The Belknap Press of Harvard University Press, 2000).

— *Consilience: The Unity of Knowledge* (New York: Alfred A. Knopf, 1998).

Wright, Robert, The Moral Animal: Why We Are, the Way We Are: The New Science of Evolutionary Psychology (New York: Routledge, 1995).

Films

Amiel, Jon (dir.) *Creation*, 2009.
Annaud, Jean Jacques (dir) *The Quest for Fire*, 1981.
Attenborough, David (prod.) *Darwin and the Tree of Life*, 2009.
Chapman, Michael (dir) *The Clan of the Cave Bear*, 1986.
Jennings, Garth (dir.) *The Hitchhiker's Guide to the Galaxy*, 2005.
Malaterre, Jacques (dir) *A Species Odyssey*, 2003.
— (dir) *Homo Sapiens*, 2005.

This book examines Darwinian fiction and non-fiction, highlighting how the theory of evolution and Darwin's life are presented in each of them. What these eleven essays share is the strong belief that comparative analysis of Darwinism-inspired texts can enrich literary studies. By studying the diverse uses of evolutionary discourse in recent literature and films, my book demonstrates how natural science influences the contemporary humanities and, conversely, how literary conventions are used in order to make scientific and popular-science texts intelligible and attractive. *Charles Darwin's Looking Glass* shows how and why today's culture gazes upon the myth of Darwin, his theory, and his life in order to find its own reflection.

Gdańsk Transatlantic Studies in British and North American Culture

Edited by Marek Wilczyński

The interdisciplinary series "GdańskTransatlantic Studies in British and North American Culture" brings together literary and cultural studies concerning literatures and cultures of the English-speaking world, particularly those of Great Britain, Ireland, the United States, and Canada. The range of topics to be addressed includes literature, theater, film, and art, considered in various twenty-first-century theoretical perspectives, such as, for example (but not exclusively), New Historicism and canon formation, cognitive narratology, gender and queer studies, performance studies, memory and trauma studies, and New Art History. The editors are leaving a broad margin for the innovative and the unpredictable, hoping to attract authors whose approaches will point to new directions of research as regards both thematic areas and methods. Comparative Polish-Anglo-American proposals will be considered, too.

Transatlantic Studies in British and North American Culture

Edited by Marek Wilczyński

www.peterlang.com